COMPUTATIONAL PHYSICS

A Practical Introduction to Computational Physics
and Scientific Computing (using C++)

Volume I

Athens, 2016

KONSTANTINOS N. ANAGNOSTOPOULOS
National Technical University of Athens

National Technical University of Athens

COMPUTATIONAL PHYSICS
A Practical Introduction to Computational Physics and Scientific Computing (C++ version)
Volume I

AUTHORED BY KONSTANTINOS N. ANAGNOSTOPOULOS
Physics Department, National Technical University of Athens, Zografou Campus, 15780 Zografou, Greece
konstant@mail.ntua.gr, www.physics.ntua.gr/~konstant/

PUBLISHED BY KONSTANTINOS N. ANAGNOSTOPOULOS
and the
NATIONAL TECHNICAL UNIVERSITY OF ATHENS

Book Website:
www.physics.ntua.gr/~konstant/ComputationalPhysics

©Konstantinos N. Anagnostopoulos 2014, 2016

First Published 2014
Second Edition 2016
Version[1] 2.0.20161206202900

Cover: Design by K.N. Anagnostopoulos. The front cover picture is a snapshot taken during Monte Carlo simulations of hexatic membranes. Work done with Mark J. Bowick. Relevant video at youtu.be/Erc7Q6YXfLk

This book and its cover(s) are subject to copyright. They are licensed under the Creative Commons Attribution-ShareAlike 4.0 International License. To view a copy of this license, visit creativecommons.org/licenses/by-sa/4.0/

The book is accompanied by software available at the book's website. All the software, unless the copyright does not belong to the author, is open source, covered by the GNU public license, see www.gnu.org/licenses/. This is explicitly mentioned at the end of the respective source files.

ISBN 978-1-365-58322-3 (lulu.com, vol. I)

ISBN 978-1-365-58338-4 (lulu.com, vol. II)

---

[1] The first number is the major version, corresponding to an "edition" of a conventional book. Versions differing by major numbers have been altered substantially. Chapter numbers and page references are not guaranteed to match between different versions. The second number is the minor version. Versions differing by a minor version may have serious errors/typos corrected and/or substantial text modifications. Versions differing by only the last number may have minor typos corrected, added references etc. When reporting errors, please mention the version number you are referring to.

Contents

Foreword to the Second Edition   **vii**

Foreword to the First Edition   **ix**

1 The Computer   **1**
    1.1 The Operating System 2
        1.1.1 Filesystem 3
        1.1.2 Commands 11
        1.1.3 Looking for Help 15
    1.2 Text Processing Tools – Filters 17
    1.3 Programming with Emacs 22
        1.3.1 Calling Emacs 23
        1.3.2 Interacting with Emacs 23
        1.3.3 Basic Editing 27
        1.3.4 Cut and Paste 29
        1.3.5 Windows 31
        1.3.6 Files and Buffers 32
        1.3.7 Modes 33
        1.3.8 Emacs Help 34
        1.3.9 Emacs Customization 36
    1.4 The C++ Programming Language 37
        1.4.1 The Foundation 37
    1.5 Gnuplot 50
    1.6 Shell Scripting 56

2 Kinematics   **69**
    2.1 Motion on the Plane 69
        2.1.1 Plotting Data 79

		2.1.2 More Examples 82
2.2	Motion in Space 95
2.3	Trapped in a Box 105
	2.3.1	The One Dimensional Box 105
	2.3.2	Errors 114
	2.3.3	The Two Dimensional Box 118
2.4	Applications 122
2.5	Problems 144

3 Logistic Map   149
- 3.1 Introduction 149
- 3.2 Fixed Points and 2^n Cycles 152
- 3.3 Bifurcation Diagrams 159
- 3.4 The Newton–Raphson Method 163
- 3.5 Calculation of the Bifurcation Points 169
- 3.6 Liapunov Exponents 174
- 3.7 Problems 189

4 Motion of a Particle   201
- 4.1 Numerical Integration of Newton's Equations 201
- 4.2 Prelude: Euler Methods 202
- 4.3 Runge–Kutta Methods 215
  - 4.3.1 A Program for the 4th Order Runge–Kutta 219
- 4.4 Comparison of the Methods 224
- 4.5 The Forced Damped Oscillator 227
- 4.6 The Forced Damped Pendulum 235
- 4.7 Appendix: On the Euler–Verlet Method 243
- 4.8 Appendix: 2nd order Runge–Kutta Method 246
- 4.9 Problems 249

5 Planar Motion   253
- 5.1 Runge–Kutta for Planar Motion 253
- 5.2 Projectile Motion 259
- 5.3 Planetary Motion 267
- 5.4 Scattering 271
  - 5.4.1 Rutherford Scattering 275
  - 5.4.2 More Scattering Potentials 283
- 5.5 More Particles 286

		5.6	Problems	299
6	**Motion in Space**			**305**
	6.1	Adaptive Stepsize Control for RK Methods	306
		6.1.1	The rksuite Suite of RK Codes............	306
		6.1.2	Interfacing C++ Programs with Fortran	311
		6.1.3	The rksuite Driver	317
	6.2	Motion of a Particle in an EM Field	322
	6.3	Relativistic Motion	324
	6.4	Problems	337

Bibliography   **340**

Index   **349**

> This book has been written assuming that the reader **executes all the commands presented in the text and follows all the instructions at the same time**. If this advice is neglected, then the book will be of little help and some parts of the text may seem incomprehensible.

The book's website is at
http://www.physics.ntua.gr/~konstant/ComputationalPhysics/
From there, you can can download the *accompanying software*, which contains, among other things, *all* the programs presented in the book.

Some conventions: Text using the font shown below refers to commands given using a shell (the "command line"), input and output of programs, code written in Fortran (or any other programming language), as well as to names of files and programs:

```
> echo Hello world
Hello world
```

When a line *starts* with the prompt

```
>
```

then the text that follows is a command, which can be given from the command line of a terminal. The second line, Hello World, is the *output* of the command.

The contents of a file with C++ code is listed below:

```
int main(){
  double x = 0.0;
  for(int i=0;i<10;i++){
    x += i;
```

```
    }
}
```

What you need in order to work on your PC:

- An operating system of the GNU/Linux family and its basic tools.

- A Fortran compiler. The gfortran compiler is freely available for all major operating systems under an open source license at http://www.gfortran.org.

- An advanced text editor, suitable for editing code in several programming languages, like Emacs[2].

- A good plotting program, suitable for data analysis, like gnuplot[3].

- The shell tcsh[4].

- The programs awk[5], grep, sort, cat, head, tail, less. Make sure that they are available in your computer environment.

If you have installed a GNU/Linux distribution on your computer, all of the above can be installed easily. For example, in a Debian like distribution (Ubuntu, ...) the commands

```
> sudo apt-get install tcsh emacs gnuplot gnuplot-doc
> sudo apt-get install gfortran gawk gawk-doc binutils
> sudo apt-get install manpages-dev coreutils liblapack3
```

install all the necessary tools.

If you don't wish to install GNU/Linux on your computer, you can try the following:

- Boot your computer using a usb/DVD live GNU/Linux, like Ubuntu[6]. This will not make any permanent changes in your hard drive but it will start and run slower. On the other hand, you may save all

---
[2] http://www.gnu.org/software/emacs/
[3] http://www.gnuplot.info
[4] http://www.tcsh.org
[5] http://www.gnu.org/software/gawk
[6] http://www.ubuntu.com

your computing environment and documents and use it on any computer you like.

- Install Cygwin[7] in your Microsoft Windows. It is a very good solution for Microsoft-addicted users. If you choose the *full* installation, then you will find all the tools needed in this book.

- Mac OS X is based on Unix. It is possible to install all the software needed in this book and follow the material as presented. Search the internet for instructions, e.g. google "gfortran for Mac", "emacs for Mac", "tcsh for Mac", etc.

---

[7]http://www.cygwin.com

Foreword to the Second Edition

This book has been out "in the wild" for more than two years. Since then, its pdf version has been downloaded 2-5000 times/month from the main server and has a few thousand hits from sites that offer science e-books for free. I have also received positive feedback from students and colleagues from all over the world and that gave me the encouragement to devote some time to create a C++ version of the book. As far as scientific programming is concerned, the material has not changed apart from some typo and error corrections[8].

I have to make it clear that by using this book you will not learn much on the advanced features of C++. Scientific computing is usually simple at its core and, since it must be made efficient and accurate, it needs to go down to the lowest levels of programming. This also partly the reason of why I chose to use Fortran for the core programming in the first edition of the book: It is a language designed for numerical programming and high performance computing in mind. It is simple and a scientist or engineer can go directly into programming her code. C++ is not designed for scientific applications[9] in mind and this reflects on some trivial omissions in its standard. Still, many scientific groups are now using C++ for programming and the C++ compilers have improved quite a lot. There is still an advantage in performance using a Fortran compiler on a supercomputer, but this is not going to last for much longer.

Still, for a scientist, the programming language is a *tool* to solve her *scientific* problems. One should not bind herself to a specific language. The treasures of today are the garbage of tomorrow, and the time scale for this happening is small in today's computing environments. What

---
[8] Check the *errata* section at the book's homepage.
[9] Object oriented languages' aim is to improve modularity, maintainability and flexibility of programs.

has really lasting value is the ability to solve problems using a computer and this is what needs to be emphasized. Consistent with this idea is that, in the course of reading this book, you will also learn how to make your C++ code interact with code written in Fortran, like in the case of the popular library Lapack. This will improve your "multilingual skills" and flexibility with interacting with legacy code.

The good news for us scientists is that numerical code usually needs simple data structures and programming is similar in *any* language. It was simple for me to "translate" my book from Fortran to C++. Unfortunately I will not touch on all this great stuff in true object oriented programming but you may be happy to know that you will most likely not need it[10].

So, I hope that you will enjoy using my book and I remind you that I love fan mail and I appreciate comments/corrections/suggestions sent to me. Now, if you want to learn about the structure and educational procedure in this book, read the foreword to the first edition, otherwise skip to the real fun of solving scientific problems numerically.

Athens, 2016.

---

[10] A lot of C++ code out there is realizing procedural and not true object oriented programming.

Foreword to the First Edition

This book is the culmination of my ten years' experience in teaching three introductory, *undergraduate level*, scientific computing/computational physics classes at the National Technical University of Athens. It is suitable mostly for junior or senior level science courses, but I am currently teaching its first chapters to sophomores without a problem. A two semester course can easily cover all the material in the book, including lab sessions for practicing.

Why another book in computational physics? Well, when I started teaching those classes there was no bibliography available in Greek, so I was compelled to write lecture notes for my students. Soon, I realized that my students, majoring in physics or applied mathematics, were having a hard time with the technical details of programming and computing, rather than with the physics concepts. I had to take them slowly by the hand through the "howto" of computing, something that is reflected in the philosophy of this book. Hoping that this could be useful to a wider audience, I decided to translate these notes in English and put them in an order and structure that would turn them into "a book".

I also decided to make the book freely available on the web. I was partly motivated by my anger caused by the increase of academic (e)book prices to ridiculous levels during times of plummeting publishing costs. Publishers play a diminishing role in academic publishing. They get an almost ready-made manuscript in electronic form by the author. They need to take no serious investment risk on an edition, thanks to print-on-demand capabilities. They have virtually zero cost ebook publishing. Moreover, online bookstores have decreased costs quite a lot. Academic books need no advertisement budget, their success is due to their academic reputation. I don't see all of these reflected on reduced book prices, quite the contrary, I'm afraid.

My main motivation, however, is the freedom that independent publishing would give me in improving, expanding and changing the book in the future. It is great to have no length restrictions for the presentation of the material, as well as not having to report to a publisher. The reader/instructor that finds the book long, can read/print the portion of the book that she finds useful for her.

This is *not* a reference book. It uses some interesting, I hope, physics problems in order to introduce the student to the fundamentals of solving a scientific problem numerically. At the same time, it keeps an eye in the direction of advanced and high performance scientific computing. The reader should follow the instructions given in each chapter, since the book *teaches by example*. Several skills are taught through the solution of a particular problem. My lectures take place in a (large) computer lab, where the students are simultaneously doing what I am doing (and more). The program that I am editing and the commands that I am executing are shown on a large screen, displaying my computer monitor and actions live. The book provides no systematic teaching of a programming language or a particular tool. A very basic introduction is given in the first chapter and then the reader learns whatever is necessary for the solution of her problem. There is more than one way to do it[11] and the problems can be solved by following a basic or a fancy way, depending on the student's computational literacy. The book provides the necessary tools for both. A bibliography is provided at the end of the book, so that the missing pieces of a puzzle can be sought in the literature.

This is also *not* a computational physics playground. Of course I hope that the reader will have fun *doing* what is in the book, but my goal is to provide an experience that will set the solid foundation for her becoming a high performance computing, number crunching, heavy duty data analysis expert in the future. This is why the programming language of the core numerical algorithms has been chosen to be Fortran, a highly optimized, scientifically oriented, programming language. The computer environment is set in a Unix family operating system, enriched by all the powerful GNU tools provided by the FSF[12]. These tools are indispensable in the complicated data manipulation needed in scientific research, which requires flexibility and imagination. Of course, Fortran

---

[11] A Perl moto!
[12] Free Software Foundation, www.fsf.org.

is not the best choice for heavy duty object oriented programming, and is not optimal for interacting with the operating system. The philosophy[13] is to let Fortran do what is best for, number crunching, and leave data manipulation and file administration to external, powerful tools. Tools, like awk, shell scripting, gnuplot, Perl and others, are quite powerful and complement all the weaknesses of Fortran mentioned before. The plotting program is chosen to be gnuplot, which provides very powerful tools to manipulate the data and create massive and complicated plots. It can also create publication quality plots and contribute to the "fun part" of the learning experience by creating animations, interactive 3d plots etc. All the tools used in the book are open source software and they are accessible to everyone for free. They can be used in a Linux environment, but they can also be installed and used in Microsoft Windows and Mac OS X.

The other hard part in teaching computational physics to scientists and engineers is to explain that the approach of solving a problem numerically is quite different from solving it analytically. Usually, students of this level are coming with a background in analysis and fundamental physics. It is hard to put them into the mode of thinking about solving a problem using only additions, multiplications and some logical operations. The hardest part is to explain the discretization of a model defined analytically, which can be done in many ways, depending on the accuracy of the approximation. Then, one has to extrapolate the numerical solution, in order to obtain a good approximation of the analytic one. This is done step by step in the book, starting with problems in simple motion and ending with discussing finite size scaling in statistical physics models in the vicinity of a continuous phase transition.

The book comes together with additional material which can be found at the web page of the book[14]. The accompanying software contains all the computer programs presented in the book, together with useful tools and programs solving some of the exercises of each chapter. Each chapter has problems complementing the material covered in the text. The student

---

[13]Java and C++ have been popular choices in computational physics courses. But object oriented programming is usually avoided in the high performance part of a computation. So, one usually uses those languages in a procedural style of programming, cheating herself that she is actually learning the advantages of object oriented programming.

[14]www.physics.ntua.gr/~konstant/ComputationalPhysics/

needs to solve them in order to obtain hands on experience in scientific computing. I hope that I have already stressed enough that, in order for this book to be useful, it is not enough to be read in a café or in a living room, but one needs to *do* what it says.

Hoping that this book will be useful to you as a student or as an instructor, I would like to ask you to take some time to send me feedback for improving and/or correcting it. I would also appreciate fan mail or, if you are an expert, a review of the book. If you use the book in a class, as a main textbook or as supplementary material, I would also be thrilled to know about it. Send me email at konstantmail.ntua.gr and let me know if I can publish, anonymously or not, (part of) what you say on the web page (otherwise I will only use it privately for my personal ego-boost). Well, nothing is given for free: As one of my friends says, some people are payed in dollars and some others in ego-dollars!

Have fun computing scientifically!

Athens, 2014.

Chapter 1

The Computer

The aim of this chapter is to lay the grounds for the development of the computational skills which are necessary in the following chapters. It is not an in depth exposition but a practical training by example. For a more systematic study of the topics discussed, we refer to the bibliography. Many of the references are freely available on the web.

The are many choices that one has to make when designing a computer project. These depend on the needs for numerical efficiency, on available programming hours, on the needs for extensibility and upgradability and so on. In this book we will get the flavor of a project that is mostly scientifically and number crunching oriented. One has to make the best of the available computing resources and have powerful tools available for a productive analysis of the data. Such an environment, found in most of today's supercomputers, that offers flexibility, dependability, simplicity, powerful tools for data analysis and effective compilers is provided by the family of the Unix operating systems. The GNU/Linux operating system is a Unix variant that is freely available and most of its utilities are open source software. The voluntary work of millions of excellent programmers worldwide has built the most stable, fastest and highest quality software available for scientific computing today. Thanks to the idea of the open source software pioneered by Richard Stallman[1] this giant collaboration has been made possible.

Another choice that we have to make is the programming language. In this edition of the book we will be programming in C++. C++ is a

---

[1] www.stallman.org

language with very high level of abstraction designed for projects where modular programming and the use of complicated data structures is of very high priority. A large and complicated project should be divided into independent programming tasks (modules), where each task contains everything that it needs and does not interfere with the functionality of other modules. Although it has not been designed for high performance numerical applications, it is becoming more and more popular in the recent years.

C++, as well as other languages like C, Java and Fortran, is a language that needs to be *compiled* by a compiler. Other languages, like python, perl, awk, shell scripting, Macsyma, Mathematica, Octave, Matlab, ..., are *interpreted* line by line. These languages can be simple in their use, but they can be prohibitively slow when it comes to a numerically demanding program. A compiler is a tool that analyzes *the whole program* and optimizes the computer instructions executed by the computer. But if programming time is more valuable, then a simple, interpreted language can lead to faster results.

Another choice that we make in this book, and we mention it because it is not the default in most Linux distributions, is the choice of shell. The shell is a program that "connects" the user to the operating system. In this book, we will teach how to use a shell[2] to "send" commands to the operating system, which is the most effective way to perform complicated tasks. We will use the shell `tcsh`, although most of the commands can be interpreted by most popular shells. Shell scripting is simpler in this shell, although shells like bash provide more powerful tools, mostly needed for complicated system administration tasks. That may cause a small inconvenience to some readers, since `tcsh` is not preinstalled in Linux distributions[3].

1.1 The Operating System

The Unix family of operating systems offer an environment where complicated tasks can be accomplished by *combining* many different tools,

---

[2]It is more popular to be called "the command line", or the "terminal", or the "console", but in fact the user interaction is through a shell.

[3]See www.tcsh.org. On Debian like systems, like Ubuntu, installation is very simple through the software center or by the command `sudo apt-get install tcsh`.

1.1. THE OPERATING SYSTEM

each of which performs a distinct task. This way, one can use the power of each tool, so that trivial but complicated parts of a calculation don't have to be programmed. This makes the life of a researcher much easier and much more productive, since research requires from us to try many things before we understand how to compute what we are looking for.

In the Unix operating system everything is a *file*, and files are organized in a unique and unified *filesystem*. Documents, pictures, music, movies, executable programs are files. But also directories or devices, like hard disks, monitors, mice, sound cards etc, are, from the point of view of the operating system, files. In order for a music file to be played by your computer, the music data needs to be written to a device file, connected by the operating system to the sound card. The characters you type in a terminal are read from a file "the keyboard", and written to a file "the monitor" in order to be displayed. Therefore, the first thing that we need to understand is the structure of the Unix filesystem.

1.1.1 Filesystem

There is at least one *path* in the filesystem associated with each file. There are two types of paths, *relative paths* and *absolute paths*. These are two examples:

```
bin/RungeKutta/rk.exe
/home/george/bin/RungeKutta/rk.exe
```

The paths shown above may refer to the same or a different file. This depends on "where we are". If "we are" in the directory /home/george, then both paths refer to the same file. If on the other way "we are" in a directory /home/john or /home/george/CompPhys, then the paths refer[4] to two different files. In the last two cases, the paths refer to the files

```
/home/john/bin/RungeKutta/rk.exe
/home/george/CompPhys/bin/RungeKutta/rk.exe
```

respectively. How can we tell the difference? An absolute path always begins with the / character, whereas a relative path does not. When we

---

[4] Some times two or more paths refer to the same file, or as we say, a file has two or more "links" in the same filesystem, but let's keep it simple for the moment.

say that "we are in a directory", we refer to a position in the filesystem called the *current directory*, or *working directory*. Every process in the operating system has a unique current directory associated with it.

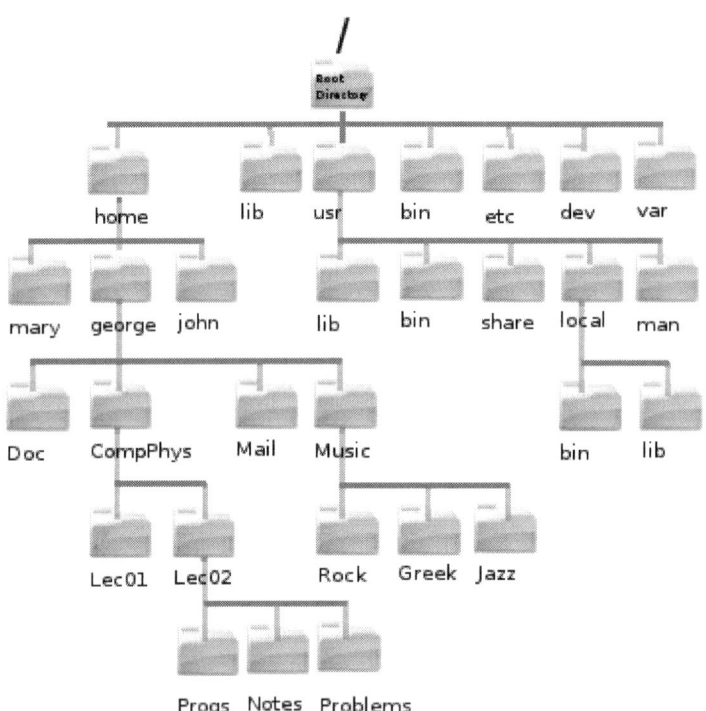

Figure 1.1: The Unix filesystem. It looks like a tree, with the *root directory* / at the top and branches that connect directories with their parents. Every directory contains files, among them other directories called its *subdirectories*. Every directory has a unique *parent directory*, noted by .. (double dots). The parent of the root directory is itself.

The filesystem is built on its *root* and looks like a tree positioned upside down. The symbol of the root is the character / The root is a *directory*. Every directory is a file that contains a list of files, and it is connected to a unique directory, its *parent directory* . Its list of files contains other directories, called its *subdirectories*, which all have it as their parent directory. All these files are the *contents* of the directory. Therefore, the filesystem is a tree of directories with the root directory at its top which branch to its subdirectories, which in their turn branch

1.1. THE OPERATING SYSTEM

into other subdirectories and so on. There is practically no limit to how large this tree can become, even for a quite demanding environment[5].

A path consists of a string of characters, with the characters / separating its *components*, and refers to a unique location in the filesystem. Every component refers to a file. All, but the last one, must be directories in a hierarchy, from parent directory to subdirectory. The only exception is a possible / in the beginning, which refers to the root directory. Such an example can be seen in figure 1.1.

In a Unix filesystem there is complete freedom in the choice of the location of the files[6]. Fortunately, there are some universally accepted conventions respected by almost everyone. One expects to find *home directories* in the directory /home, configuration files in the directory /etc, application executables in directories with names such as /bin, /usr/bin, /usr/local/bin, software libraries in directories with names such as /lib, /usr/lib etc.

There are some important conventions in the naming of the paths. A single dot "." refers to the current directory and a double dot ".." to the parent directory. Similarly, a tilde "~" refers to the home directory of the user. Assume, e.g., that we are the user george running a process with a current directory /home/george/Music/Rock (see figure 1.1). Then, the following paths refer to the same file /home/george/Doc/lyrics.doc:

```
../../Doc/lyrics.doc
~/Doc/lyrics.doc
~george/Doc/lyrics.doc
./../../Doc/lyrics.doc
```

Notice that ~ and ~george refer to the home directory of the user george (ourselves), whereas ~mary refer to the home directory of another user, mary.

---

[5]Of course, the capacity of the filesystem is finite, issue the command "df -i ." in order to see the number of inodes available in your filesystem. Every file corresponds to one and only one inode of the filesystem. Every path is mapped to a unique inode, but an inode maybe pointed to by more than one paths.

[6]This gives a great sense of freedom, but historically this was a important factor that led the Unix operating systems, although superior in quality, not to win a fair share of the market! The Linux family tries to keep things simple and universal to a large extent, but one should be aware that because of this freedom files in different version of Linuxes or Unices can be in different places.

We are now going to introduce the basic commands for filesystem navigation and manipulation[7]. The command cd (=change directory) changes the current directory, whereas the command pwd (=print working directory) prints the current directory:

```
> cd /usr/bin
> pwd
/usr/bin
> cd /usr/local/lib
> pwd
/usr/local/lib
> cd
> pwd
/home/george
> cd -
> pwd
/usr/local/lib
> cd ../../
> pwd
/usr
```

The argument of the command cd is an absolute or a relative path. If the path is correct and we have the necessary permissions, the command changes the current directory to this path. If no path is given, then the current directory changes to the home directory of the user. If the character - is given instead of a path, then the command changes the current directory to the previous current directory.

The command mkdir creates new directories, whereas the command rmdir removes empty directories. Try:

```
> mkdir new
> mkdir new/01
> mkdir new/01/02/03
mkdir: cannot create directory 'new/01/02/03': No such file or
      directory
> mkdir -p new/01/02/03
> rmdir new
rmdir: 'new': Directory not empty
> rmdir new/01/02/03
```

---

[7]Remember that lines that begin with the > character are commands. All other lines refer to the output of the commands.

1.1. THE OPERATING SYSTEM

```
> rmdir new/01/02
> rmdir new/01
> rmdir new
```

Note that the command `mkdir` cannot create directories more than one level down the filesystem, whereas the command `mkdir -p` can. The "switch" -p makes the behavior of the command different than the default one.

In order to list the contents of a directory, we use the command `ls` (=list):

```
> ls
BE.eps    Byz.eps   Programs        srBE_xyz.eps   srB_xyz.eps
B.eps     Bzy.eps   srBd_xyz.eps    srB_xy.eps
> ls Programs
Backup              rk3_Byz.cpp   rk3.cpp
plot-commands       rk3_Bz.cpp    rk3_g.cpp
```

The first command is given without an argument and it lists the contents of the current directory. The second one, lists the contents of the subdirectory of the current directory Programs. If the argument is a list of paths pointing to regular files, then the command prints the names of the paths. Another way of giving the command is

```
total 252
-rw-r--r--  1 george users 24284 May  1 12:08 BE.eps
-rw-r--r--  1 george users 22024 May  1 11:53 B.eps
-rw-r--r--  1 george users 29935 May  1 13:02 Byz.eps
-rw-r--r--  1 george users 48708 May  1 12:41 Bzy.eps
drwxr-xr-x  4 george users  4096 May  1 23:38 Programs
-rw-r--r--  1 george users 41224 May  1 22:56 srBd_xyz.eps
-rw-r--r--  1 george users 23187 May  1 21:13 srBE_xyz.eps
-rw-r--r--  1 george users 24610 May  1 20:29 srB_xy.eps
-rw-r--r--  1 george users 23763 May  1 20:29 srB_xyz.eps
```

The switch -l makes `ls` to list the contents of the current directory together with useful information on the files in 9 columns. The first column lists the *permissions* of the files (see below). The second one lists the number of links of the files[8]. The third one lists the user who is the *owner* of

---

[8]For a directory it means the number of its subdirectories plus 2 (the parent directory

each file. The fourth one lists the group that is assigned to the files. The fifth one lists the size of the file in bytes (=8 bits). The next three ones list the modification time of the file and the last one the paths of the files.

File permissions[9] are separated in three classes: owner permissions, group permissions and other permissions. Each class is given three specific permissions, r=**r**ead, w=**w**rite and x=e**x**ecute. For regular files, read permission effectively means access to the file for reading/copying, write permission means permission to modify the contents of the file and execute permission means permission to execute the file as a command[10]. For directories, read permission means that one is able to read the names of the files in the directory (but not make it as current directory with the cd command), write permission means to be able to modify its contents (i.e. create, delete, and rename files) and execute permission grants permission to access/modify the contents of the files (but not list the names of the files, this is granted by the read permission).

The command ls -l lists permissions in three groups. The owner (positions 2-4), the group (positions 5-7) and the rest of the world (others - positions 8-10). For example

```
-rw-r--r--
-rwxr-----
drwx--x--x
```

In the first case, the owner has read and write but not execute permissions and the group+others have only read permissions. In the second case, the user has read, write and execute permissions, the group has read permissions and others have no permissions at all. In the last case, the user has read, write and execute permissions, whereas the group and the world have only execute permissions. The first character d indicates a special file, which in this case is a **d**irectory. All special files have this position set to a character, while regular files have it set to -.

File permissions can be modified by using the command chmod:

---

and itself). For a regular file, it shows how many paths in the filesystem point to this file.

[9]See the "File system permissions" entry in en.wikipedia.org.

[10]Of course it is the user's responsibility to make sure the file with execute permission is actually a program that is possible to execute. An error results if this is not the case.

1.1. THE OPERATING SYSTEM

```
> chmod u+x   file
> chmod og-w  file1 file2
> chmod a+r   file
```

Using the first command, the owner (u≡ user) obtains (+) permission to execute (x) the file named `file`. Using the second one, the rest of the world (o≡ others) and the group (g≡group) loose (-) the write (w) permission to the files named `file1` and `file2`. Using the third one, everyone (a≡all) obtain read (r) permission on the file named `file`.

We will close this section by discussing some commands which are used for administering files in the filesystem. The command cp (copy) copies the contents of files into other files:

```
> cp file1.cpp file2.cpp
> cp file1.cpp file2.cpp file3.cpp Programs
```

If the file `file2.cpp` does not exist, the first command copies the contents of `file1.cpp` to a new file `file2.cpp`. If it already exists, it replaces its contents by the contents of the file `file2.cpp`. In order for the second command to be executed, `Programs` needs to be a directory. Then, the contents of the files `file1.cpp`, `file2.cpp`, `file3.cpp` are copied to indentical files in the directory `Programs`. Of course, we assume that the user has the appropriate privileges for the command to be executed successfully.

The command mv "moves", or renames, files:

```
> mv file1.cpp file2.cpp
> mv file1.cpp file2.cpp file3.cpp Programs
```

The first command renames the file `file1.cpp` to `file2.cpp`. The second one moves files `file1.cpp`, `file2.cpp`, `file3.cpp` into the directory `Programs`.

The command rm (remove) deletes files[11]. Beware, the command is unforgiving: after deletion, a file cannot be restored into the filesystem[12].

---

[11]Actually it removes "links" from files. A file may have more than one links in the same partition of a filesystem. A file is deleted when its last link is removed.

[12]This does not mean that its contents have been deleted from the disk. Deletion means marking for overwriting. Until the data is overwritten it can be recovered by the

Therefore, after executing successfully the following commands

```
> ls
file1.cpp   file2.cpp   file3.cpp   file4.csh
> rm file1.cpp file2.cpp file3.cpp
> ls
file4.csh
```

the files file1.cpp, file2.cpp, file3.cpp do not exist in the filesystem anymore. A more prudent use of the command demands the flag -i. Then, before deletion we are asked for confirmation:

```
> rm -i *
rm: remove regular file 'file1.cpp'? y
rm: remove regular file 'file2.cpp'? y
rm: remove regular file 'file3.cpp'? y
rm: remove regular file 'file4.csh'? n
> ls
file4.csh
```

When we type y, the file is deleted, when we type n, the file is not deleted.

We cannot remove directories the same way. It is possible to use the command *rmdir* in order to remove *empty* directories. In order to delete directories together with their contents (including subdirectories and their contents) use the command[13] rm -r. For example, assume that the contents of the directories dir1 and dir1/dir2 are the files:

```
./dir1
./dir1/file2.cpp
./dir1/file1.cpp
./dir1/dir2
./dir1/dir2/file3.cpp
```

Then the results of the following commands are:

```
> rm dir1
rm: cannot remove 'dir1': Is a directory
> rm dir1/dir2
```

---

use of special tools. Shredding sensitive data can be tricky business...

[13] A small mistake, like rm -rf * and your data is ... history!

1.1. THE OPERATING SYSTEM

```
rm: cannot remove 'dir1/dir2': Is a directory
> rmdir dir1
rmdir: dir1: Directory not empty
> rmdir dir1/dir2
rmdir: dir1/dir2: Directory not empty
> rm -r dir1
```

The last command removes all files (assuming that we have write permissions for all directories and subdirectories). Alternatively, we can empty the contents of all directories first, and then remove them with the command rmdir:

```
> cd dir1/dir2 ; rm file3.cpp
> cd .. ; rmdir dir2
> rm file1.cpp file2.cpp
> cd .. ; rmdir dir1
```

Note that by using a semicolon, we can execute two or more commands on the same line.

1.1.2 Commands

Commands in a Unix operating system are files with execute permission. When we write a sentence on the command line, like

```
> ls -l test.cpp test.dat
```

the shell reads its and interprets it. The shell is a program that creates a interface between a user and the operating system. The first word (ls) of the sentence is interpreted as a command. The rest of the words are the *arguments* of the command and the program can use them (or not) at the discretion of its programmer. There is a special convention for arguments that begin with a - (e.g. -l, --help, --version, -O3). They are called *options* or *switches*, and they act as virtual switches that make the program act in a particular way. We have already seen that the program ls gives a different output with the switch -l.

In order for a command to be executed, the shell looks for a file that has the same name as the command (here a file named ls). In order to understand where the shell looks for such a file, we should digress a little bit and explain the use of *shell variables* and *environment variables*. These have a name, which is a string of permissible characters, and their values are obtained by preceding their name with the $ character. For example the variable PATH has value $PATH. The values of the environment variables can be set with the command[14] setenv and of the shell variables with the command set:

```
> setenv MYVAR test-env
> set myvar = test-shell
> echo $MYVAR $myvar
test-env test-shell
```

Two special variables are the variables PATH and path:

```
>echo $PATH
/usr/local/bin:/usr/bin:/bin:/usr/X11/bin
>echo $path
/usr/local/bin /usr/bin /bin /usr/X11/bin
```

The first one is an environment variable and the second one is a shell variable. Their values are set by the shell, and we don't need to worry about them, unless we want to change them. Their value is a string of characters whose components should be valid paths to directories. In the first case, the components are separated by a :, while in the second case, by one or more spaces. In the example shown above, the shell searches each component of the path or PATH variables (in this order) until it finds a file ls in their contents. If it succeeds and the file has execute permissions, then the program in this file is executed. If it fails, then it prints an error message. Try the commands:

```
> which ls
/bin/ls
> ls -l /bin/ls
-rwxr-xr-x 1 root root 93560 Sep 28  2006 /bin/ls
```

We see that the program that the ls command executes the program in the file /bin/ls.

---

[14]The command setenv is special to the tcsh shell. For example the bash shell uses the syntax MYVAR=test-env in order to set the value of an environment variable.

1.1. THE OPERATING SYSTEM

The arguments of a command are passed on to the program that the command executes for possible interpretation. For example:

```
> ls -l test.cpp test.dat
```

The argument -l is the switch that results in a long listing of the files. The arguments test.cpp and test.dat are interpreted by the program ls as paths that it will look up for file information.

You can use the * (wildcard) character as a shorthand notation for a group of files. For example, in the command shown below

```
> ls -l *.cpp *.dat
```

the shell will expand *.cpp and *.dat to a list of all files whose names end with .cpp or .dat. Therefore, if the current directory contains the files test.cpp, test1.cpp, myprog.cpp, test.dat, hello.dat, the arguments that will be passed on to the command ls are

```
> ls -l myprog.cpp test1.cpp test.cpp hello.dat test.dat
```

For each command there are three special files associated with it. The first one is the *standard input* (stdin), the second one is the *standard output* (stdout) and the third one the *standard error* (stderr). These are files where the program can print or read data from. By default, these files are the terminal that the user uses to execute the command. In this case, when the program reads data from the stdin, then it reads the data that we type to the terminal using the keyboard. When the program writes data to the stdout or to the stderr, then the data is written to the terminal.

The advantage of using these special files in order to read/write data is that the user can *redirect* the input/output to these files to any file she wants. Using the character > at the end of a command redirects the stdout to the file whose name is written after >. For example:

```
> ls
file1.cpp   file2.cpp   file3.cpp   file4.csh
> ls > results
> ls
```

```
file1.cpp   file2.cpp   file3.cpp   file4.csh   results
```

The first of the above commands, prints the contents of the current working directory to the terminal. The second command redirects data written to the stdout to the file results. After executing the command, the file results is created and its contents are the names of the files file1.cpp file2.cpp file3.cpp file4.csh. If the file results does not exist (as in the above example), the file is created. If it already exists, it is *truncated* and its contents replaced by the data written to the stdout of the command. If we want to *append* data without erasing the existing contents, then we should use the string of characters >>. Therefore, if we give the command

```
> ls >> results
```

after executing the previous commands, then the contents of the file results will be

```
file1.cpp   file2.cpp   file3.cpp   file4.csh
file1.cpp   file2.cpp   file3.cpp   file4.csh   results
```

The redirection of the stdin is accomplished by the use of the character < while that of the stderr by the use of the string of characters[15] >&. We will see more examples in section 1.2.

It is possible to redirect the stdout of a command to be the stdin of another command. This is very useful for creating *filters*. A filter is a command that creates a flow of data between two or more programs. This process is called *piping*. Pipes are creating by using the character |

```
> cmd1 | cmd2 | cmd3 | ... | cmdN
```

Using the syntax shown above, the stdout of the command cmd1 is redirected to the stdin of the command cmd2, the stdout of the command cmd2 is redirected to the stdin of the command cmd3 etc. More examples will be presented in section 1.2.

---
[15]This syntax is particular to the tcsh shell. For other shells (bash, sh, ...) read their documentation.

1.1. THE OPERATING SYSTEM

1.1.3 Looking for Help

Unix got itself a reputation for not being user friendly. This is far from the truth. Although there is a steep learning curve, detailed documentation for almost everything is available online.

The key for a comfortable ride is to learn how to use the help system available on your computer and on the internet. Most of the commands are self documented. A simple test, like the one shown below, will help you with the basic usage of most of the commands:

```
> cmd --help
> cmd -h
> cmd -help
> cmd -\?
```

For example, try the command `ls --help`. For a window application, start from the menu "Help". You should not be afraid and/or lazy and you should proceed with careful searching and reading.

For example, let's assume that you have heard about a command that sounds like `printf`, or something like that. The first level of online help is the `man` (=manual) command that searches the "man pages". Read the output of the command

```
> man printf
```

The command `info` usually provides more detailed and user friendly documentation. It has basic browsing capabilities like the browsers you use to read pages from the internet. Try the command

```
> info printf
```

Furthermore, the commands

```
> man -k printf
> whatis printf
```

will inform you that there are other, possibly related, commands with names like `fprintf`, `fwprintf`, `wprintf`, `sprintf`...:

```
> whatis printf
printf                  (1)   - format and print data
printf                  (1p)  - write formatted output
printf                  (3)   - formatted output conversion
printf                  (3p)  - print formatted output
printf [builtins]       (1)   - bash built-in commands, see bash↵
    (1)
```

The second column printed by the whatis command is the "section" of the man pages. In order to gain access to the information in a particular section, you have to give it as an argument to the man command:

```
> man 1  printf
> man 1p printf
> man 3  printf
> man 3p printf
> man bash
```

Section 1 of the man pages contains information of ordinary command line commands, section 3 contains information on functions in libraries of the C language. Section 2 contains information on commands used for system administration. You may browse the directory /usr/share/man, or read the man page of the man command (use the command man man for that!).

By using the command

```
> printf --help
```

we obtain plenty of memory refreshing information. The command

```
> locate printf
```

shows us many files related to the command printf. The commands

```
> which printf
> where printf
```

give information on the location of the executable(s) of the command printf.

1.2. TEXT PROCESSING TOOLS – FILTERS

Another useful feature of the shell is the *command* or it filename completion. This means that we can write only the first characters of the name of a command or filename and then press simultaneously the keys [Ctrl-d][16] (i.e. press the key Ctrl and the key of the letter d at the same time). Then the shell will complete the name of the command up to the point that is is unique with the given string of characters[17]:

```
> pri[Ctrl-d]
printafm     printf  printenv  printnodetest
```

Try to type an x on the command line and then type [Ctrl-d]. You will learn all the commands that are available and whose name begins with an x: xterm, xeyes, xclock, xcalc, ...

Finally, the internet contains a wealth of information. Google your blues... and you will be rewarded!

1.2 Text Processing Tools – Filters

For doing data analysis, we will need powerful tools for manipulating data in text files. These are files that consist solely of printable characters. Some tools that can be used in order to construct complicated and powerful filters are the programs cat, less, head, tail, grep, sort and awk.

Suppose that we have data in a file named data[18] which contains information on the contents of a food warehouse and their prices:

```
bananas    100 pieces  1.45
apples     325 boxes   1.18
pears       34 kilos   2.46
bread       62 kilos   0.60
ham         85 kilos   3.56
```

---

[16] If you use the bash shell press [Tab] once or twice.

[17] Use the same procedure to auto-complete the names of files in the arguments of commands.

[18] The particular file, as well as most of the files in this section, can be found in the accompanying software of the chapter. It is highly recommended that you try all the commands in this section by using all the provided files.

The command

```
> cat data
```

prints the contents of the file data to the stdout. In general, this command prints the contents of all files given in its arguments or the stdin if none is given. Since the stdin and the stdout can be redirected, the command

```
> cat < data > data1
```

takes the contents of the file data from the stdin and prints them to the stdout, which in this case is the file data1. This command has the same result as the command:

```
> cp data data1
```

The command

```
> cat data data1 > data2
```

prints the contents of the file data and then the contents of the file data1 to the stdout. Since the stdout is redirected to the file data2, data2 contains the data of both files.

By giving the command

```
> less gfortran.txt
```

you can browse the data contained in the file gfortran.txt one page at a time. Press [space] in order to "turn" a page, [b] to turn back a page. Press the up and down arrows to move one line backwards/forward. Press [g] in order to jump to the beginning of the file and press [G] in order to jump to the end. Press [h] in order to get a help message and press [q] in order to quit.

The commands

```
> head -n 1 data
```

1.2. TEXT PROCESSING TOOLS – FILTERS

```
bananas    100 pieces  1.45
> tail -n 2 data
bread       62 kilos   0.60
ham         85 kilos   3.56
> tail -n 2 data | head -n 1
bread       62 kilos   0.60
```

print the first line, the last two lines and the second to the last line of the file data to the stdout respectively. Note that, by piping the stdout of the command tail to the stdin of the command head, we are able to construct the filter "print the line before the last one".

The command sort sorts the contents of a file by comparing each line of its text with all others. The sorting is alphabetical, unless otherwise set by using options. For example

```
> sort data
apples     325 boxes   1.18
bananas    100 pieces  1.45
bread       62 kilos   0.60
ham         85 kilos   3.56
pears       34 kilos   2.46
```

For reverse sorting, try sort -r data. We can also sort by comparing specific *fields* of each line. By default, fields are words separated by one or more spaces. For example, in order to sort w.r.t. the second column of the file data, we can use the switch -k 2 (=second field). Furthermore, we can use the switch -n for numerical sorting:

```
> sort -k 2 -n data
pears       34 kilos   2.46
bread       62 kilos   0.60
ham         85 kilos   3.56
bananas    100 pieces  1.45
apples     325 boxes   1.18
```

If we omit the switch -n, the comparison of the lines is performed based on character sorting of the second field and the result is

```
> sort -k 2 data
bananas    100 pieces  1.45
apples     325 boxes   1.18
```

```
pears       34 kilos   2.46
bread       62 kilos   0.60
ham         85 kilos   3.56
```

The last column contains floating point numbers (not integers). In order to sort by the values of such numbers we should use the switch -g:

```
> sort -k 4 -g data
bread       62 kilos   0.60
apples     325 boxes   1.18
bananas    100 pieces  1.45
pears       34 kilos   2.46
ham         85 kilos   3.56
```

The command grep processes a text file line by line, searching for a given string of characters. When this string is found anywhere in a line, this line is printed to the stdout. The command

```
> grep kilos data
pears       34 kilos   2.46
bread       62 kilos   0.60
ham         85 kilos   3.56
```

prints each line containing the string "kilos". If we want to search for all line *not* containing the string "kilos", then we add the switch -v:

```
> grep -v kilos data
bananas    100 pieces  1.45
apples     325 boxes   1.18
```

We can use a *regular expression* for searching a whole family of strings of characters. These monsters need a full book for discussing them in detail! But it is not hard to learn how to use some simple forms of regular expressions. Here are some examples:

```
> grep ^b data
bananas    100 pieces  1.45
bread       62 kilos   0.60
> grep '0$' data
bread       62 kilos   0.60
> grep '3[24]' data
```

1.2. TEXT PROCESSING TOOLS – FILTERS 21

```
apples    325 boxes   1.18
pears      34 kilos   2.46
```

The first one, prints each line whose *first* character is a b. The second one, prints each line that *ends* with a 0. The third one, prints each line contaning the strings 32 *or* 34.

By far, the strongest tool in our toolbox is the awk program. By default, awk analyzes a text file line by line. Each word (or *field* in the awk jargon) of these lines is stored in a set of variables with names $1, $2, ... The variable $0 contains the full line currently processed, whereas the variable NF counts the number of fields in the current line. The variable NR counts the number of lines of the file processed so far by awk.

An awk program can be written in the command line. A set of commands within { ... } is executed for each line of input. The constructs BEGIN{ ... } and END{ ... } contain commands executed, only once, before and after the processing of the file respectively. For example, the command

```
> awk '{print $1,"total value= ",$2*$4}' data
bananas total value=  145
apples  total value=  383.5
pears   total value=  83.64
bread   total value=  37.2
ham     total value=  302.6
```

prints the name of the product (1st column = $1) and the total value stored in the warehouse (2nd column = $2) × (4th column = $4). More examples are given below:

```
> awk '{value += $2*$4}END{print "Total= ",value}' data
Total=  951.94
> awk '{av += $4}END{print "Average Price= ",av/NR}' data
Average Price=  1.85
> awk '{print $2^2 * sin($4) + exp($4)}' data
```

The first one calculates the total value of all products: The processing of each line results in the increment (+=) of the variable value by the product of the second and fourth fields. In the end (END{ ... }), the string Total= is printed, together with the final value of the variable value. This is an easy way for computing the sum of the values calculated for each line. The second command, calculates and prints an average. The sum is calculated in each line and stored in the variable av. In the end, we print the quotient of the sum of all values by the number of lines that have been processed (NR). The last command shows a (crazy) mathematical expression based on numerical values found in each line of the file data: It computes the square of the second field times the sine of the fourth field plus the exponential of the fourth field.

There is much more potential in the commands presented above. Reading the documentation and getting experience by using them will provide you with very strong tools in order to accomplish complicated tasks.

1.3 Programming with Emacs

For a programmer that spends many hours programming every day, the environment and the tools available for editing the commands of a large and complicated program determine, to a large extent, the quality of her life! An editor edits the contents of a text file, that consists solely of printable characters. Such editors, available in most Linux environments, are the programs gedit, vim, pico, nano, zile... They provide basic functionality such as adding, removing or changing text within a file as well as more complicated functions, such as copying, pasting, searching and replacing text etc. There are many functions that are particularly useful to a programmer, such as detecting and formatting keywords of a particular programming language, pretty printing, closing scopes etc, which can be very useful for comfortable programming and for spotting errors. A very powerful and "knowledgeable" editor, offering many such functions for several programming languages, is the GNU Emacs editor[19]. Emacs is open source software, it is available for free and can be used in most available operating systems. It is programmable[20] and the user

---

[19] http://www.gnu.org/software/emacs/ (main site), http://www.emacswiki.org/ (expert tips), http://en.wikipedia.org/wiki/Emacs (general info)

[20] Emacs is written in a dialect of the programming language Lisp, called Elisp. There is no need of an in-depth knowledge of the language in order to program simple functions, just see how others are doing it...

1.3. PROGRAMMING WITH EMACS

can automate most of her everyday repeated tasks and configure it to her liking. There is a full interaction with the operating system, in fact Emacs has been built with the ambition of becoming an operating system. For example, a programmer can edit a C++ file, compile it, debug it and run it, everything done with Emacs commands.

1.3.1 Calling Emacs

In the command line type

```
> emacs &
```

Note the character & at the end of the line. This makes the particular command to run in the *background*. Without it, the shell waits until a command exits in order to return the prompt.

In a desktop environment, Emacs starts in its own window. For a quick and dirty editing session, or in the case that a windows environment is not available[21], we can run Emacs in a terminal mode. Then, we omit the & at the end of the line and we run the command

```
> emacs -nw
```

The switch -nw forces Emacs to run in terminal mode.

1.3.2 Interacting with Emacs

We can interact with Emacs in various ways. Newbies will prefer buttons and menus that offer a simple and intuitive interface. For advanced usage, however, we recommend that you make an effort to learn the keyboard shortcuts. There are also thousands of functions available to be used interactively. They are called from a "command line", called *the minibuffer* in the Emacs jargon.

Keyboard shortcuts are usually combinations of keystrokes that consist of the simultaneous pressing of the Ctrl or Alt keys together with other keys. Our convention is that a key sequence starting with a C- means that the characters that follow are keys simultaneously pressed

---
[21]Quite handy when we edit files in a remote computer.

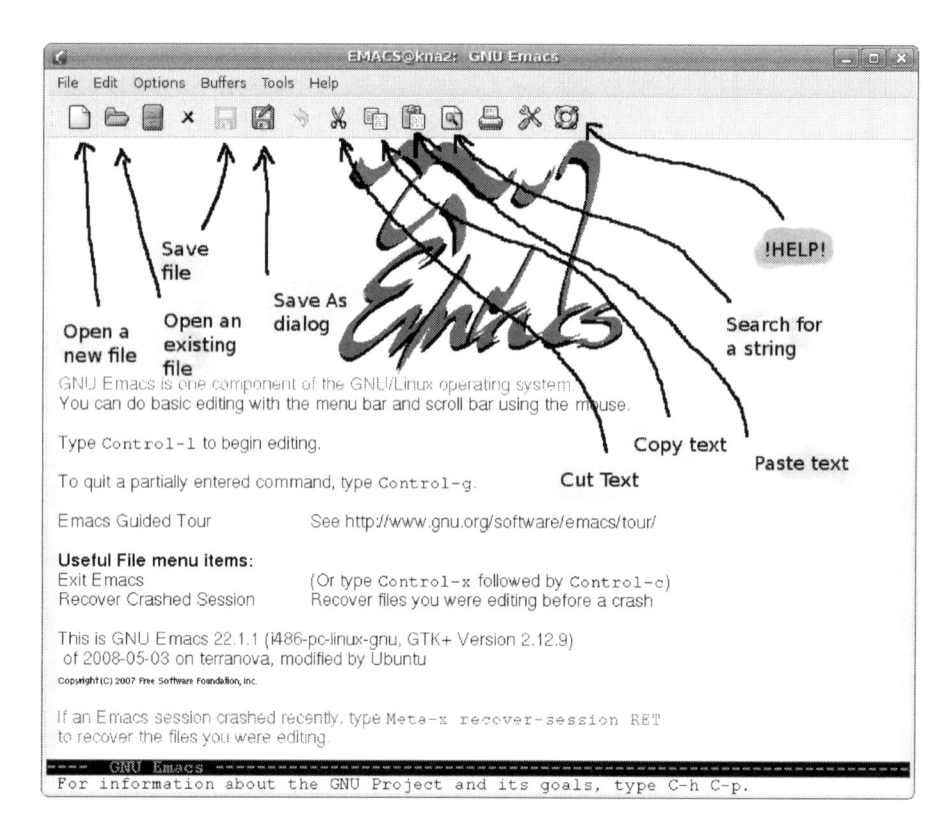

Figure 1.2: The Emacs window in a windows environment. The buttons of very basic functions found on its toolbar are shown and explained.

with the `Ctrl` key. A key sequence starting with a `M-` means that the characters that follow are keys simultaneously pressed with the `Alt` key[22]. Some commands have shortcuts consisting of two or more composite keystrokes. For example by `C-x C-c` we mean that we have to press simultaneously the `Ctrl` key together with x and *then* press simultaneously the `Ctrl` key together with c. This sequence is a shortcut to the command that exits Emacs. Another example is `C-x 2` which means to press the `Ctrl` key together with x and then press only the key 2. This is a shortcut to the command that splits a window horizontally to two equal parts.

The most useful shortcuts are `M-x` (press the `Alt` key siumutaneously

---

[22]Actually, `M-` is the so called `Meta` key, usually bound to the `Alt` key. It is also bound to the `Esc` and `C-[` keys. The latter can be our *only* choices available in dumb terminals.

1.3. PROGRAMMING WITH EMACS

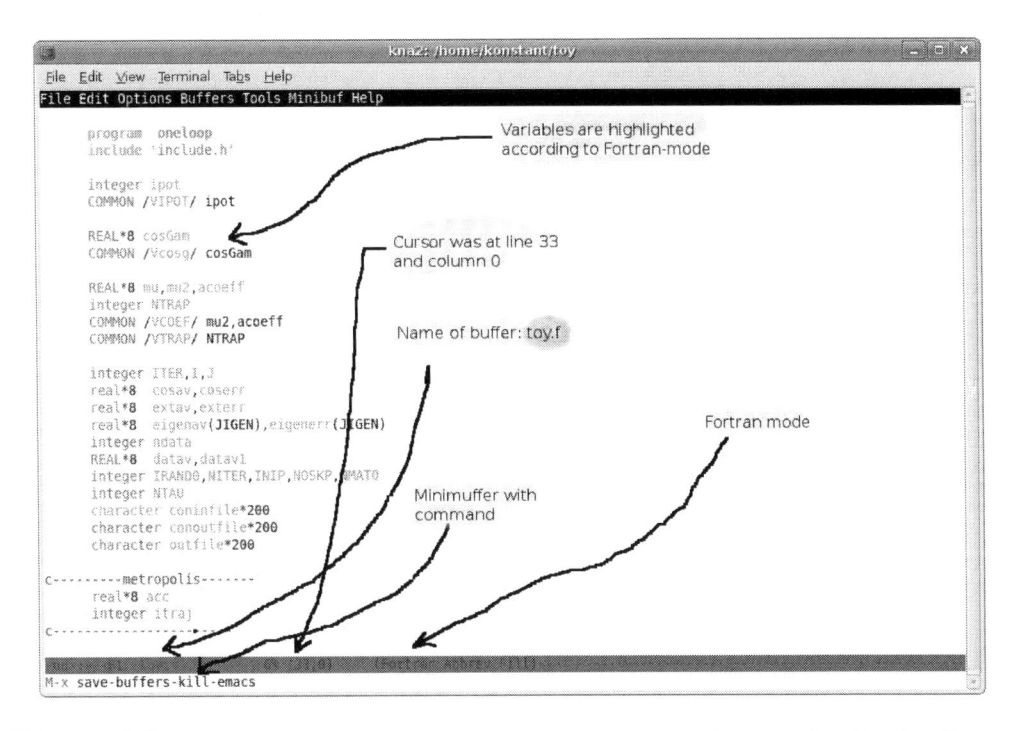

Figure 1.3: Emacs in a non-window mode running on the console. In this figure, we have typed the command save-buffers-kill-emacs in the minibuffer, a command that exits Emacs after saving edited data from all buffers. The same command can be given using the keyboard shortcut C-x C-c. We can see the *mode line* and the name of the buffer toy.f written on it, the percentage of the buffer (6%) shown in the window, the line and columns (33,0) where the point lies and the editing *mode* which is active on the buffer (Fortran mode (Fortran), Abbreviation mode (Abbrev), Auto Fill mode (Fill)).

with the x key) and C-g. The first command takes us to the minibuffer where we can give a command by typing its name. For example, type M-x and then type save-buffers-kill-emacs in the minibuffer (this will terminate Emacs). The second one is an "SOS button" that interrupts anything Emacs does and returns control to the working buffer. This can be pretty handy when a command hangs or destroys our work and we need to interrupt it.

The conventions for the mouse events are as follows: With Mouse-1, Mouse-2 and Mouse-3 we denote a simple click with the left, middle and right buttons of the mouse respectively. With Drag-Mouse-1 we mean to press the left button of the mouse and at the same time drag the mouse.

Figure 1.4: The basic menus found in Emacs when run in a desktop environment. We can see the basic commands and the keyboard shortcut reminders in the parentheses. E.g. the command File → Visit New File can be given by typing C-x C-f. Note the commands File → Visit New File (open a file), File→Save (write contents of a buffer to a file), File→Exit Emacs, File → Split Window (split window in two), File→New Frame (open a new Emacs desktop window) and of course the well known commands Cut, Copy, Paste, Undo from the Edit menu. We can choose different buffers from the menu Buffers, which contain the contents of other files that we have opened for editing. We recommend trying the Emacs Tutorial and Read Emacs Manual in the Help menu.

We summarize the possible ways of giving a command in Emacs with the following examples that have the same effect: Open a file and put its contents in a buffer for editing.

- By pressing the toolbar button that looks like a white sheet of paper (see figure 1.2).

- By choosing the File→Visit New File menu entry.

- By typing the keyboard shortcut C-x C-f.

- By typing the name of the command in the minibuffer: M-x find-file

1.3. PROGRAMMING WITH EMACS

The number of available commands increases from the top to the bottom of the above list.

1.3.3 Basic Editing

In order to edit a file, Emacs places the *contents* of a file in a *buffer*. Such a buffer is a chunk of computer memory where the contents of the file are copied and it is not the file itself. When we make changes to the contents of a buffer, the file remains intact. For our changes to take effect and be written to the file, we have to save the buffer. Then, the contents of the buffer are written back to the file. It is important to understand the following cycle of events:

- Read a file's contents to a buffer.

- Edit buffer contents.

- Write (save) buffer's contents back into the file.

Emacs may have more than one buffers open for editing simultaneously. By default, the name of the buffer is the same as the name of the file that is edited, although this is not necessary[23]. The name of a buffer is written in the modeline of the window of the buffer, as can be seen in figure 1.3.

If Emacs crashes or exits before we save our edits, it is possible to recover (part of) them. There is a command M-x recover-file that will guide us through the necessary recovery steps, or we can look for a file that has the same name as the buffer we were editing surrounded by two #. For example, if we were editing the file file.cpp, the automatically saved changes can be found in the file #file.cpp#. Auto saving is done periodically by Emacs and its frequency can be controlled by the user.

The point where we insert text while editing is called "the point". This is right before the blinking cursor[24]. Each buffer has another position marked by "the mark". A point and the mark define a "region"

---

[23]The user can change the name of the buffer without affecting the name of the file it edits. Also, if we open more than one files with the same name, emacs gives each buffer a unique name. E.g. if we edit more than one files named index.html then the corresponding buffers are named index.html, index.html<2>, index.html<3>, ...

[24]Strictly speaking, the point lies between two characters and not on top of a character. The cursor lies on the character immediately to the right of the point. A point is assigned

in the buffer. This is a part of the text in the buffer where the functions of Emacs can act (e.g. copy, cut, change case, spelling etc.). We can set the region by setting a point and then press C-SPC[25] or give the command M-x set-mark-command. This defines the current point to be the mark. Then we can move the cursor to another point which will define a region together with the mark that we set. Alternatively we can use Drag-Mouse-1 (hold the left mouse button and drag the mouse) and mark a region. The mark can be set with Mouse-3, i.e. with a simple click of the right button of the mouse. Therefore by Mouse-1 at a point and then Mouse-3 at a different point will set the region between the two points.

We can open a file in a buffer with the command C-x C-f, and then by typing its path. If the file already exists, its contents are copied to a buffer, otherwise a new buffer is created. Then:

- We can browse the buffer's contents with the Up/Down/Left/Right arrows. Alternatively, by using the commands C-n, C-p, C-f and C-b.

- If the buffer is large, we can browse its contents one page at a time by using the Page Up/Page Dn keys. Alternatively, by using the commands C-v, M-v.

- Enter text at the points simply by typing it.

- Delete characters before the point by using the Backspace key and after the point by using the Delete key. The command C-d deletes a forward character.

- Erase all the characters in a line that lie ahead of the point by using the command C-k.

- Open a new line by using Enter or C-o.

- Go to the first character of a line by using Home and the last one by using End. Alternatively, by using the commands C-a and C-e, respectively.

---

to every *window*, therefore a buffer can have multiple points, one for each window that displays its contents.

[25]Press the Ctrl and spacebar keys simultanesouly.

1.3. PROGRAMMING WITH EMACS

- Go to the first character of the buffer with the key C-Home and the last one with the key C-End. Alternatively, with M-x beginning-of-buffer and M-x end-of-buffer.

- Jump to any line we want: Type M-x goto-line and then the line number.

- Search for text after the point: Press C-s and then the text you are looking for. This is an incremental search and the point jumps immediately to the first string that matches the search. The same search can be repeated by pressing C-s repeatedely.

When we finish editing (or frequently enough so that we don't loose our work due to an unfortunate event), we save the changes in the buffer, either by pressing the save icon on the toolbar, or by pressing the keys C-s, or by giving the command M-x save-buffer.

1.3.4 Cut and Paste

Use the instructions below for slightly more advanced editing:

- Undo! Some of the changes described below can be catastrophic. Emacs has a great Undo function that keeps in its memory many of the changes inflicted by our editing commands. By repeatedely pressing C-/, we undo the changes we made. Alternatively, we can use C-x u or the menu entry Edit→Undo. Remember that C-g interrupts any Emacs process currently running in the buffer.

- Cut text by using the mouse: Click with Mouse-1 at the point before the beginning of the text and then Mouse-3 at the point after the end. A second Mouse-3 and the region is ... gone (in fact it is written in the "kill ring" and it is available for pasting)!

- Cut text by using a keyboard shortcut: Set the mark by C-SPC at the point before the beginning of the text that you want to cut. Then move the cursor after the last character of the text that you want to cut and type C-w.

- Copy text by using the mouse: Drag the mouse Drag-Mouse-1 and mark the region that you want to copy. Alternatively, Mouse-1 at the point before the beginning of the text and then Mouse-3 at the point after the end.

- Copy text by using a keyboard shortcut: Set the mark at the beginning of the text with C-SPC and then move the cursor after the last character of the text. Then type M-w.

- Pasting text with the mouse: We click the middle button[26] Mouse-2 at the point that we want to insert the text from the kill ring (the copied text).

- Pasting text with a keyboard shortcut: We move the point to the desired insertion point and type C-y.

- Pasting text from previous copying: A fast choice is the menu entry Edit→Paste from kill manu and then select from the copied texts. The keyboard shortcut is to first type C-y and then M-y repeatedly, until the text that we want is yanked.

- Insert the contents of a file: Move the point to the desired place and type C-x i and the path of the file. Alternatively, give the command M-x insert-file.

- Insert the contents of a buffer: We can insert the contents of a whole buffer at a point by giving the command M-x insert-buffer.

- Replace text: We can replace text interactively with the command M-x query-replace, then type the string we want to replace, and then the replacement string. Then, we will be asked whether we want the change to be made and we can answer by typing y (yes), n (no), q (quit the replacements). A , (comma) makes only one replacement and quits (useful if we know that this is the last change that we want to make). If we are confident, we can change all string in a buffer, no questions asked, by giving the command M-x replace-string.

- Change case: We can change the case in the words of a region with the commands M-x upcase-region, M-x capitalize-region and M-x downcase-region. Try it.

---

[26]If it is a two button mouse, try clicking the left and right buttons simultaneously.

1.3. PROGRAMMING WITH EMACS

We note that cutting and pasting can be made between different windows of the same or different buffers.

1.3.5 Windows

Sometimes it is very convenient to edit one or more different buffers in two or more windows. The term "windows" in Emacs refers to regions of the same Emacs desktop window. In fact, a desktop window running an Emacs session is referred to as a *frame* in the Emacs jargon. Emacs can split a frame in two or more windows, horizontally or/and vertically. Study figure 1.5 on page 63 for details. We can also open a new frame and edit several buffers simultaneously[27]. We can manipulate windows and frames as follows:

- Position the point at the center of the window and clear the screen from garbage: C-l (careful: l not 1).

- Split a window in two, horizontally: C-x 2.

- Split a window in two, vertically: C-x 3.

- Delete all other windows (remain only with the current one): C-x 1.

- Delete the current windows (the others remain): C-x 0.

- Move the cursor to the other window: Mouse-1 or C-x o.

- Change the size of window: Use Drag-Mouse-1 on the line separating two windows (the mode line). Use C-^, C-} for making a change of the horizontal/vertical size of a window respectively.

- Create a new frame: C-x 5 2.

- Delete a frame: C-x 5 0.

- Move the cursor to a different frame: With Mouse-1 or with C-x 5 o.

---

[27] Be careful not to start a new Emacs session each time that all you need is a new frame. A new Emacs process takes time to start, binds computer resources and does not communicate with a different Emacs process.

You can have many windows in a dumb terminal. This is a blessing when a dekstop environment is not available. Of course, in that case you cannot have many frames.

1.3.6 Files and Buffers

- Open a file: `C-x C-f` or `M-x find-file`.

- Save a buffer: `C-x C-s` or `M-x save buffer`. With `C-x C-c` or `M-x save-buffers-kill-emacs` we can also exit Emacs. From the menu: File→Save. From the toolbar: click on the save icon.

- Save buffer contents to a different file: `C-x C-w` or `M-x write-file`. From the menu: File→Save As. From the toolbar: click on the "save as" icon.

- Save all buffers: `C-x s` or `M-x save-some-buffers`.

- Connect a buffer to a different file: `M-x set-visited-filename`.

- Kill a buffer: `C-x k`.

- Change the buffer of the current window: `C-x b`. Also, use the menu `Buffers`, then choose the name of the buffer.

- Show the list of all buffers: `C-x C-b`. From the menu: `Buffers → List All Buffers`. By typing `Enter` next to the name of the buffer, we make it appear in the window. There are several buffer administration commands. Learn about them by typing `C-h m` when the cursor is in the Bufer List window.

- Recover data from an edited buffer: If Emacs crashed, do not despair. Start a new Emacs and type `M-x recover-file` and follow the instructions. The command `M-x recover-session` recovers all unsaved buffers.

- Backup files: When you save a buffer, the previous contents of the file become a backup file. This is a file whose path is the same as the original's file with a ~ appended in the end. For example a file `test.cpp` will have as a backup the file `test.cpp~`. Emacs has

1.3. PROGRAMMING WITH EMACS

version control, and you can configure it to keep as many versions of your edits as you want.

- Directory browsing and directory administration commands: C-x d or M-x dired. You can act on the files of a directory (open, delete, rename, copy etc) by giving appropriate commands. When the cursor is in the dired window, type C-h m to read the relevant documentation.

1.3.7 Modes

Each buffer can be in different *modes*. Each mode may activate different commands or editing environment. For example each mode can color keywords relevant to the mode and/or bind keys to different commands. There exist *major modes*, and a buffer can be in only one of them. There are also *minor modes*, and a buffer can be in one or more of them. Emacs activates major and minor modes by default for each file. This usually depends on the filename but there are also other ways to control this. The user can change both major and minor modes at will, using appropriate commands.

Active modes are shown in a parenthesis on the mode line (see figures 1.3 and 1.5.

- M-x c++-mode: This mode is of special interest in this book since we will edit a lot of C++ code. We need it activated in buffers that contain a C++ program and its most useful characteristics are automatic code alignment by pressing the key TAB, the coloring of C++ statements, variables and other structural constructs (classes, if statements, for loops, variable declarations, comments etc). Another interesting function is the one that comments out a whole region of code, as well as the inverse function.

- M-x c-mode: For files containing programs written in the C language. Related modes are the java-mode, perl-mode, awk-mode, python-mode, makefile-mode, octave-mode, gnuplot-mode and others.

- latex-mode: For files containing LaTeX text formatting commands.

- text-mode: For editing simple text files (.txt).

- `fundamental-mode`: The basic mode, when one that fits better doesn't exist...

Some interesting minor modes are:

- `M-x auto-fill-mode`: When a line becomes too long, it is wrapped automatically. A related command to do that for the whole region is `M-x fill-region`, and for a paragraph `M-x fill-paragraph`.

- `M-x overwrite-mode`: Instead of inserting characters at the point, overwrite the existing ones. By giving the command several times, we toggle between activating and deactivating the mode.

- `M-x read-only mode`: When visiting a file with valuable data that we don't want to change by mistake, we can activate this mode so that changes will not be allowed by Emacs. When we open a file with the command `C-x C-r` or `M-x find-file-read-only` this mode is activated. We can toggle this mode on and off with the command `C-x C-q` (`M-x toggle-read-only`). See the mode line of the buffer `jack.c` in figure 1.5 which contains a string %%. By clicking on the %% we can toggle the `read-only` mode on and off.

- `flyspell-mode`: Spell checking as we type.

- `font-lock-mode`: Colors the structural elements of the buffer which are defined by the major mode (e.g. the commands of a C++ program).

In a desktop environment, we can choose modes from the menu of the mode line. By clicking with `Mouse-3` on the name of a mode we are offered options for (de)activating minor modes. With a `Mouse-1` we can (de)activate the `read-only` mode with a click on :%% or :-- respectively. See figure 1.5.

1.3.8 Emacs Help

Emacs' documentation is impressive. For newbies, we recommend to follow the mini course offered by the Emacs tutorial. You can start the tutorial by typing `C-h t` or select Help → Emacs Tutorial from the menu. Enjoy... The Emacs man page (give the man emacs command in

1.3. PROGRAMMING WITH EMACS

the command line) will give you a summary of the basic options when calling Emacs from the command line.

A quite detailed manual can be found in the Emacs info pages[28]. Using `info` needs some training, but using the Emacs interface is quite intuitive and similar to using a web browser. Type the command C-h r (or choose Help→Emacs Tutorial from the menu) and you will open the front page of the emacs manual in a new window. By using the keys SPC and Backspace we can read the documentation page by page. When you find a link (similar to web page hyperlinks), you can click on it in order to open to read the topic it refers to. Using the navigation icons on the toolbar, you can go to the previous or to the next pages, go up one level etc. There are commands that can be given by typing single characters. For example, type d in order to jump to the main `info` directory. There you can find all the available manuals in the `info` system installed on your computer. Type g (emacs) and go to the top page of the Emacs manual. Type g (info) and read the `info` manual.

Emacs is structured in an intuitive and user friendly way. You will learn a lot from the names of the commands: Almost all names of Emacs commands consist of whole words, separated by a hyphen "-", which almost form a full sentence. These make them quite long sometimes, but by using auto completion of their names this does not pose a grave problem.

- auto completion: The names of the commands are auto completed by typing a TAB one or more times. E.g., type M-x in order to go to the minibuffer. Type `capi[TAB]` and the command autocompletes to `capitalize-`. By typing [TAB] for a second time, a new window opens and offers the options for completing to two possible commands: `capitalize-region` and `capitalize-word`. Type an extra `r[TAB]` and the command auto completes to the only possible choice `capitalize-region`. You can see all the commands that start with an s by typing M-x `s[TAB][TAB]`. Sure, there are many... Click on the *Completions* buffer and browse the possibilities. A lot will become clear just by reading the names of the commands. By typing M-x [TAB][TAB], all available commands will appear in your buffer!

---

[28] If you prefer books in the form of PDF visit the page www.gnu.org/software/emacs and click on Documentation. You will find a 600 page book that has almost everything!

- keyboard shortcuts: If you don't remember what happens when you type C-s, no problem: Type C-h k and then the ... forgotten key sequence C-s. Conversely, have you forgotten what is the keyboard shortcut of the command save-buffer? Type C-h w and then the command.

- functions: Are you looking for a command, e.g. save-something-I-forgot? Type C-h f and then save-[TAB] in order to browse over different choices. Use Mouse-2 in order to select the command you are interested in, or type and complete the rest of its name (you may use [TAB] again). Read about the function in the *Help* buffer that opens.

- variables: Do the same after typing C-h v in order to see a variable's value and documentation.

- command apropos: Have you forgotten the exact name of a command? No problem... Type C-h a and a keyword. All commands related to the keyword you typed will appear in a buffer. Use C-h d for even more information.

- modes: When in a buffer, type C-h m and read information about the active modes of the buffer.

- info: Type C-h i

- Have you forgotten everything mentioned above? Just type C-h ?

1.3.9 Emacs Customization

You can customize everything in Emacs. From key bindings to programming your own functions in the Elisp language. The most common way for a user to customize her Emacs sessions, is to put all her customization commands in the file ~/.emacs in her home directory. Emacs reads and executes all these commands just before starting a session. Such a .emacs file is given below:

```
; Define F1 key to save the buffer
(global-set-key [f1]    'save-buffer)
```

```
; Define Control-c s to save the buffer
(global-set-key "\C-cs" 'save-some-buffers)
; Define Meta-s (Alt-s) to interactively search forward
(global-set-key "\M-s" 'isearch-forward)
; Define M-x is         to interactively search forward
(defalias       'is     'isearch-forward)
; Define M-x cm         to set c++-mode for the buffer
(defun cm()     (interactive) (c++-mode))
; Define M-x sign       to sign my name
(defun sign()   (interactive) (insert "K. N. Anagnostopoulos"))
```

Everything after a ; is a comment. Functions/commands are enclosed in parentheses. The first three ones bind the keys F1, C-c s and M-s to the commands save-buffer, save-some-buffers and isearch-forward respectively. The next one defines an *alias* of a command. This means that, when we give the command M-x is in the minibuffer, then the command isearch-forward will be executed. The last two commands are the definitions of the functions (fm) and (sign), which can be called interactively from the minibuffer.

For more complicated examples google "emacs .emacs file" and you will see other users' .emacs files. You may also customize Emacs from the menu commands Options→Customize Emacs. For learning the Elisp language, you can read the manual "Emacs Lisp Reference Manual" found at the address
www.gnu.org/software/emacs/manual/elisp.html

1.4 The C++ Programming Language

In this section, we give a very basic introduction to the C++ programming language. This is not a systematic exposition and you are expected to learn what is needed in this book *by example*. So, please, if you have not already done it, get in front of a computer and *do* what you read. You can find many good tutorials and books introducing C++ in a more complete way in the bibliography.

1.4.1 The Foundation

The first program that one writes when learning a new programming language is the "Hello World!" program. This is the program that prints

"Hello World!" on your screen:

```
#include <iostream>
using namespace std;

int main(){

  // This is a comment.
  cout << "Hello World!\n";

}
```

Commands, or *statements*, in C++ are strings of characters separated by blanks ("words") and end with a semicolon (;). We can put more than one command on each line by separating them with a semicolon.

Everything after two slashes (//) is a comment. Proliferation of comments is necessary for documenting our code. Good documentation of our code is an integral part of programming. If we plan to have our code read by others (or by us) at a later time, we have to make sure to explain in detail what each line is supposed to do. You and your collaborators will save a lot of time in the process of debugging, improving and extending your code.

The first line of the code shown above is a *preprocessor directive*. These lines start with a # and are interpreted by a separate program, the preprocessor. The #include directive, inserts the contents of a file replacing the line where the directive is. This acts like an editor! Actually, the code that will be compiled is not the one shown above, but the result of adding the contents of a file whose name is iostream[29]. iostream is an example of a *header* file that has many definitions of functions and symbols used by the program. The particular header has the necessary definitions in order to perform standard input and standard output operations.

The execution of a C++ program starts by calling a *function* whose name is main(). Therefore, the line int main(){ shows how to actually define a function in C++. Its name is the word before the parentheses () and the keyword int specifies that the function returns a value of integer

---

[29]The path to the file is determined by the compiler. If you are curious to see the file, search for it with the command locate iostream. In order to see the result of adding the contents of the file (and, actually several other files added by preprocessor directives in iostream), call the preprocessor with the command cpp hello.cpp

1.4. THE C++ PROGRAMMING LANGUAGE

type[30]. Within the parentheses placed after the name of the function, we put the *arguments* that we pass to the function. In our case the parentheses contain nothing, showing how to define a function without arguments.

The curly brackets { ... } define the *scope* or the *body* of the function and contain the statements to be executed when the function is called.

The line

```
cout << "Hello World!\n";
```

is the only line that contains an executable statement that actually *does* something. Notice that it ends with a semicolon. This statement performs an output operation printing a *string* to the standard output. The sentence Hello World!\n is a constant string and contains a sequence of printable characters enclosed in double quotes. The last character \n is a newline character, that prints a new line to the stdout.

cout identifies the standard **c**haracter **out**put device, which gives access to the stdout. The characters << indicate that we write *to* cout the expression to the right. In order to make cout accessible to our program, we need both the inclusion of the header file iostream and the statement using namespace std[31].

Statements in C++ end with a semicolon. Splitting them over a number of lines is only a matter of making the code legible. Therefore, the following parts of the code have equivalent effect as the one written above:

```
int main()
{
  cout <<
    "Hello World!\n";
}
```

---

[30]The value returned by main() is useless within our program, but it is used by the operating system in order to inform us about the successful (or not) termination of the program. Of course, functions returning values is a very useful feature in general!

[31]Omitting using namespace std does not make cout inaccessible. One can use its "full name" std::cout instead. Remove the statement and try it. cout is part of the C++ *Standard Library*. All elements of the library belong to the std *namespace* and their names can be prefixed by std:: Using the using namespace std statement, the prefix may be omitted.

```
int main(){cout <<"Hello World!\n";}
```

Finally notice that, for C++, uppercase and lowercase characters are different. Therefore main(), Main() and MAIN() are names of *different* functions.

In order to execute the commands in a program, it is necessary to *compile* it. This is a job done by a program called the *compiler* that translates the human language programming statements into binary commands that can be loaded to the computer memory for execution. There are many C++ compilers available, and you should learn which compilers are available for use in your computing environment. Typical names for C++ compilers are g++, c++, icc, ... You should find out which compiler is best suited for your program and spend time reading its documentation carefully. It is important to learn how to use a new compiler so that you can finely tune it to optimize the performance of *your* program.

We are going to use the open source and freely available compiler g++, which can be installed on most popular operating systems[32]. The compilation command is:

```
> g++ hello.cpp -o hello
```

The extension .cpp to the file name hello.cpp is important and instructs the compiler that the file contains source code in C++. Use your editor and edit a file with the name hello.cpp with the program shown above before executing the above command.

The switch -o defines the name of the executable file, which in our case is hello. If the compilation is successful, the program runs with the command:

```
> ./hello
Hello world!
```

The ./ is not a special symbol for running programs. The dot is the current working directory and ./hello is the full path to the file hello.

---

[32]g++ is a *front end* to the GNU collection of compilers gcc. By installing gcc, you obtain a collection of compilers for several languages, like C, C++, Fortran, Java and others. See http://gcc.gnu.org/

1.4. THE C++ PROGRAMMING LANGUAGE 41

Now, we will try a simple calculation. Given the radius of a circle we will compute its length and area. The program can be found in the file area_01.cpp:

```
#include <iostream>
using namespace std;

int main(){
  double PI = 3.1415926535897932;
  double R  = 4.0;

  cout << "Perimeter= " << 2.0*PI*R << "\n";
  cout << "Area=      " << PI*R*R   << "\n";
}
```

The first two statements in main() declare the values of the *variables* PI and R. These variables are of type double, which are *floating point* numbers[33].

The following two commands have two effects: Computing the length $2\pi R$ and the area πR^2 of the circle and printing the results. The expressions 2.0*PI*R and PI*R*R are evaluated before being printed to the stdout. Note the explicit decimal points at the constants 2.0 and 4.0. If we write 2 or 4 instead, then these are going to be constants of the int type and by using them the wrong way we may obtain surprising results[34]. We compile and run the program with the commands:

```
> g++ area_01.cpp -o area
> ./area
Perimeter= 25.1327
Area=      50.2655
```

Now we will try a process that repeats itself for many times. We will calculate the length and area of 10 circles of different radii $R_i = 1.28 + i$,

---

[33]Don't confuse double variables with the real numbers. double variables take values that are finite approximations of real numbers and take values that are a subset of the rational numbers. This approximation becomes better by increasing the amount of memory allocated to store them. In most computing environments, doubles are allocated 8 bytes of memory, in which case they approximate real numbers with, more or less, 17 significant digits.

[34]Try adding the command cout << 2/4 << 2.0/4.0; and check the results.

$i = 1, 2, \ldots, 10$. We will store the values of the radii in an *array* R[10] of the double type. The code can be found in the file area_02.cpp:

```
#include <iostream>
using namespace std;

int main(){
  double PI = 3.1415926535897932;
  double R[10];
  double area, perimeter;
  int i;

  R[0] = 2.18;
  for(i=1;i<10;i++){
    R[i] = R[i-1] + 1.0;
  }

  for(i=0;i<10;i++){
    perimeter = 2.0*PI*R[i];
    area      = PI*R[i]*R[i];
    cout << (i+1) << ") R= " << R[i] << " perimeter= "
         << perimeter << '\n';
    cout << (i+1) << ") R= " << R[i] << " area      = "
         << area      << '\n';
  }

}
```

The declaration double R[10] defines an array of length 10. This way, the elements of the array are referred to by an index that takes values from 0 to 9. For example, R[0] is the first, R[3] is the fourth and R[9] is the last element of the array.

Between the lines

```
for(i=1;i<10;i++){
  ...
}
```

we can write commands that are repeatedly executed while the int variable i takes values from 1 to 9 with increasing step equal to 1. The way it works is the following: In the round brackets after the keyword for, there exist three statements separated by semicolons. The first, i=1, is

1.4. THE C++ PROGRAMMING LANGUAGE

the statement executed once before the loop starts. The second, i<10, is a statement that is evaluated each time before the loop repeats itself. If it is true, then the statements in the loop are executed. If it is false, the control of the program is transferred after the end of the loop. The last statement, i++, is evaluated each time after the last statement in the loop has been executed. The operator ++ is the *increment* operator, and its effect is equivalent to the statement:

```
i = i + 1;
```

The value of i is increased by one. The command:

```
R[i] = R[i-1] + 1.0;
```

defines the i-th radius from the value R[i-1]. For the loop to work correctly, we must define the initial value of R[0], otherwise[35] R[0]=0.0. The second loop uses the defined R-values in order to do the computation and print of the results.

Now we will write an interactive version of the program. Instead of *hard coding* the values of the radii, we will interact with the user asking her to give her own values. The program will read the 10 values of the radii from the standard input (stdin). We will also see how to write the results directly to a file instead of the standard output (stdout). The program can be found in the file area_03.cpp:

```
#include <iostream>
#include <fstream>
using namespace std;

int main(){
  const int    N  = 10;
  const double PI = 3.1415926535897932;
  double R[N];
  double area,perimeter;
  int    i;

  for(i=0;i<N;i++){
    cout << "Enter radius of circle: ";
```

---
[35] Arrays in C++ are zero-initialized.

```
    cin  >> R[i];
    cout << "i= " << (i+1) << " R(i)= " << R[i] << '\n';
 }

 ofstream myfile ("AREA.DAT");
 for(i=0;i<N;i++){
   perimeter = 2.0*PI*R[i];
   area      = PI*R[i]*R[i];
   myfile << (i+1) << ") R= " << R[i]
          << " perimeter= "   << perimeter << '\n';
   myfile << (i+1) << ") R= " << R[i]
          << " area      = "  << area      << '\n';
 }

 myfile.close();

}
```

In the above program, the size of the array R is defined by a const int. A const declares a variable to be a parameter whose value does not change during the execution of the program and, if it is of int type, it can be used to declare the size of an array.

The array elements R[i] are read using the command:

```
    cin  >> R[i];
```

cin is the standard input stream, the same way that cout is the standard output stream[36]. We read input using the >> operator, which indicates that input is written *to* the variable on the right.

In order to interact with ordinary files, we need to include the header

```
#include <fstream>
```

In this header, the C++ class ofstream is defined and it can be used in order to write to files (output stream). An object in this class, like myfile, is defined ("instantiated") by the statement:

```
ofstream myfile ("AREA.DAT");
```

---

[36] And cerr is the standard error stream.

1.4. THE C++ PROGRAMMING LANGUAGE

This object's constructor is called by placing the parentheses ("AREA.DAT"), and then the output stream myfile is directed to the file AREA.DAT. Then we can write output to the file the same way we have already done with cout:

```
myfile << (i+1) << ") R= " << R[i]
       << " perimeter= "   << perimeter << '\n';
```

When we are done writing to the file, we can close the stream with the statement:

```
myfile.close();
```

Reading from files is done in a similar way by using the class ifstream instead of ofstream.

The next step will be to learn how to define and use functions. The program below shows how to define a function area_of_circle(), which computes the length and area of a circle of given radius. The following program can be found in the file area_04.cpp:

```
#include <iostream>
#include <fstream>
using namespace std;

const double PI = 3.1415926535897932;

void area_of_circle(const double& R, double& L, double& A);

int main(){
  const int    N = 10;
  double R[N];
  double area,perimeter;
  int    i;

  for(i=0;i<N;i++){
    cout << "Enter radius of circle: ";
    cin  >> R[i];
    cout << "i= " << (i+1) << " R(i)= " << R[i] << '\n';
  }

  ofstream myfile ("AREA.DAT");
```

```
  for(i=0;i<N;i++){
    area_of_circle(R[i],perimeter,area);
    myfile << (i+1) << ") R= " << R[i] << " perimeter= "
           << perimeter << '\n';
    myfile << (i+1) << ") R= " << R[i] << " area     = "
           << area      << '\n';
  }

  myfile.close();

}
//----------------------------------------------------------------
void area_of_circle(const double& R, double& L, double& A){
  L = 2.0*PI*R;
  A =     PI*R*R;
}
```

The calculation of the length and the area of the circle is performed by the function

```
area_of_circle(R[i],perimeter,area);
```

Calling a function, transfers the control of the program to the statements within the *body* of the function. The above function has *arguments* (R[i], perimeter, area). The argument R[i] is intended to be only an input variable whose value is not going to change during the calculation. The arguments perimeter and area are intended for output. Upon return of the function to the main program, they store the result of the computation. The user of a function must learn how to use its arguments in order to be able to call it in her program. These must be documented carefully by the programmer of the function.

In order to use a function, we need to declare it the same way we do with variables or, as we say, to provide its *prototype*. The prototype of a function can be declared without providing the function's definition. We may provide just enough details that determine the types of its arguments and the value returned. In our program this is done on the line:

```
void area_of_circle(const double& R, double& L, double& A);
```

This is the same syntax used later in the definition of the function, but replacing the body of the function with a semicolon. The argument list

1.4. THE C++ PROGRAMMING LANGUAGE

does not need to include the argument names, only their types. We could have also used the following line in order to declare the function's prototype:

```
void area_of_circle(const double& , double& , double& );
```

We could also have used different names for the arguments, if we wished so. Including the names is a matter of style that improves legibility of the code.

The argument R is intended to be left unchanged during the function execution. This is why we used the keyword const in its declaration. The arguments L and R, however, will return a different value to the calling program. This is why const is not used for them.

The actual program executed by the function is between the lines:

```
void area_of_circle(const double& R, double& L, double& A){
  L = 2.0*PI*R;
  A =     PI*R*R;
}
```

The type of the value returned by a function is declared by the keyword before its name. In our case, this is void which declares that the function does not return a value.

The arguments (R,L,A) must be declared in the function and need not have the same names as the ones that we use when we call it. All arguments are declared to be of type double. The character & indicates that they are passed to the function *by reference*. This makes possible to change their values from within the function.

If & is omitted, then the arguments will be passed *by value* and a statement like L = 2.0*PI*R *will not* change the value of the variable passed by the calling program. This happens because, in this case, only the *value* of the variable L of the calling program is *copied* to a local variable which is used only within the function. This is important to understand and you are encouraged to run the program with and without the & and check the difference in the computed results.

The names of variables in a function are only valid within the *scope* of the function, i.e. between the curly brackets that contain the body of the function. Therefore the variable const int N is valid only within the scope of main(). You may use any name you like, even if it is already used outside the scope of the function. The names of arguments need not be the same as the ones used in the calling program. Only their types have to match.

Variables in the *global scope* are accessible by all functions in the same file[37]. An example of such a variable is PI, which is accessible by main(), as well as by area_of_circle().

We summarize all of the above in a program trionymo.cpp, which computes the roots of a second degree polynomial:

```cpp
// ================================================================
// Program to compute roots of a 2nd order polynomial
// Tasks: Input from user, logical statements,
//        use of functions, exit
//
// Tests: a,b,c= 1   2    3      D=  -8
//        a,b,c= 1  -8   16      D=   0      x1=  4
//        a,b,c= 1  -1   -2      D=   9.     x1=  2.    x2=  -1.
//        a,b,c= 2.3 -2.99 -16.422           x1=  3.4   x2=  -2.1
// ================================================================
#include <iostream>
#include <cstdlib>
#include <cmath>
using namespace std;

double Discriminant(double a, double b, double c);
void   roots(double a, double b, double c, double& x1,
             double& x2);

int main(){
  double a,b,c,D;
  double x1,x2;

  cout << "Enter a,b,c: ";
  cin  >> a >> b >> c;
  cout << a << " " << b << " " << c << " " << '\n';

  // Test if we have a well defined polynomial of 2nd degree:
  if( a == 0.0 ){
```

---

[37]If the code is spread over multiple files, then all files must use the keyword external in order to make the variable accessible to the functions that they contain. In *one* of the files, the variable must be defined without the word external. More on that later...

1.4. THE C++ PROGRAMMING LANGUAGE

```cpp
    cerr << "Trionymo: a=0\n";
    exit(1);
  }

  // Compute the discriminant
  D = Discriminant(a,b,c);
  cout << "Discriminant: D= " << D << '\n';

  // Compute the roots in each case: D>0, D=0, D<0 (no roots)
  if     (D > 0.0) {
    roots(a,b,c,x1,x2);
    cout << "Roots:   x1= " << x1 << " x2= "<< x2 << '\n';
  }
  else if(D == 0.0) {
    roots(a,b,c,x1,x2);
    cout << "Double Root:  x1= " << x1 << '\n';
  }
  else{
    cout << "No real roots\n";
    exit(1);
  }

}
// =====================================================
// This is the function that computes the discriminant
// A function returns a value. This value is returned using
// the return statement
// =====================================================
double Discriminant(double a,double b,double c){
  return b * b - 4.0 * a * c;
}
// =====================================================
// The function that computes the roots.
// a,b,c are passed by value: Their values cannot change
//                            within the function
// x1,x2 are passed by reference: Their values DO change
//                            within the function
// =====================================================
void  roots(double a,double b,double c, double& x1,
            double& x2){
  double D;

  D = Discriminant(a,b,c);
  if(D >= 0.0){
    D = sqrt(D);
```

```
}else{
   cerr << "roots: Sorry, cannot compute roots, D<0="
        << D << '\n';
}

x1 = (-b + D)/(2.0*a);
x2 = (-b - D)/(2.0*a);
}
```

The program reads the coefficients of the polynomial $ax^2 + bx + c$. After a check whether $a \neq 0$, it computes the discriminant $D = b^2 - 4ac$ by calling the Discriminant(a,b,c).

The type of the value returned must be declared at the function's prototype

```
double Discriminant(double a, double b, double c);
```

and at the function's definition

```
double Discriminant(double a, double b, double c){
   return b * b - 4.0 * a * c;
}
```

The value returned to the calling program is the value of the expression given as an argument to the return statement. return has also the effect of transferring the control of the program back to the calling statement.

1.5 Gnuplot

Plotting data is an indispensable tool for their qualitative, but also quantitative, analysis. Gnuplot is a high quality, open source, plotting program that can be used for generating publication quality plots, as well as for heavy duty analysis of a large amount of scientific data. Its great advantage is the possibility to use it from the command line, as well as from shell scripts and other programs. Gnuplot is programmable and it is possible to call external programs in order manipulate data and create complicated plots. There are many mathematical functions built in gnuplot and a fit command for non linear fitting of data. There exist

1.5. GNUPLOT

interactive *terminals* where the user can transform a plot by using the mouse and keyboard commands.

This section is brief and only the features, necessary for the following chapters, are discussed. For more information visit the official page of gnuplot http://gnuplot.info. Try the rich demo gallery at http://gnuplot.info/screenshots/, where you can find the type of graph that you want to create and obtain an easy to use recipe for it. The book [16] is an excellent place to look for many of gnuplot's secrets[38].

You can start a gnuplot session with the gnuplot command:

```
> gnuplot

  G N U P L O T
  Version X.XX
  ....
  The gnuplot FAQ is available from www.gnuplot.info/faq/
  ....
gnuplot>
```

There is a welcome message and then a prompt gnuplot> is issued waiting for your command. Type a command an press [Enter]. Type quit in order to quit the program. In the following, when we show a prompt gnuplot>, it is assumed that the command after the prompt is executed from within gnuplot.

Plotting a function is extremely easy. Use the command plot and x as the independent variable of the function[39]. The command

```
gnuplot> plot x
```

plots the function $y = f(x) = x$ which is a straight line with slope 1. In order to plot many functions simultaneously, you can write all of them in one line:

```
gnuplot> plot [-5:5][-2:4] x, x**2, sin(x), besj0(x)
```

---

[38]A the time of the writing of this book, there was a very nice site www.gnuplotting.org which shows how to create many beautiful and complicated plots.

[39]You can change the symbol of the independent variable. For example, the command set dummy t sets the independent variable to be t.

The above command plots the functions x, x^2, $\sin x$ and $J_0(x)$. Within the square brackets [:], we set the limits of the x and y axes, respectively. The bracket [-5:5] sets $-5 \leq x \leq 5$ and the bracket [-2:4] sets $-2 \leq y \leq 4$. You may leave the job of setting such limits to gnuplot, by omitting some, or all of them, from the respective positions in the brackets. For example, typing [1:] [:5] changes the lower and upper limits of x and y and leaves the upper and lower limits unchanged[40].

In order to plot data points (x_i, y_i), we can read their values from files. Assume that a file data has the following numbers recorded in it:

```
#  x    y1    y2
0.5  1.0  0.779
1.0  2.0  0.607
1.5  3.0  0.472
2.0  4.0  0.368
2.5  5.0  0.287
3.0  6.0  0.223
```

The first line is taken by gnuplot as a *comment line*, since it begins with a #. In fact, gnuplot ignores everything after a #. In order to plot the second column as a function of the first, type the command:

```
gnuplot> plot "data" using 1:2 with points
```

The name of the file is within double quotes. After the keyword using, we instruct gnuplot which columns to use as the x and y coordinates, respectively. The keywords with points instructs gnuplot to add each pair (x_i, y_i) to the plot with points.

The command

```
gnuplot> plot "data" using 1:3 with lines
```

---

[40] By default, the x and y ranges are determined automatically. In order to *force* them to be automatic, you can insert a * in the brackets at the corresponding position(s). For example plot [1:*] [*:5] sets the upper and lower limits of x and y to be determined automatically.

1.5. GNUPLOT

plots the third column as a function of the first, and the keywords with lines instruct gnuplot to connect each pair (x_i, y_i) with a straight line segment.

We can combine several plots together in one plot:

```
gnuplot> plot   "data" using 1:3 with points, exp(-0.5*x)
gnuplot> replot "data" using 1:2
gnuplot> replot 2*x
```

The first line plots the 1st and 3rd columns in the file data together with the function $e^{-x/2}$. The second line *adds* the plot of the 1st and 2nd columns in the file data and the third line adds the plot of the function $2x$.

There are many powerful ways to use the keyword using. Instead of column numbers, we can put mathematical expressions enclosed inside brackets, like using (...):(...). Gnuplot evaluates each expression within the brackets and plots the result. In these expressions, the values of each column in the file data are represented as in the awk language. $i are variables that expand to the number read from columns i=1,2,3,.... Here are some examples:

```
gnuplot> plot   "data" using 1:($2*sin($1)*$3) with points
gnuplot> replot 2*x*sin(x)*exp(-x/2)
```

The first line plots the 1st column of the file data together with the value $y_i sin(x_i) z_i$, where y_i, x_i and z_i are the numbers in the 2nd, 1st and 3rd columns respectively. The second line adds the plot of the function $2x \sin(x) e^{-x/2}$.

```
gnuplot> plot   "data" using (log($1)):(log($2**2))
gnuplot> replot 2*x+log(4)
```

The first line plots the logarithm of the 1st column together with the logarithm of the square of the 2nd column.

We can plot the data written to the standard output of *any* command. Assume that there is a program called area that prints the perimeter and area of a circle to the stdout in the form shown below:

```
> ./area
R=      3.280000        area=       33.79851
R=      6.280000        area=       123.8994
R=      5.280000        area=       87.58257
R=      4.280000        area=       57.54895
```

The interesting data is at the second and fourth columns. These can be plotted directly with the gnuplot command:

```
gnuplot> plot "< ./area" using 2:4
```

All we have to do is to type the full command after the < within the double quotes. We can create complicated filters using pipes as in the following example:

```
gnuplot> plot \
"< ./area|sort -g -k 2|awk '{print log($2),log($4)}'" \
using 1:2
```

The filter produces data to the stdout, by combining the action of the commands area, sort and awk. The data printed by the last program is in two columns and we plot the results using 1:2.

In order to save plots in files, we have to change the *terminal* that gnuplot outputs the plots. Gnuplot can produce plots in several languages (e.g. PDF, postscript, SVG, LaTeX, jpeg, png, gif, etc), which can be interpreted and rendered by external programs. By redirecting the output to a file, we can save the plot to the hard disk. For example:

```
gnuplot> plot "data" using 1:3
gnuplot> set terminal jpeg
gnuplot> set output "data.jpg"
gnuplot> replot
gnuplot> set output
gnuplot> set terminal qt
```

The first line makes the plot as usual. The second one sets the output to be in the JPEG format and the third one sets the name of the file to which the plot will be saved. The fourth lines repeats all the previous plotting commands and the fifth one closes the file data.jpg. The last line chooses the interactive terminal qt to be the output of the next plot. High

1.5. GNUPLOT

quality images are usually saved in the PDF, encapsulated postcript or SVG format. Use set terminal pdf, postscript eps or svg, respectively.

And now a few words for 3-dimensional (3d) plotting. The next example uses the command splot in order to make a 3d plot of the function $f(x, y) = e^{-x^2-y^2}$. After you make the plot, you can use the mouse in order to rotate it and view it from a different perspective:

```
gnuplot> set pm3d
gnuplot> set hidden3d
gnuplot> set size ratio 1
gnuplot> set isosamples 50
gnuplot> splot [-2:2][-2:2] exp(-x**2-y**2)
```

If you have data in the form (x_i, y_i, z_i) and you want to create a plot of $z_i = f(x_i, y_i)$, write the data in a file, like in the following example:

```
-1 -1 2.000
-1  0 1.000
-1  1 2.000

 0 -1 1.000
 0  0 0.000
 0  1 1.000

 1 -1 2.000
 1  0 1.000
 1  1 2.000
```

Note the empty line that follows the change of the value of the first column. If the name of the file is data3, then you can plot the data with the commands:

```
gnuplot> set pm3d
gnuplot> set hidden3d
gnuplot> set size ratio 1
gnuplot> splot "data3" with lines
```

We close this section with a few words on parametric plots. A parametric plot on the plane (2-dimensions) is a curve $(x(t), y(t))$, where t is a parameter. A parametric plot in space (3-dimensions) is a surface $(x(u, v), y(u, v), z(u, v))$, where (u, v) are parameters. The following commands plot the circle $(\sin t, \cos t)$ and the sphere $(\cos u \cos v, \cos u \sin v, \sin u)$:

```
gnuplot> set parametric
gnuplot> plot sin(t),cos(t)
gnuplot> splot cos(u)*cos(v),cos(u)*sin(v),sin(u)
```

1.6 Shell Scripting

A typical GNU/Linux environment offers very powerful tools for complicated system administration tasks. They are much more simple to use than to incorporate them into your program. This way, the programmer can concentrate on the high performance and scientific computing part of the project and leave the administration and trivial data analysis tasks to other, external, programs.

One can avoid repeating the same sequence of commands by coding them in a file. An example can be found in the file script01.csh:

```
#!/bin/tcsh -f
g++ area_01.cpp -o area
./area
g++ area_02.cpp -o area
./area
g++ area_03.cpp -o area
./area
g++ area_04.cpp -o area
./area
```

This is a very simple shell script. The first line instructs the operating system that the lines that follow are to be *interpreted* by the program /bin/tcsh[41]. This can be any program in the system, which in our case is the tcsh shell. The following lines are valid commands for the shell, one in each line. They compile the C++ programs found in the files that we created in section 1.4 with g++, and then they run the executable ./area. In order to execute the commands in the file, we have to make

---
[41] Use #!/bin/bash if you prefer the bash shell.

1.6. SHELL SCRIPTING

sure that the file has the appropriate execute permissions. If not, we have to give the command:

```
> chmod u+x script01.csh
```

Then we simply type the path to the file script01.csh

```
> ./script01.csh
```

and the above commands are run the one after the other. Some of the versions of the programs that we wrote are asking for input from the stdin, which, normally, you have to type on the terminal. Instead of interacting directly with the program, we can write the input data to a file Input, and run the command

```
./area < Input
```

A more convenient solution is to use the, so called, "Here Document". A "Here Document" is a section of the script that is treated as if it were a separate file. As such, it can be used as input to programs by sending its "contents" to the stdin of the command that runs the program[42]. The "Here Document" does not appear in the filesystem and we don't need to administer it as a regular file. An example of using a "Here Document" can be found in the file script02.csh:

```
#!/bin/tcsh -f
g++ area_04.cpp -o area
./area <<EOF
1.0
2.0
3.0
4.0
5.0
6.0
7.0
8.0
9.0
```

---

[42]Their great advantage is that we can use variable and command substitution in them, therefore sending this information to the program that we want to run.

```
10.0
EOF
```

The stdin of the command ./area is redirected to the contents between the lines

```
./area <<EOF
...
EOF
```

The string EOF marks the beginning and the end of the "Here Document", and can be any string you like. The last EOF has to be placed exactly in the beginning of the line.

The power of shell scripting lies in its programming capabilities: Variables, arrays, loops and conditionals can be used in order to create a complicated program. Shell variables can be used as discussed in section 1.1.2: The value of a variable name is $name and it can be set with the command set name = value. An array is defined, for example, by the command

```
set R = (1.0 2.0 3.0 4.0 5.0 6.0 7.0 8.0 9.0 10.0)
```

and its data can be accessed using the syntax $R[1] ... $R[10].

Lets take a look at the following script:

```
#!/bin/tcsh -f

set files = (area_01.cpp area_02.cpp area_03.cpp area_04.cpp)
set R     = (1.0 2.0 3.0 4.0 5.0 6.0 7.0 8.0 9.0 10.0)

echo "Hello $USER Today is " `date`
foreach file ($files)
 echo "# ——————— Working on file $file "
 g++ $file -o area
 ./area <<EOF
$R[1]
$R[2]
$R[3]
$R[4]
$R[5]
$R[6]
```

1.6. SHELL SCRIPTING

```
$R[7]
$R[8]
$R[9]
$R[10]
EOF
  echo "# ———————— Done "
  if( -f AREA.DAT ) cat AREA.DAT
end
```

The first two lines of the script define the values of the arrays `files` (4 values) and R (10 values). The command echo echoes its argument to the stdin. $USER is the name of the user running the script. `date` is an example of *command substitution*: When a command is enclosed between backquotes and is part of a string, then the command is executed and its stdout is pasted back to the string. In the example shown above, `date` is replaced by the current date and time in the format produced by the date command.

The foreach loop

```
foreach file ($files)
  ...
end
```

is executed once for each of the 4 values of the array `files`. Each time the value of the variable `file` is set equal to one of the values area_01.cpp, area_02.cpp, area_03.cpp, area_04.cpp. These values can be used by the commands in the loop. Therefore, the command g++ $file -o area compiles a different file each time that it is executed by the loop.

The last line in the loop

```
if( -f AREA.DAT ) cat AREA.DAT
```

is a conditional. It executes the command cat AREA.DAT if the condition -f AREA.DAT is true. In this case, -f constructs a logical expression which is true when the file AREA.DAT exists.

We close this section by presenting a more complicated and advanced script. It only serves as a demonstration of the shell scripting capabilities. For more information, the reader is referred to the bibliography [18, 19, 20, 21, 22]. Read carefully the commands, as well as the comments which follow the # mark. Then, write the commands to a file script04.csh[43], make it an executable file with the command chmod u+x script04.csh and give the command

```
> ./script04.csh This is my first serious tcsh script
```

The script will run with the words "This is my first serious tcsh script" as its arguments. Notice how these arguments are manipulated by the script. Then, the script asks for the values of the radii of ten or more circles interactively, so that it will compute their perimeter and area. Type them on the terminal and then observe the script's output, so that you understand the function of each command. You will not regret the time investment!

```
#!/bin/tcsh -f
# Run this script as:
# ./script04.csh Hello this is a tcsh script
#
# 'command' is command substitution: it is replaced by stdout of command
set now = `date` ; set mypc = `uname -a`
# Print information: variables are expanded within double quotes
echo "I am user $user working on the computer $HOST" #HOST is predefined
echo "Today the date is        : $now"              #now  is defined above
echo "My home directory is     : $home"             #home is predefined
echo "My current directory is: $cwd"                #cwd changes with cd
echo "My computer runs         : $mypc"             #mypc is defined above
echo "My process id is         : $$ "               #$$   is predefined
# Manipulate the command line: ($#argv is number of elements in array argv)
echo "The command line has $#argv arguments"
echo "The name of the command I am running is: $0"
echo "Arguments 3rd to last of the command   : $argv[3-]"    #third to last
echo "The last argument is                   : $argv[$#argv]" #last element
echo "All arguments                          : $argv"

# Ask user for input: enter radii of circles
echo -n "Enter radii of circles: "  # variable $< stores one line of input
set Rs = ($<) #Rs is now an array with all words entered by user
if($#Rs < 10 )then #make a test, need at least 10 of them
  echo "Need more than 10 radii. Exiting...."
  exit(1)
endif
echo "You entered $#Rs radii, the first is $Rs[1] and the last $Rs[$#Rs]"
echo "Rs= $Rs"
# Now, compute the perimeter of each circle:
foreach R ($Rs)
 # -v rad=$R set the awk variable rad equal to $R. pi=atan2(0,-1)=3.14...
  set l = `awk -v rad=$R 'BEGIN{print 2*atan2(0,-1)*rad}'`
```

---

[43]You will find it also in the accompanying software

1.6. SHELL SCRIPTING

```
  echo "Circle with R= $R has perimeter $l"
end
# alias defines a command to do what you want: use awk as a calculator
alias acalc  'awk "BEGIN{ print \!* }"' # \!* substitutes args of acalc
echo "Using acalc to compute        2+3=" `acalc 2+3`
echo "Using acalc to compute cos(2*pi)=" `acalc cos(2*atan2(0,-1))`
# Now do the same loop over radii as above in a different way
# while( expression ) is executed as long as "expression" is true
while($#Rs > 0) #executed as long as $Rs contains radii
  set R = $Rs[1] #take first element of $Rs
  shift Rs        #now $Rs has one less element: old $Rs[1] has vanished
  set a = `acalc atan2(0,-1)*${R}*${R}` # =pi*R*R calculated by acalc
  # construct a filename to save the result from the value of R:
  set file = area${R}.dat
  echo "Circle with R= $R has area $a" > $file #save result in a file
end              #end while
# Now look for our files: save their names in an array files:
set files = (`ls -1 area*.dat`)
if( $#files == 0) echo "Sorry, no area files found"
echo "_____"
echo "files: $files"
ls -l $files
echo "_____"
echo "And the results for the area are:"
foreach f ($files)
  echo -n "file ${f}: "
  cat $f
end
# now play a little bit with file names:
echo "_____"
set f = $files[1] # test permissions on first file
# -f, -r, -w, -x, -d test existence of file, rwxd permissions
# the ! negates the expression (true -> false, false -> true)
echo "testing permissions on files:"
if(  -f $f     ) echo "$file exists"
if(  -r $f     ) echo "$file is readable by me"
if(  -w $f     ) echo "$file is writable by be"
if(! -w /bin/ls) echo "/bin/ls is NOT writable by me"
if(! -x $f     ) echo "$file is NOT an executable"
if(  -x /bin/ls) echo "/bin/ls is executable by me"
if(! -d $f     ) echo "$file is NOT a directory"
if(  -d /bin   ) echo "/bin is a directory"
echo "_____"
# transform the name of a file
set f = $cwd/$f       # add the full path in $f
set filename  = $f:r  # removes extension .dat
set extension = $f:e  # gets     extension .dat
set fdir      = $f:h  # gets     directory of $f
set base      = `basename $f` # removes directory name
echo "file      is: $f"
echo "filename  is: $filename"
echo "extension is: $extension"
echo "directory is: $fdir"
echo "basename  is: $base"
# now transform the name to one with different extension:
set newfile = ${filename}.jpg
echo "jpeg name is: $newfile"
```

```
echo "jpeg base is:" `basename $newfile`
if($newfile:e == jpg)echo `basename $newfile` " is a picture"
echo "_____"
# Now save all data in a file using a "here document"
# A here document starts with <<EOF and ends with a line
# starting exactly with EOF (EOF can be any string as below)
# In a "here document" we can use variables and command
# substitution:
cat <<AREAS >> areas.dat
# This file contains the areas of circle of given radii
# Computation done by ${user} on ${HOST}. Today is `date`
`cat $files`
AREAS
# now see what we got:
if( -f areas.dat ) cat areas.dat
# You can use a "here document" as standard input to any command:
# use gnuplot to save a plot: gnuplot does the job and exits...
gnuplot <<GNU
set terminal jpeg
set output    "areas.jpg"
plot "areas.dat" using 4:7 title "areas.dat",\
     pi*x*x                title "pi*R^2"
set output
GNU
# check our results: display the jpeg file using eog
if( -f areas.jpg) eog areas.jpg &
```

1.6. SHELL SCRIPTING

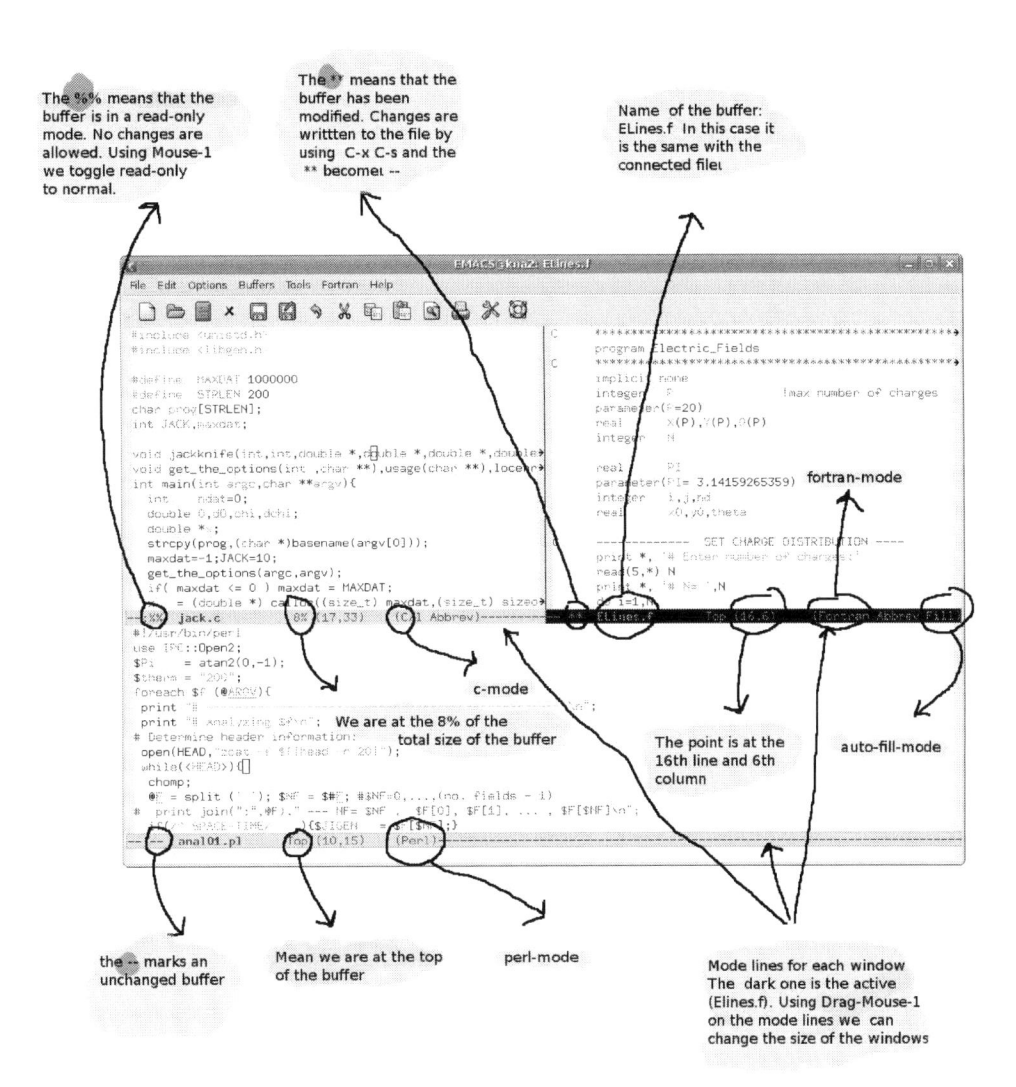

Figure 1.5: In this figure, the Emacs window has been split in three *windows*. The splitting was done horizontally first (C-x 2), and then vertically (C-x 3). By dragging the mouse (Drag-Mouse-1) on the horizontal *mode lines* and vertical lines that separate the windows, we can change window sizes. Notice the useful information diplayed on the mode lines. Each window has one *point* and the *cursor* is on the active window (in this case the window of the buffer named ELines.f). A buffer with no active changes in its contents is marked by a --, an edited buffer is marked by ** and a buffer in *read only* mode with (%%). With a mouse click on a %%, we can change them to -- (so that we can edit) and vice versa. With Mouse-3 on the name of a *mode* we can activate a choice of *minor modes*. With Mouse-1 on the name of a mode we ca have access to commands relevant to the mode. The numbers (17,31), (16,6) and (10,15) on the mode lines show the (line,column) of the point location on the respective windows.

awk	search for and process patterns in a file,
cat	display, or join, files
cd	change working directory
chmod	change the access mode of a file
cp	copy files
date	display current time and date
df	display the amount of available disk space
diff	display the differences between two files
du	display information on disk usage
echo	echo a text string to output
find	find files
grep	search for a pattern in files
gzip	compress files in the gzip (.gz) format (gunzip to uncompress)
head	display the first few lines of a file
kill	send a signal (like KILL) to a process
locate	search for files stored on the system (faster than find)
less	display a file one screen at a time
ln	create a link to a file
lpr	print files
ls	list information about files
man	search information about command in man pages
mkdir	create a directory
mv	move and/or rename a file
ps	report information on the processes run on the system
pwd	print the working directory
rm	remove (delete) files
rmdir	remove (delete) a directory
sort	sort and/or merge files
tail	display the last few lines of a file
tar	store or retrieve files from an archive file
top	dynamic real-time view of processes
wc	counts lines, words and characters in a file
whatis	list man page entries for a command
where	show where a command is located in the path (alternatively: whereis)
which	locate an executable program using "path"
zip	create compressed archive in the zip format (.zip)
unzip	get/list contents of zip archive

Table 1.1: Basic Unix commands.

1.6. SHELL SCRIPTING

Table 1.2: Basic Emacs commands.

Leaving Emacs	
suspend Emacs (or iconify it under X)	C-z
exit Emacs permanently	C-x C-c

Files	
read a file into Emacs	C-x C-f
save a file back to disk	C-x C-s
save **all** files	C-x s
insert contents of another file into this buffer	C-x i
toggle read-only status of buffer	C-x C-q

Getting Help	
The help system is simple. Type C-h (or F1) and follow the directions. If you are a first-time user, type C-h t for a **tutorial**.	
remove help window	C-x 1
apropos: show commands matching a string	C-h a
describe the function a key runs	C-h k
describe a function	C-h f
get mode-specific information	C-h m

Error Recovery	
abort partially typed or executing command	C-g
recover files lost by a system crash	M-x recover-session
undo an unwanted change	C-x u, C-_ or C-/
restore a buffer to its original contents	M-x revert-buffer
redraw garbaged screen	C-l

Incremental Search	
search forward	C-s
search backward	C-r
regular expression search	C-M-s
abort current search	C-g
Use C-s or C-r again to repeat the search in either direction. If Emacs is still searching, C-g cancels only the part not matched.	

Motion

entity to move over	backward	forward
character	C-b	C-f
word	M-b	M-f
line	C-p	C-n

Continued...

Table 1.2: Continued...

go to line beginning (or end)	C-a	C-e
go to buffer beginning (or end)	M-<	M->
scroll to next screen	C-v	
scroll to previous screen	M-v	
scroll left	C-x <	
scroll right	C-x >	
scroll current line to center of screen	C-u C-l	

Killing and Deleting

entity to kill	backward	forward
character (delete, not kill)	DEL	C-d
word	M-DEL	M-d
line (to end of)	M-0 C-k	C-k
kill **region**	C-w	
copy region to kill ring	M-w	
yank back last thing killed	C-y	
replace last yank with previous kill	M-y	

Marking

set mark here	C-@ or C-SPC	
exchange point and mark	C-x C-x	
mark **paragraph**	M-h	
mark entire **buffer**	C-x h	

Query Replace

interactively replace a text string	M-% or M-x query-replace
using regular expressions	M-x query-replace-regexp

Buffers

select another buffer	C-x b
list all buffers	C-x C-b
kill a buffer	C-x k

Multiple Windows

When two commands are shown, the second is a similar command for a frame instead of a window.

delete all other windows	C-x 1	C-x 5 1
split window, above and below	C-x 2	C-x 5 2
delete this window	C-x 0	C-x 5 0

Continued...

1.6. SHELL SCRIPTING

Table 1.2: Continued...

split window, side by side	C-x 3	
switch cursor to another window	C-x o	C-x 5 o
grow window taller	C-x ^	
shrink window narrower	C-x {	
grow window wider	C-x }	

Formatting

indent current **line** (indent code etc)	TAB
insert newline after point	C-o
fill paragraph	M-q

Case Change

uppercase word	M-u
lowercase word	M-l
capitalize word	M-c
uppercase region	C-x C-u
lowercase region	C-x C-l

The Minibuffer

The following keys are defined in the minibuffer.

complete as much as possible	TAB
complete up to one word	SPC
complete and execute	RET
abort command	C-g

Type C-x ESC ESC to edit and repeat the last command that used the minibuffer.
Type F10 to activate menu bar items on text terminals.

Spelling Check

check spelling of current word	M-$
check spelling of all words in region	M-x ispell-region
check spelling of entire buffer	M-x ispell-buffer
On the fly spell checking	M-x flyspell-mode

Info – Getting Help Within Emacs

enter the Info documentation reader	C-h i
scroll forward	SPC
scroll reverse	DEL
next node	n

Continued...

Table 1.2: Continued...

previous node	p
move **up**	u
select menu item by name	m
return to last node you saw	l
return to directory node	d
go to top node of Info file	t
go to any node by name	g
quit Info	q

Chapter 2

Kinematics

In this chapter we show how to program simple kinematic equations of motion of a particle and how to do basic analysis of numerical results. We use simple methods for plotting and animating trajectories on the two dimensional plane and three dimensional space. In section 2.3 we study numerical errors in the calculation of trajectories of freely moving particles bouncing off hard walls and obstacles. This will be a prelude to the study of the integration of the *dynamical* equations of motion that we will introduce in the following chapters.

2.1 Motion on the Plane

When a particle moves on the plane, its position can be given in Cartesian coordinates $(x(t), y(t))$. These, as a function of time, describe the particle's trajectory. The position vector is $\vec{r}(t) = x(t)\,\hat{x} + y(y)\,\hat{y}$, where $\hat{x}$ and $\hat{y}$ are the unit vectors on the x and y axes respectively. The velocity vector is $\vec{v}(t) = v_x(t)\,\hat{x} + v_y(t)\,\hat{y}$ where

$$\vec{v}(t) = \frac{d\vec{r}(t)}{dt}$$
$$v_x(t) = \frac{dx(t)}{dt} \qquad v_y(t) = \frac{dy(t)}{dt}, \qquad (2.1)$$

The acceleration $\vec{a}(t) = a_x(t)\,\hat{x} + a_y(t)\,\hat{y}$ is given by

$$\vec{a}(t) = \frac{d\vec{v}(t)}{dt} = \frac{d^2\vec{r}(t)}{dt^2}$$

$$a_x(t) = \frac{dv_x(t)}{dt} = \frac{d^2 x(t)}{dt^2} \qquad a_y(t) = \frac{dv_y(t)}{dt} = \frac{d^2 y(t)}{dt^2}. \qquad (2.2)$$

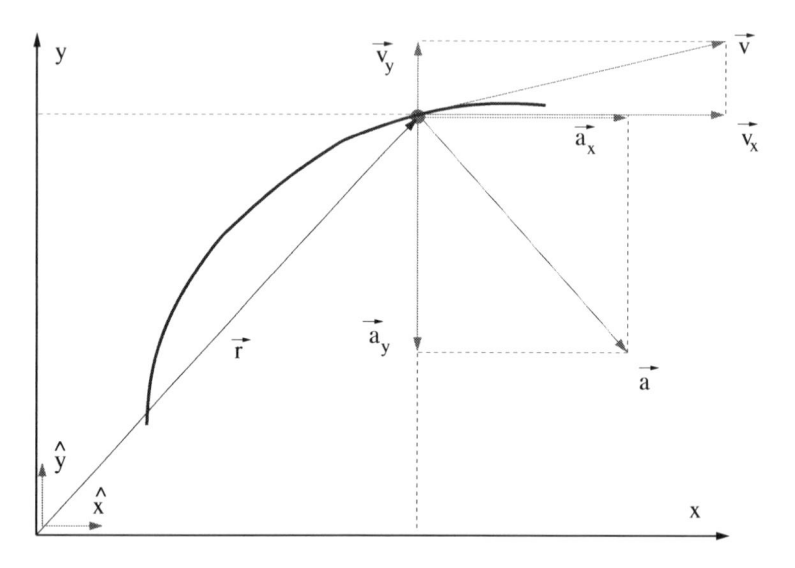

Figure 2.1: The trajectory of a particle moving in the plane. The figure shows its position vector $\vec{r}$, velocity $\vec{v}$ and acceleration $\vec{a}$ and their Cartesian components in the chosen coordinate system at a point of the trajectory.

In this section we study the kinematics of a particle trajectory, therefore we assume that the functions $(x(t), y(t))$ are known. By taking their derivatives, we can compute the velocity and the acceleration of the particle in motion. We will write simple programs that compute the values of these functions in a time interval $[t_0, t_f]$, where t_0 is the initial and t_f is the final time. The continuous functions $x(t), y(t), v_x(t), v_y(t)$ are approximated by a discrete sequence of their values at the times $t_0, t_0 + \delta t, t_0 + 2\delta t, t_0 + 3\delta t, \ldots$ such that $t_0 + n\delta t \leq t_f$.

We will start the design of our program by forming a generic template to be used in all of the problems of interest. Then we can study each problem of particle motion by programming only the equations of motion without worrying about the less important tasks, like input/output,

2.1. MOTION ON THE PLANE

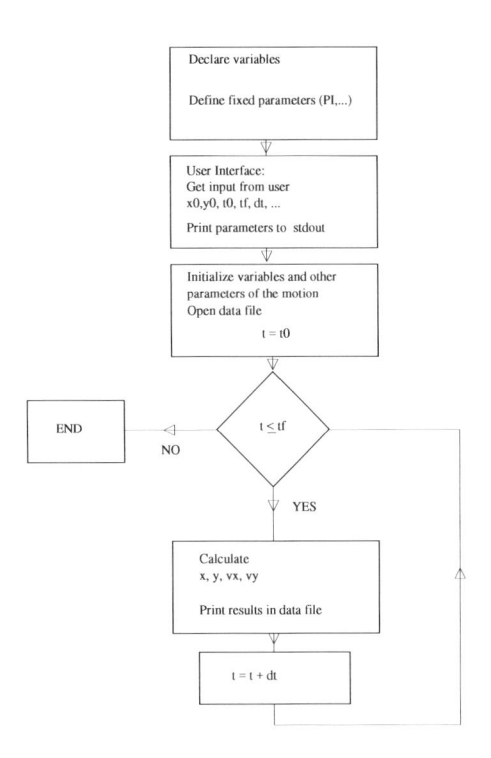

Figure 2.2: The flowchart of a typical program computing the trajectory of a particle from its (kinematic) equations of motion.

user interface etc. Figure 2.2 shows a *flowchart* of the basic steps in the algorithm. The first part of the program declares variables and defines the values of the fixed parameters (like $\pi = 3.1459\ldots$, $g = 9.81$, etc). The program starts by interacting with the user ("user interface") and asks for the values of the variables x_0, y_0, t_0, t_f, $\delta t \ldots$. The program prints these values to the stdout so that the user can check them for correctness and store them in her data.

The main calculation is performed in a *loop* executed while $t \leq t_f$. The values of the positions and the velocities $x(t), y(t), v_x(t), v_y(t)$ are calculated and printed in a file together with the time t. At this point we fix the format of the program output, something that is very important to do it in a consistent and convenient way for easing data analysis. We choose to print the values t, x, y, vx, vy in five columns in each line of the output file.

The specific problem that we are going to solve is the computation of the trajectory of the circular motion of a particle on a circle with center (x_0, y_0) and radius R with constant angular velocity ω. The position on the circle can be defined by the angle θ, as can be seen in figure 2.3. We define the initial position of the particle at time t_0 to be $\theta(t_0) = 0$.

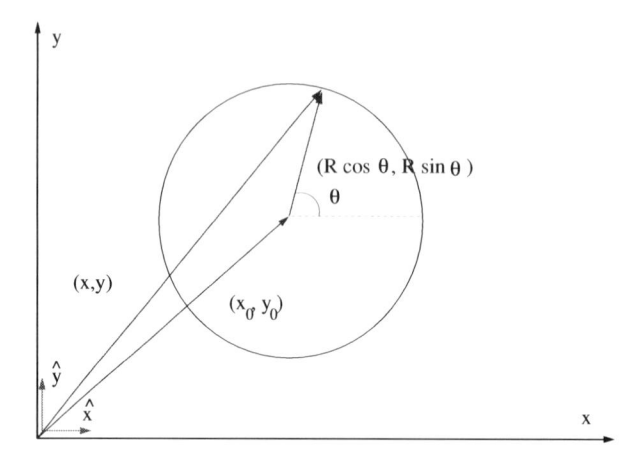

Figure 2.3: The trajectory of a particle moving on a circle with constant angular velocity calculated by the program Circle.cpp.

The equations giving the position of the particle at time t are

$$\begin{aligned} x(t) &= x_0 + R\cos\left(\omega(t-t_0)\right) \\ y(t) &= y_0 + R\sin\left(\omega(t-t_0)\right). \end{aligned} \qquad (2.3)$$

Taking the derivative w.r.t. t we obtain the velocity

$$\begin{aligned} v_x(t) &= -\omega R\sin\left(\omega(t-t_0)\right) \\ v_y(t) &= \omega R\cos\left(\omega(t-t_0)\right), \end{aligned} \qquad (2.4)$$

and the acceleration

$$\begin{aligned} a_x(t) &= -\omega^2 R\cos\left(\omega(t-t_0)\right) = -\omega^2(x(t)-x_0) \\ a_y(t) &= -\omega^2 R\sin\left(\omega(t-t_0)\right) = -\omega^2(y(t)-y_0). \end{aligned} \qquad (2.5)$$

We note that the above equations imply that $\vec{R}\cdot\vec{v} = 0$ ($\vec{R} \equiv \vec{r} - \vec{r}_0$, $\vec{v} \perp \vec{R}$, $\vec{v}$ tangent to the trajectory) and $\vec{a} = -\omega^2 \vec{R}$ ($\vec{R}$ and $\vec{a}$ anti-parallel, $\vec{a} \perp \vec{v}$).

The data structure is quite simple. The constant angular velocity ω is stored in the double variable omega. The center of the circle (x_0, y_0), the

2.1. MOTION ON THE PLANE

radius R of the circle and the angle θ are stored in the `double` variables `x0`, `y0`, `R`, `theta`. The times at which we calculate the particle's position and velocity are defined by the parameters $t_0, t_f, \delta t$ and are stored in the `double` variables `t0`, `tf`, `dt`. The current position $(x(t), y(t))$ is calculated and stored in the `double` variables `x`, `y` and the velocity $(v_x(t), v_y(t))$ in the `double` variables `vx`, `vy`. The declarations of the variables are put in the beginning of the program:

```
double x0,y0,R,x,y,vx,vy,t,t0,tf,dt;
double theta,omega;
```

The user interface of the program is the interaction of the program with the user and, in our case, it is the part of the program where the user enters the parameters `omega`, `x0`, `y0`, `R`, `t0`, `tf`, `dt`. The program issues a prompt with the names the variables expected to be read. The variables are read from the `stdin` by reading from the stream `cin` and the values entered by the user are printed to the `stdout` using the stream `cout`[1]:

```
cout << "# Enter omega:\n";
cin  >> omega;            getline(cin,buf);
cout << "# Enter center of circle (x0,y0) and radius R:\n";
cin  >> x0 >> y0 >> R;    getline(cin,buf);
cout << "# Enter t0,tf,dt:\n";
cin  >> t0 >> tf >> dt;   getline(cin,buf);
cout <<"# omega= " << omega << endl;
cout <<"# x0= "    << x0    << " y0= " << y0
     << " R= "     << R     << endl;
cout <<"# t0= "    << t0    << " tf= " << tf
     << " dt= "    << dt    << endl;
```

There are a couple of things to explain. Notice that after reading each variable from the standard input stream `cin`, we call the function `getline`. By calling `getline(cin,buf)`, a whole line is read from the input stream `cin` into the *string* `buf`[2]. Then the statement

---
[1] This is done so that the used can check for typos and see the actual value read by the program. By redirecting the `stdout` of a file on the hard disk, the parameters can be saved for future reference and used in data analysis.

[2] In fact it is possible to call `getline(cin,buf,char)` and read a line until the character `char` is encountered.

```
cin >> x0 >> y0 >> R;   getline(cin,buf);
```

has the effect of reading three doubles from the stdin and put the rest of the line in the string buf. Since we never use buf, this is a mechanism to discard the rest of the line of input. The reason for doing so will become clear later.

Objects of type string in C++ store character sequences. In order to use them you have to include the header

```
#include <string>
```

and, e.g., declare them like

```
string buf,buf1,buf2;
```

Then you can store data in the obvious way, like buf="Hello World!", manipulate string data using operators like buf=buf1 (assign buf1 to buf), buf=buf1+buf2 (concatenate buf1 and buf2 and store the result in buf), buf1==buf2 (compare strings) etc.

Finally, endl is used to end all the cout statements. This has the effect of adding a newline to the output stream and flush the output[3].

Next, the program initializes the state of the computation. This includes checking the validity of the parameters entered by the user, so that the computation will be possible. For example, the program computes the expression 2.0*PI/omega, where it is assumed that omega has a non zero value. We will also demand that $R > 0$ and $\omega > 0$. An if statement will make those checks and if the parameters have illegal values, the exit statement[4] will stop the program execution and print an

---

[3] When *buffered* output is used, it is not written out immediately but stored in a temporary memory location (a *buffer*). When the buffer fills, it is automatically flushed to the output stream. If we want to force flushing before the buffer is full, then we have to *flush* the buffer. There are several methods to flush an output stream os (like os.flush()).

[4] The exit(1) statement returns 1 as exit code for the program. This is the int that main() returns. exit(0) is conventionally used for a normal exit and a non zero value is used when an error occurs. In order to use exit() you must include the header cstdlib.

2.1. MOTION ON THE PLANE

informative message to the standard error stream cerr[5]. The program opens the file Circle.dat for writing the calculated values of the position and the velocity of the particle.

```
if(R     <=0.0){cerr <<"Illegal value of R    \n";exit(1);}
if(omega<=0.0){cerr <<"Illegal value of omega\n";exit(1);}
cout     << "# T= "  << 2.0*PI/omega              << endl;
ofstream myfile("Circle.dat");
myfile.precision(17);
```

The line myfile.precision(17) sets the precision of the floating point numbers (like double) printed to myfile to 17 significant digits accuracy. The default is 6 which is a pity, because doubles have up to 17 significant digits accuracy.

If $R \leq 0$ or $\omega \leq 0$ the corresponding exit statements are executed which end the program execution. The optional error messages are included after the stop statements which are printed to the stderr. The value of the period $T = 2\pi/\omega$ is also calculated and printed for reference.

The main calculation is performed within the loop

```
t = t0;
while( t <= tf ){
   .........
   t = t + dt;
}
```

The first statement sets the initial value of the time. The statements between within the scope of the while(*condition*) are executed as long as *condition* has a true value. The statement t=t+dt increments the time and this is necessary in order not to enter into an infinite loop. The statements put in place of the dots .... calculate the position and the velocity and print them to the file Circle.dat:

```
#include <cmath>
.................
    theta = omega * (t-t0);
    x     = x0+R*cos(theta);
```

---

[5]Note that there are more assumptions that need to be checked by the program. We leave this as an exercise for the reader.

```
y   =   y0+R*sin(theta);
vx  =  -omega*R*sin(theta);
vy  =   omega*R*cos(theta);
myfile << t << " "
       << x  << " " << y  << " "
       << vx << " " << vy << endl;
```

Notice the use of the functions sin and cos that calculate the sine and cosine of an angle expressed in radians. The header cmath is necessary to be included.

The program is stored in the file Circle.cpp and can be found in the accompanied software. The *extension* .cpp is used to inform the compiler that the file contains source code written in the C++ language. Compilation and running can be done using the commands:

```
> g++ Circle.cpp -o cl
> ./cl
```

The switch -o cl forces the compiler g++ to write the binary commands executed by the program to the file[6] cl. The command ./cl loads the program instructions to the computer memory for execution. When the programs starts execution, it first asks for the parameter data and then performs the calculation. A typical session looks like:

```
> g++ Circle.cpp -o cl
> ./cl
# Enter omega:
1.0
# Enter center of circle (x0,y0) and radius R:
1.0 1.0 0.5
# Enter t0,tf,dt:
0.0 20.0 0.01
# omega= 1
# x0= 1 y0= 1 R=  0.5
# t0= 0 tf= 20 dt= 0.01
# T= 6.28319
```

The lines shown above that start with a # character are printed by the program and the lines without # are the values of the parameters entered

---

[6]If omitted, the executable file has the default name a.out.

2.1. MOTION ON THE PLANE

interactively by the user. The user types in the parameters and then presses the Enter key in order for the program to read them. Here we have used $\omega = 1.0$, $x_0 = y_0 = 1.0$, $R = 0.5$, $t_0 = 0.0$, $t_f = 20.0$ and $\delta t = 0.01$.

You can execute the above program many times for different values of the parameters by writing the parameter values in a file using an editor. For example, in the file Circle.in type the following data:

```
1.0              omega
1.0  1.0  0.5    (x0, y0) , R
0.0  20.0 0.01   t0 tf dt
```

Each line has the parameters that we want to pass to the program with each call to cout. The rest of the line consists of comments that explain to the user what each number is there for. We want to discard these characters during input and this is the reason for using getline to complete reading the rest of the line. The program can read the above values of the parameters with the command:

```
> ./cl < Circle.in > Circle.out
```

The command ./cl runs the commands found in the executable file ./cl. The < Circle.in redirects the contents of the file Circle.in to the standard input (stdin) of the command ./cl. This way the program reads in the values of the parameters from the contents of the file Circle.in. The > Circle.out redirects the standard output (stdout) of the command ./cl to the file Circle.out. Its contents can be inspected after the execution of the program with the command cat:

```
> cat Circle.out
# Enter omega:
# Enter center of circle (x0,y0) and radius R:
# Enter t0,tf,dt:
# omega= 1
# x0= 1 y0= 1 R=  0.5
# t0= 0 tf= 20 dt= 0.01
# T= 6.28319
```

We list the full program in Circle.cpp below:

```cpp
//========================================================
//File Circle.cpp
//Constant angular velocity circular motion
//Set (x0,y0) center of circle, its radius R and omega.
//At t=t0, the particle is at theta=0
//--------------------------------------------------------
#include <iostream>
#include <fstream>
#include <cstdlib>
#include <string>
#include <cmath>
using namespace std;

#define PI 3.1415926535897932

int main(){
//--------------------------------------------------------
//Declaration of variables
  double x0,y0,R,x,y,vx,vy,t,t0,tf,dt;
  double theta,omega;
  string buf;
//--------------------------------------------------------
//Ask user for input:
  cout << "# Enter omega:\n";
  cin  >> omega;              getline(cin,buf);
  cout << "# Enter center of circle (x0,y0) and radius R:\n";
  cin  >> x0 >> y0 >> R;   getline(cin,buf);
  cout << "# Enter t0,tf,dt:\n";
  cin  >> t0 >> tf >> dt;  getline(cin,buf);
  cout <<"# omega= " << omega << endl;
  cout <<"# x0= "    << x0    << " y0= " << y0
       << " R= "     << R     << endl;
  cout <<"# t0= "    << t0    << " tf= " << tf
       << " dt= "    << dt    << endl;
//--------------------------------------------------------
//Initialize
  if(R    <=0.0){cerr <<"Illegal value of R    \n";exit(1);}
  if(omega<=0.0){cerr <<"Illegal value of omega\n";exit(1);}
  cout    << "# T= " << 2.0*PI/omega              << endl;
  ofstream myfile("Circle.dat");
  // Set precision for numeric output to myfile to 17 digits
  myfile.precision(17);
//--------------------------------------------------------
//Compute:
  t = t0;
```

2.1. MOTION ON THE PLANE

```
while( t <= tf ){
  theta = omega * (t-t0);
  x  = x0+R*cos(theta);
  y  = y0+R*sin(theta);
  vx = -omega*R*sin(theta);
  vy =  omega*R*cos(theta);
  myfile << t << " "
     << x << " " << y << " "
     << vx << " " << vy << endl;
  t = t + dt;
  }
} //main()
```

2.1.1 Plotting Data

We use gnuplot for plotting the data produced by our programs. The file Circle.dat has the time t and the components x, y, vx, vy in five columns. Therefore we can plot the functions $x(t)$ and $y(t)$ by using the gnuplot commands:

```
gnuplot> plot   "Circle.dat" using 1:2 with lines title "x(t)"
gnuplot> replot "Circle.dat" using 1:3 with lines title "y(t)"
```

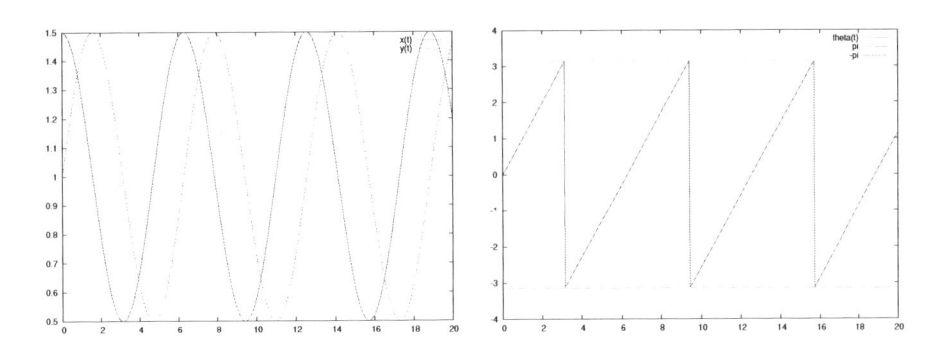

Figure 2.4: The plots $(x(t), y(t))$ (left) and $\theta(t)$ (right) from the data in Circle.dat for $\omega = 1.0$, $x_0 = y_0 = 1.0$, $R = 0.5$, $t_0 = 0.0$, $t_f = 20.0$ and $\delta t = 0.01$.

The second line puts the second plot together with the first one. The results can be seen in figure 2.4.

Let's see now how we can make the plot of the function $\theta(t)$. We can do that using the raw data from the file Circle.dat within gnuplot, without having to write a new program. Note that $\theta(t) = \tan^{-1}\left((y-y_0)/(x-x_0)\right)$. The function atan2 is available in gnuplot[7] as well as in C++. Use the online help system in gnuplot in order to see its usage:

```
gnuplot> help atan2
 The 'atan2(y,x)' function returns the arc tangent (inverse
 tangent) of the ratio of the real parts of its arguments.
 'atan2' returns its argument in radians or degrees, as
 selected by 'set angles', in the correct quadrant.
```

Therefore, the right way to call the function is atan2(y-y0,x-x0). In our case x0=y0=1 and x, y are in the 2nd and 3rd columns of the file Circle.dat. We can construct an *expression* after the using command as in page 53, where $2 is the value of the second and $3 the value of the third column:

```
gnuplot> x0 = 1 ; y0 = 1
gnuplot> plot "Circle.dat" using 1:(atan2($3-y0,$2-x0)) \
              with lines title "theta(t)",pi,-pi
```

The second command is broken in two lines by using the character \ so that it fits conveniently in the text[8]. Note how we defined the values of the variables x0, y0 and how we used them in the expression atan2($3-x0,$2-y0). We also plot the lines which graph the constant functions $f_1(t) = \pi$ and $f_2(t) = -\pi$ which mark the limit values of $\theta(t)$. The gnuplot variable[9] pi is predefined and can be used in formed expressions. The result can be seen in the left plot of figure 2.4.

The velocity components $(v_x(t), v_y(t))$ as function of time as well as the trajectory $\vec{r}(t)$ can be plotted with the commands:

```
gnuplot> plot    "Circle.dat" using 1:4 title "v_x(t)" \
                         with lines
gnuplot> replot "Circle.dat" using 1:5 title "v_y(t)" \
```

---

[7]The command help functions will show you all the available functions in gnuplot.
[8]This can be done on the gnuplot command line as well.
[9]Use the command show variables in order to see the current/default values of gnuplot variables.

2.1. MOTION ON THE PLANE

```
gnuplot> plot     "Circle.dat" using 2:3 title "x-y"
                                 with lines
                                 with lines
```

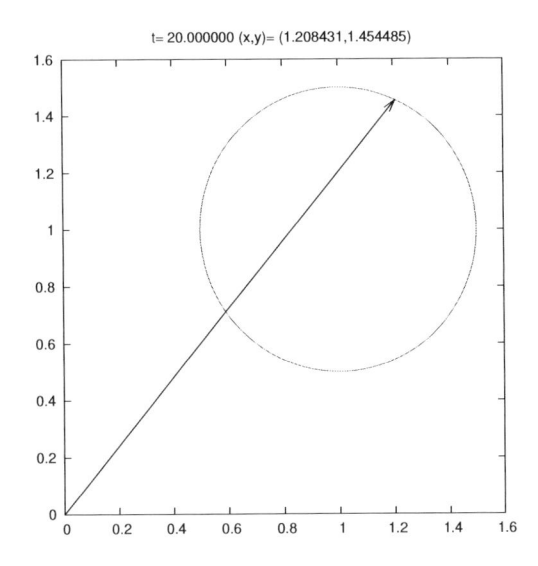

Figure 2.5: The particle trajectory plotted by the gnuplot program in the file animate2D.gnu of the accompanied software. The position vector is shown at a given time t, which is marked on the title of the plot together with the coordinates (x,y). The data is produced by the program Circle.cpp described in the text.

We close this section by showing how to do a simple animation of the particle trajectory using gnuplot. There is a file animate2D.gnu in the accompanied software which you can copy in the directory where you have the data file Circle.dat. We are not going to explain how it works[10] but how to use it in order to make your own animations. The final result is shown in figure 2.5. All that you need to do is to define the data file[11], the initial time t0, the final time tf and the time step dt. These times can be different from the ones we used to create the data in Circle.dat. A full animation session can be launched using the commands:

---

[10] You are most welcome to study the commands in the script and guess how it works of course!

[11] It can be *any* file that has (t,x,y) in the 1st, 2nd and 3rd columns respectively.

```
gnuplot> file = "Circle.dat"
gnuplot> set xrange [0:1.6]; set yrange [0:1.6]
gnuplot> t0   = 0; tf = 20 ; dt = 0.1
gnuplot> load "animate2D.gnu"
```

The first line defines the data file that animate2D.gnu reads data from. The second line sets the range of the plots and the third line defines the time parameters used in the animation. The final line launches the animation. If you want to rerun the animation, you can repeat the last two commands as many times as you want using the same or different parameters. E.g. if you wish to run the animation at "half the speed" you should simply redefine dt=0.05 and set the initial time to t0=0:

```
gnuplot> t0   = 0; dt = 0.05
gnuplot> load "animate2D.gnu"
```

2.1.2 More Examples

We are now going to apply the steps described in the previous section to other examples of motion on the plane. The first problem that we are going to discuss is that of the small oscillations of a simple pendulum. Figure 2.6 shows the single oscillating degree of freedom $\theta(t)$, which is the small angle that the pendulum forms with the vertical direction. The motion is periodic with angular frequency $\omega = \sqrt{g/l}$ and period

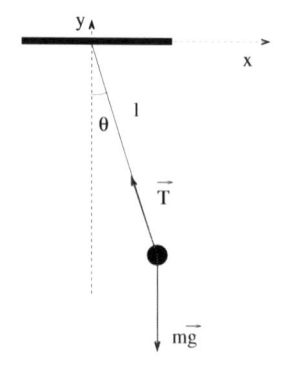

Figure 2.6: The simple pendulum whose motion for $\theta \ll 1$ is described by the program SimplePendulum.cpp.

2.1. MOTION ON THE PLANE

$T = 2\pi/\omega$. The angular velocity is computed from $\dot{\theta} \equiv d\theta/dt$ which gives

$$\begin{aligned} \theta(t) &= \theta_0 \cos(\omega(t - t_0)) \\ \dot{\theta}(t) &= -\omega\theta_0 \sin(\omega(t - t_0)) \end{aligned} \quad (2.6)$$

We have chosen the initial conditions $\theta(t_0) = \theta_0$ and $\dot{\theta}(t_0) = 0$. In order to write the equations of motion in the Cartesian coordinate system shown in figure 2.6 we use the relations

$$\begin{aligned} x(t) &= l\sin(\theta(t)) \\ y(t) &= -l\cos(\theta(t)) \\ v_x(t) &= \frac{dx(t)}{dt} = l\dot{\theta}(t)\cos(\theta(t)) \\ v_y(t) &= \frac{dy(t)}{dt} = l\dot{\theta}(t)\sin(\theta(t))\,. \end{aligned} \quad (2.7)$$

These are similar to the equations (2.3) and (2.4) that we used in the case of the circular motion of the previous section. Therefore the structure of the program is quite similar. Its final form, which can be found in the file SimplePendulum.cpp, is:

```cpp
//==========================================================
// File SimplePendulum.cpp
// Set pendulum original position at theta0
// with no initial speed
//----------------------------------------------------------
#include <iostream>
#include <fstream>
#include <cstdlib>
#include <string>
#include <cmath>
using namespace std;

#define PI 3.1415926535897932
#define g  9.81

int main(){
//----------------------------------------------------------
// Declaration of variables
    double l,x,y,vx,vy,t,t0,tf,dt;
    double theta,theta0,dtheta_dt,omega;
```

```
  string buf;
//-----------------------------------------------
//Ask user for input:
  cout << "# Enter l:\n";
  cin  >> l;           getline(cin,buf);
  cout << "# Enter theta0:\n";
  cin  >> theta0;      getline(cin,buf);
  cout << "# Enter t0,tf,dt:\n";
  cin  >> t0 >> tf >> dt;  getline(cin,buf);
  cout <<"# l= "    << l  << " theta0= " << theta0 << endl;
  cout <<"# t0= "   << t0 << " tf= " << tf
       << " dt= "   << dt << endl;
//-----------------------------------------------
//Initialize
  omega = sqrt(g/l);
  cout << "# omega= " << omega
       << " T= "      << 2.0*PI/omega << endl;
  ofstream myfile("SimplePendulum.dat");
  myfile.precision(17);
//-----------------------------------------------
//Compute:
  t   = t0;
  while( t <= tf ){
    theta    =         theta0*cos(omega*(t-t0));
    dtheta_dt = -omega*theta0*sin(omega*(t-t0));
    x   =  l*sin(theta);
    y   = -l*cos(theta);
    vx  =  l*dtheta_dt*cos(theta);
    vy  =  l*dtheta_dt*sin(theta);
    myfile << t      << " "
           << x      << " " << y  << " "
           << vx     << " " << vy << " "
           << theta  << " " << dtheta_dt
           << endl;
    t = t + dt;
  }
} //main()
```

We note that the acceleration of gravity g is hard coded in the program and that the user can only set the length l of the pendulum. The data file SimplePendulum.dat produced by the program, contains two extra columns with the current values of $\theta(t)$ and the angular velocity $\dot\theta(t)$.

A simple session for the study of the above problem is shown below[12]:

---

[12]Notice that we replaced the command "using 1:2 with lines title" with "u

2.1. MOTION ON THE PLANE

```
> g++ SimplePendulum.cpp -o sp
> ./sp
# Enter l:
1.0
# Enter theta0:
0.314
# Enter t0,tf,dt:
0 20 0.01
# l= 1 theta0= 0.314
# t0= 0 tf= 20 dt= 0.01
# omega= 3.13209 T= 2.00607
> gnuplot
gnuplot> plot    "SimplePendulum.dat" u 1:2 w l t "x(t)"
gnuplot> plot    "SimplePendulum.dat" u 1:3 w l t "y(t)"
gnuplot> plot    "SimplePendulum.dat" u 1:4 w l t "v_x(t)"
gnuplot> replot  "SimplePendulum.dat" u 1:5 w l t "v_y(t)"
gnuplot> plot    "SimplePendulum.dat" u 1:6 w l t "theta(t)"
gnuplot> replot  "SimplePendulum.dat" u 1:7 w l t "theta'(t)"
gnuplot> plot    [-0.6:0.6][-1.1:0.1] "SimplePendulum.dat" \
                  u 2:3 w l t "x-y"
gnuplot> file = "SimplePendulum.dat"
gnuplot> t0=0;tf=20.0;dt=0.1
gnuplot> set xrange  [-0.6:0.6]; set yrange [-1.1:0.1]
gnuplot> load "animate2D.gnu"
```

The next example is the study of the trajectory of a particle shot near the earth's surface[13] when we consider the effect of air resistance to be negligible. Then, the equations describing the trajectory of the particle and its velocity are given by the parametric equations

$$
\begin{aligned}
x(t) &= v_{0x}t \\
y(t) &= v_{0y}t - \frac{1}{2}gt^2 \\
v_x(t) &= v_{0x} \\
v_y(t) &= v_{0y} - gt\,,
\end{aligned}
\quad (2.8)
$$

where t is the parameter. The initial conditions are $x(0) = y(0) = 0$, $v_x(0) = v_{0x} = v_0 \cos\theta$ and $v_y(0) = v_{0y} = v_0 \sin\theta$, as shown in figure 2.7.

1:2 w lines t". These abbreviations can be done with every gnuplot command if an abbreviation uniquely determines a command.

[13]I.e. $\vec{g} =$ const. and the Coriolis force can be ignored.

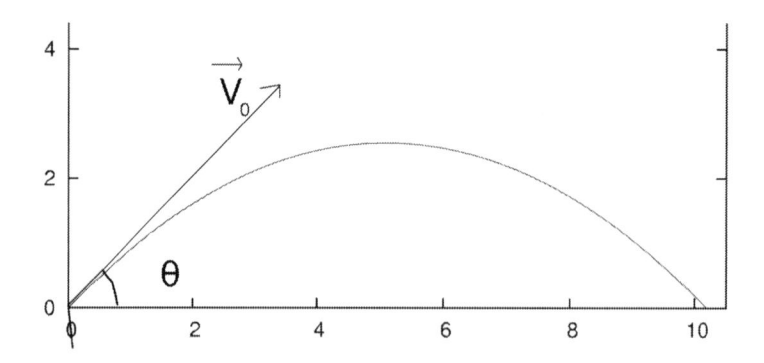

Figure 2.7: The trajectory of a particle moving under the influence of a constant gravitational field. The initial conditions are set to $x(0) = y(0) = 0$, $v_x(0) = v_{0x} = v_0 \cos\theta$ and $v_y(0) = v_{0y} = v_0 \sin\theta$.

The structure of the program is similar to the previous ones. The user enters the magnitude of the particle's initial velocity and the shooting angle θ in *degrees*. The initial time is taken to be $t_0 = 0$. The program calculates v_{0x} and v_{0y} and prints them to the stdout. The data is written to the file Projectile.dat. The full program is listed below and it can be found in the file Projectile.cpp in the accompanied software:

```cpp
//=========================================================
//File Projectile.cpp
//Shooting a progectile near the earth surface.
//No air resistance.
//Starts at (0,0), set (v0,theta).
//---------------------------------------------------------
#include <iostream>
#include <fstream>
#include <cstdlib>
#include <string>
#include <cmath>
using namespace std;

#define PI 3.1415926535897932
#define g  9.81

int main(){
//---------------------------------------------------------
// Declaration of variables
```

2.1. MOTION ON THE PLANE

```cpp
  double x0,y0,R,x,y,vx,vy,t,tf,dt;
  double theta,v0x,v0y,v0;
  string buf;
//----------------------------------------
//Ask user for input:
  cout << "# Enter v0,theta (in degrees):\n";
  cin >> v0 >> theta;        getline(cin,buf);
  cout << "# Enter tf,dt:\n";
  cin >> tf >> dt;           getline(cin,buf);
  cout <<"# v0= "     << v0
       << " theta= "<< theta << "o (degrees)" << endl;
  cout <<"# t0= "     << 0.0  << " tf= "       << tf
       << " dt= "    << dt    << endl;
//----------------------------------------
//Initialize
  if(v0   <= 0.0)
    {cerr <<"Illegal value of v0   <= 0\n";exit(1);}
  if(theta<= 0.0)
    {cerr <<"Illegal value of theta<= 0\n";exit(1);}
  if(theta>=90.0)
    {cerr <<"Illegal value of theta>=90\n";exit(1);}
  theta    = (PI/180.0)*theta; //convert to radians
  v0x      = v0*cos(theta);
  v0y      = v0*sin(theta);
  cout    << "# v0x= "  << v0x
       << " v0y= "  << v0y  << endl;
  ofstream myfile("Projectile.dat");
  myfile.precision(17);
//----------------------------------------
//Compute:
  t = 0.0;
  while( t <= tf ){
    x  = v0x * t;
    y  = v0y * t - 0.5*g*t*t;
    vx = v0x;
    vy = v0y       -     g*t;
    myfile << t << " "
           << x << " " << y << " "
           << vx << " " << vy << endl;
    t  = t + dt;
  }
} //main()
```

A typical session for the study of this problem is shown below:

```
> g++ Projectile.cpp -o pj
> ./pj
# Enter v0,theta (in degrees):
10 45
# Enter tf,dt:
1.4416 0.001
# v0= 10 theta= 45o (degrees)
# t0= 0 tf= 1.4416 dt= 0.001
# v0x= 7.07107 v0y= 7.07107
> gnuplot
gnuplot> plot    "Projectile.dat" using 1:2 w l t "x(t)"
gnuplot> replot  "Projectile.dat" using 1:3 w l t "y(t)"
gnuplot> plot    "Projectile.dat" using 1:4 w l t "v_x(t)"
gnuplot> replot  "Projectile.dat" using 1:5 w l t "v_y(t)"
gnuplot> plot    "Projectile.dat" using 2:3 w l t "x-y"
gnuplot> file = "Projectile.dat"
gnuplot> set xrange [0:10.3];set yrange [0:10.3]
gnuplot> t0=0;tf=1.4416;dt=0.05
gnuplot> load "animate2D.gnu"
```

Next, we will study the effect of air resistance of the form $\vec{F} = -mk\vec{v}$. The solutions to the equations of motion

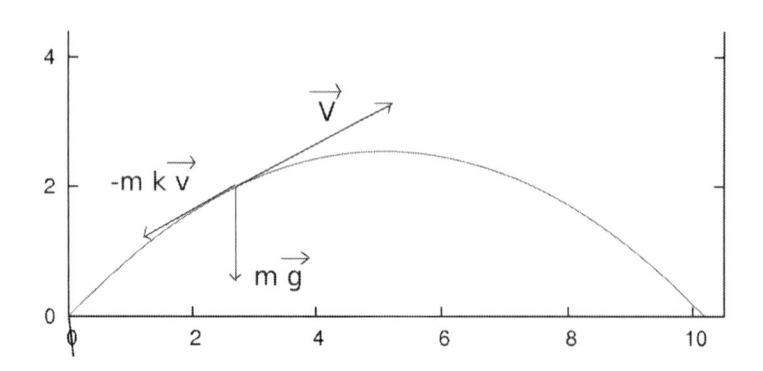

Figure 2.8: The forces that act on the particle of figure 2.7 when we assume air resistance of the form $\vec{F} = -mk\vec{v}$.

$$\begin{aligned} a_x &= \frac{dv_x}{dt} = -kv_x \\ a_y &= \frac{dv_y}{dt} = -kv_y - g \end{aligned} \quad (2.9)$$

2.1. MOTION ON THE PLANE

with initial conditions $x(0) = y(0) = 0$, $v_x(0) = v_{0x} = v_0 \cos\theta$ and $v_y(0) = v_{0y} = v_0 \sin\theta$ are[14]

$$\begin{aligned} v_x(t) &= v_{0x} e^{-kt} \\ v_y(t) &= \left(v_{0y} + \frac{g}{k}\right) e^{-kt} - \frac{g}{k} \\ x(t) &= \frac{v_{0x}}{k}\left(1 - e^{-kt}\right) \\ y(t) &= \frac{1}{k}\left(v_{0y} + \frac{g}{k}\right)\left(1 - e^{-kt}\right) - \frac{g}{k}t \end{aligned} \qquad (2.10)$$

Programming the above equations is as easy as before, the only difference being that the user needs to provide the value of the constant k. The full program can be found in the file ProjectileAirResistance.cpp and it is listed below:

```
//==========================================================
//File ProjectileAirResistance.cpp
//Shooting a progectile near the earth surface.
//No air resistance.
//Starts at (0,0), set k, (v0,theta).
//----------------------------------------------------------
#include <iostream>
#include <fstream>
#include <cstdlib>
#include <string>
#include <cmath>
using namespace std;

#define PI 3.1415926535897932
#define g  9.81

int main(){
//----------------------------------------------------------
//Declaration of variables
   double x0,y0,R,x,y,vx,vy,t,tf,dt,k;
   double theta,v0x,v0y,v0;
   string buf;
//----------------------------------------------------------
//Ask user for input:
   cout << "# Enter k,v0,theta (in degrees):\n";
```

---

[14]The proof of equations (2.10) is left as an exercise for the reader.

```cpp
  cin    >> k >> v0 >> theta; getline(cin,buf);
  cout << "# Enter tf,dt:\n";
  cin    >> tf >> dt;          getline(cin,buf);
  cout <<"# k = "    << k     << endl;
  cout <<"# v0= "    << v0
       << " theta= "<< theta << "o (degrees)" << endl;
  cout <<"# t0= "    << 0.0   << " tf= "       << tf
       << " dt= "    << dt    << endl;
//-----------------------------------------------------------
// Initialize
  if(v0   <= 0.0)
    {cerr <<"Illegal value of v0   <= 0\n";exit(1);}
  if(k    <= 0.0)
    {cerr <<"Illegal value of k    <= 0\n";exit(1);}
  if(theta<= 0.0)
    {cerr <<"Illegal value of theta<= 0\n";exit(1);}
  if(theta>=90.0)
    {cerr <<"Illegal value of theta>=90\n";exit(1);}
  theta   = (PI/180.0)*theta; //convert to radians
  v0x     = v0*cos(theta);
  v0y     = v0*sin(theta);
  cout    << "# v0x= " << v0x
       << " v0y= " << v0y << endl;
  ofstream myfile("ProjectileAirResistance.dat");
  myfile.precision(17);
//-----------------------------------------------------------
//Compute:
  t = 0.0;
  while( t <= tf ){
    x  = (v0x/k)*(1.0-exp(-k*t));
    y  = (1.0/k)*(v0y+(g/k))*(1.0-exp(-k*t))-(g/k)*t;
    vx = v0x*exp(-k*t);
    vy = (v0y+(g/k))*exp(-k*t)-(g/k);
    myfile << t << " "
        << x  << " " << y  << " "
        << vx << " " << vy << endl;
    t = t + dt;
  }
} //main()
```

We also list the commands of a typical session of the study of the problem:

```
> g++ ProjectileAirResistance.cpp -o pja
```

2.1. MOTION ON THE PLANE

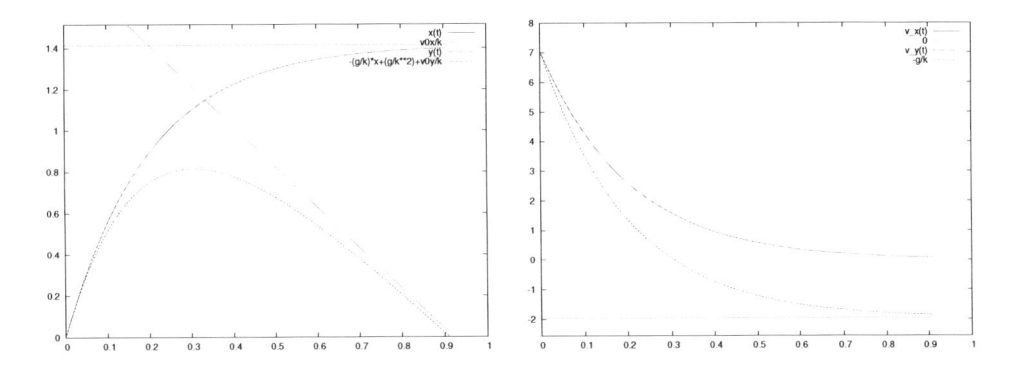

Figure 2.9: The plots of $x(t),y(t)$ (left) and $v_x(t),v_y(t)$ (right) from the data produced by the program `ProjectileAirResistance.cpp` for $k=5.0$, $v_0=10.0$, $\theta=\pi/4$, $t_f=0.91$ and $\delta t=0.001$. We also plot the asymptotes of these functions as $t\to\infty$.

```
# Enter k,v0,theta (in degrees):
5.0 10.0 45
# Enter tf,dt:
0.91 0.001
# k = 5
# v0= 10 theta= 45o (degrees)
# t0= 0  tf= 0.91  dt= 0.001
# v0x= 7.07107   v0y= 7.07107
> gnuplot
gnuplot> v0x = 10*cos(pi/4) ; v0y = 10*sin(pi/4)
gnuplot> g = 9.81 ; k = 5
gnuplot> plot [:][:v0x/k+0.1]  "ProjectileAirResistance.dat" \
         using 1:2 with lines title "x(t)",v0x/k
gnuplot> replot                "ProjectileAirResistance.dat" \
         using 1:3 with lines title "y(t)",\
         -(g/k)*x+(g/k**2)+v0y/k
gnuplot> plot [:][-g/k-0.6:]   "ProjectileAirResistance.dat" \
         using 1:4 with lines title "v_x(t)",0
gnuplot> replot                "ProjectileAirResistance.dat" \
         using 1:5 with lines title "v_y(t)",-g/k
gnuplot> plot                  "ProjectileAirResistance.dat" \
         using 2:3 with lines title "With air resistance k=5.0"
gnuplot> replot                "Projectile.dat"              \
         using 2:3 with lines title "No air resistance k=0.0"
gnuplot> file = "ProjectileAirResistance.dat"
gnuplot> set xrange [0:1.4];set yrange [0:1.4]
gnuplot> t0=0;tf=0.91;dt=0.01
gnuplot> load "animate2D.gnu"
```

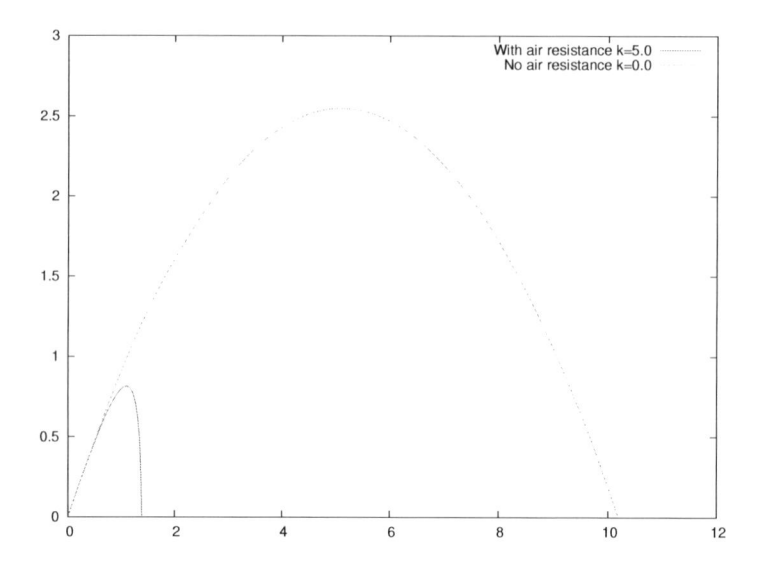

Figure 2.10: Trajectories of the particles shot with $v_0 = 10.0$, $\theta = \pi/4$ in the absence of air resistance and when the air resistance is present in the form $\vec{F} = -mk\vec{v}$ with $k = 5.0$.

Long commands have been continued to the next line as before. We defined the gnuplot variables v0x, v0y, g and k to have the values that we used when running the program. We can use them in order to construct the asymptotes of the plotted functions of time. The results are shown in figures 2.9 and 2.10.

The last example of this section will be that of the anisotropic harmonic oscillator. The force on the particle is

$$F_x = -m\omega_1^2 x \qquad F_y = -m\omega_2^2 y \qquad (2.11)$$

where the "spring constants" $k_1 = m\omega_1^2$ and $k_2 = m\omega_2^2$ are different in the directions of the axes x and y. The solutions of the dynamical equations of motion for $x(0) = A$, $y(0) = 0$, $v_x(0) = 0$ and $v_y(0) = \omega_2 A$ are

$$\begin{aligned} x(t) &= A\cos(\omega_1 t) \qquad y(t) = A\sin(\omega_2 t) \\ v_x(t) &= -\omega_1 A\sin(\omega_1 t) \qquad v_y(t) = \omega_2 A\cos(\omega_2 t)\,. \end{aligned} \qquad (2.12)$$

If the angular frequencies ω_1 and ω_2 satisfy certain relations, the trajectories of the particle are closed and self intersect at a given number of

2.1. MOTION ON THE PLANE

points. The proof of these relations, as well as their numerical confirmation, is left as an exercise for the reader. The program listed below is in the file `Lissajoux.cpp`:

```cpp
//================================================================
// File Lissajous.cpp
// Lissajous curves (special case)
// x(t)= cos(o1 t), y(t)= sin(o2 t)
//----------------------------------------------------------------
#include <iostream>
#include <fstream>
#include <cstdlib>
#include <string>
#include <cmath>
using namespace std;

#define PI 3.1415926535897932

int main(){
//----------------------------------------------------------------
// Declaration of variables
  double x0,y0,R,x,y,vx,vy,t,t0,tf,dt;
  double o1,o2,T1,T2;
  string buf;
//----------------------------------------------------------------
// Ask user for input:
  cout << "# Enter omega1 and omega2:\n";
  cin  >> o1     >> o2;getline(cin,buf);
  cout << "# Enter tf,dt:\n";
  cin  >> tf     >> dt;getline(cin,buf);
  cout <<"# o1= " << o1 << " o2= " << o2 << endl;
  cout <<"# t0= " << 0.0 << " tf= " << tf
       << " dt= " << dt  << endl;
//----------------------------------------------------------------
// Initialize
  if(o1 <=0.0){cerr <<"Illegal value of o1\n";exit(1);}
  if(o2 <=0.0){cerr <<"Illegal value of o2\n";exit(1);}
  T1    = 2.0*PI/o1;
  T2    = 2.0*PI/o2;
  cout    << "# T1= "   << T1 << " T2= " << T2 << endl;
  ofstream myfile("Lissajous.dat");
  myfile.precision(17);
//----------------------------------------------------------------
// Compute:
```

```
    t = t0;
    while( t <= tf ){
      x  =  cos(o1*t);
      y  =  sin(o2*t);
      vx = -o1*sin(o1*t);
      vy =  o2*cos(o2*t);
      myfile << t  << " "
             << x  << " " << y  << " "
             << vx << " " << vy << endl;
      t  =  t + dt;
    }
} //main()
```

We have set $A = 1$ in the program above. The user must enter the two angular frequencies ω_1 and ω_2 and the corresponding times. A typical session for the study of the problem is shown below:

```
> g++ Lissajous.cpp -o lsj
> ./lsj
# Enter omega1 and omega2:
3 5
# Enter tf,dt:
10.0 0.01
# o1= 3 o2= 5
# t0= 0 tf= 10 dt= 0.01
# T1= 2.0944 T2= 1.25664
>gnuplot
gnuplot> plot   "Lissajous.dat" using 1:2 w l t "x(t)"
gnuplot> replot "Lissajous.dat" using 1:3 w l t "y(t)"
gnuplot> plot   "Lissajous.dat" using 1:4 w l t "v_x(t)"
gnuplot> replot "Lissajous.dat" using 1:5 w l t "v_y(t)"
gnuplot> plot   "Lissajous.dat" using 2:3 w l t "x-y for 3:5"
gnuplot> file = "Lissajous.dat"
gnuplot> set xrange [-1.1:1.1]; set yrange [-1.1:1.1]
gnuplot> t0=0;tf=10;dt=0.1
gnuplot> load "animate2D.gnu"
```

The results for $\omega_1 = 3$ and $\omega_2 = 5$ are shown in figure 2.11.

2.2. MOTION IN SPACE

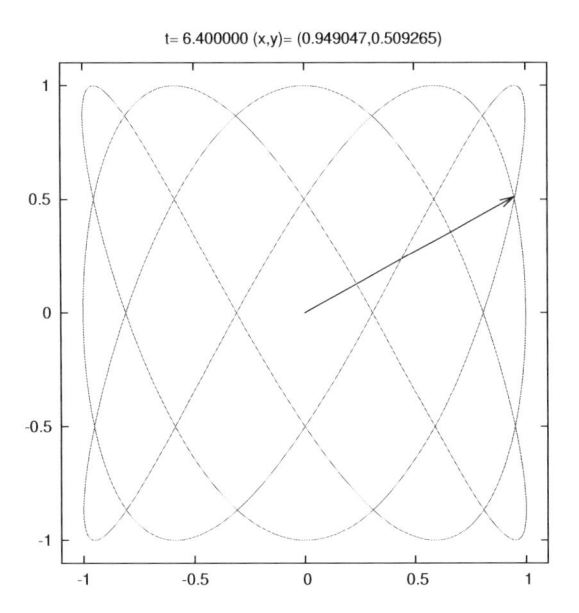

Figure 2.11: The trajectory of the anisotropic oscillator with $\omega_1 = 3$ and $\omega_2 = 5$.

2.2 Motion in Space

By slightly generalizing the methods described in the previous section, we will study the motion of a particle in three dimensional space. All we have to do is to add an extra equation for the coordinate $z(t)$ and the component of the velocity $v_z(t)$. The structure of the programs will be exactly the same as before.

The first example is the conical pendulum, which can be seen in figure 2.12. The particle moves on the xy plane with constant angular velocity ω. The equations of motion are derived from the relations

$$T_z = T\cos\theta = mg \qquad T_{xy} = T\sin\theta = m\omega^2 r, \qquad (2.13)$$

where $r = l\sin\theta$. Their solution[15] is

$$\begin{aligned} x(t) &= r\cos\omega t \\ y(t) &= r\sin\omega t \\ z(t) &= -l\cos\theta, \end{aligned} \qquad (2.14)$$

---

[15]One has to choose appropriate initial conditions. Exercise: find them!

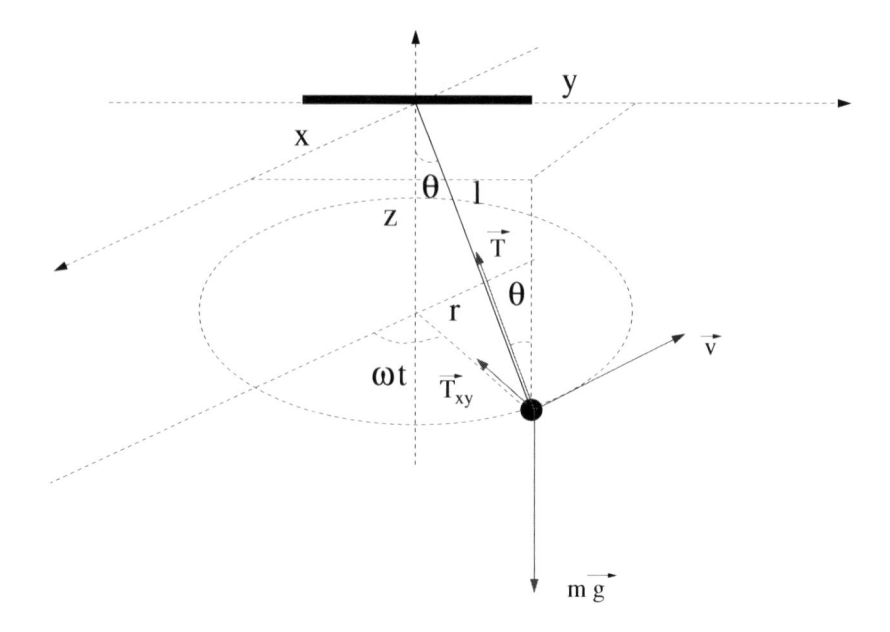

Figure 2.12: The conical pendulum of the program `ConicalPendulum.cpp`.

where we have to substitute the values

$$\cos\theta = \frac{g}{\omega^2 l}$$
$$\sin\theta = \sqrt{1-\cos^2\theta}$$
$$r = \frac{g}{\omega^2}\frac{\sin\theta}{\cos\theta}. \qquad (2.15)$$

For the velocity components we obtain

$$v_x = -r\omega\sin\omega t$$
$$v_y = r\omega\cos\omega t$$
$$v_z = 0. \qquad (2.16)$$

Therefore we must have

$$\omega \geq \omega_{\min} = \sqrt{\frac{g}{l}}, \qquad (2.17)$$

and when $\omega \to \infty$, $\theta \to \pi/2$.

2.2. MOTION IN SPACE

In the program that we will write, the user must enter the parameters l, ω, the final time t_f and the time step δt. We take $t_0 = 0$. The convention that we follow for the output of the results is that they should be written in a file where the first 7 columns are the values of t, x, y, z, v_x, v_y and v_z. The full program is listed below:

```cpp
//========================================
// File ConicalPendulum.cpp
// Set pendulum angular velocity omega and display motion in 3D
//----------------------------------------
#include <iostream>
#include <fstream>
#include <cstdlib>
#include <string>
#include <cmath>
using namespace std;

#define PI 3.1415926535897932
#define g  9.81

int main(){
//----------------------------------------
// Declaration of variables
  double l,r,x,y,z,vx,vy,vz,t,tf,dt;
  double theta,cos_theta,sin_theta,omega;
  string buf;
//----------------------------------------
// Ask user for input:
  cout << "# Enter l,omega:\n";
  cin  >> l  >> omega;          getline(cin,buf);
  cout << "# Enter tf,dt:\n";
  cin  >> tf >> dt;             getline(cin,buf);
  cout << "# l= "    << l    << " omega= " << omega << endl;
  cout << "# T=  "   << 2.0*PI/omega
       << " omega_min= " << sqrt(g/l) << endl;
  cout <<"# t0= "    << 0.0  << " tf= "    << tf
       << " dt= "    << dt   << endl;
//----------------------------------------
// Initialize
  cos_theta = g/(omega*omega*l);
  if( cos_theta >= 1.0){
    cerr << "cos(theta)>= 1\n";
    exit(1);
```

```
}
   sin_theta = sqrt(1.0-cos_theta*cos_theta);
   z  = -g/(omega*omega); //they remain constant throught;
   vz = 0.0;              //the motion
   r  = g/(omega*omega)*sin_theta/cos_theta;
   ofstream myfile("ConicalPendulum.dat");
   myfile.precision(17);
//------------------------------------------------------------
//Compute:
   t   = 0.0;
   while( t <= tf ){
     x  =  r*cos(omega*t);
     y  =  r*sin(omega*t);
     vx = -r*sin(omega*t)*omega;
     vy =  r*cos(omega*t)*omega;
     myfile << t     << " "
            << x  << " " << y  << " " << z  << " "
            << vx << " " << vy << " " << vz << " "
            << endl;
     t  = t + dt;
   }
} //main()
```

In order to compile and run the program we can use the commands shown below:

```
> g++  ConicalPendulum.cpp -o cpd
> ./cpd
# Enter l,omega:
1.0  6.28
# Enter tf,dt:
10.0 0.01
# l= 1 omega= 6.28
# T= 1.00051 omega_min= 3.13209
# t0= 0 tf= 10 dt= 0.01
```

The results are recorded in the file ConicalPendulum.dat. In order to plot the functions $x(t)$, $y(t)$, $z(t)$, $v_x(t)$, $v_y(t)$, $v_z(t)$ we give the following gnuplot commands:

```
> gnuplot
gnuplot> plot    "ConicalPendulum.dat" u 1:2 w l t "x(t)"
gnuplot> replot  "ConicalPendulum.dat" u 1:3 w l t "y(t)"
```

2.2. MOTION IN SPACE

```
gnuplot> replot "ConicalPendulum.dat" u 1:4 w l t "z(t)"
gnuplot> plot   "ConicalPendulum.dat" u 1:5 w l t "v_x(t)"
gnuplot> replot "ConicalPendulum.dat" u 1:6 w l t "v_y(t)"
gnuplot> replot "ConicalPendulum.dat" u 1:7 w l t "v_z(t)"
```

The results are shown in figure 2.13. In order to make a three dimen-

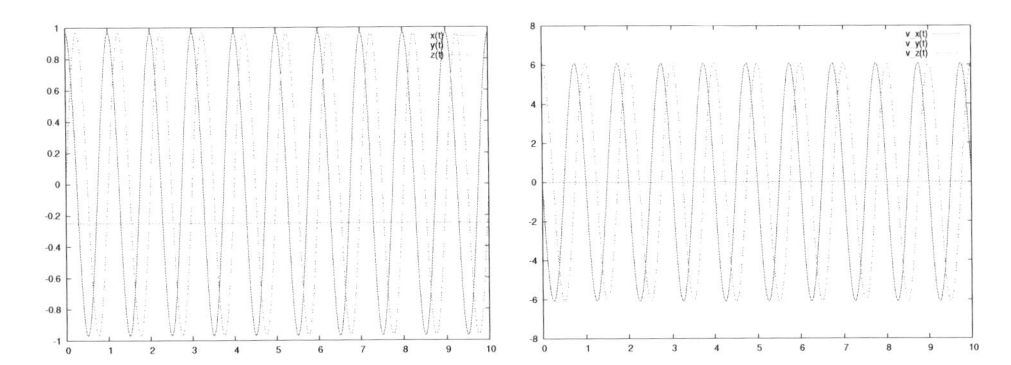

Figure 2.13: The plots of the functions $x(t), y(t), z(t), v_x(t), v_y(t), v_z(t)$ of the program ConicalPendulum.cpp for $\omega = 6.28$, $l = 1.0$.

sional plot of the trajectory, we should use the gnuplot command splot:

```
gnuplot> splot "ConicalPendulum.dat" u 2:3:4 w l t "r(t)"
```

The result is shown in figure 2.14. We can click on the trajectory and rotate it and view it from a different angle. We can change the plot limits with the command:

```
gnuplot> splot [-1.1:1.1][-1.1:1.1][-0.3:0.0] \
         "ConicalPendulum.dat" using 2:3:4 w l t "r(t)"
```

We can animate the trajectory of the particle by using the file animate3D.gnu from the accompanying software. The commands are similar to the ones we had to give in the two dimensional case for the planar trajectories when we used the file animate2D.gnu:

```
gnuplot> file = "ConicalPendulum.dat"
gnuplot> set xrange [-1.1:1.1]; set yrange [-1.1:1.1]
gnuplot> set zrange [-0.3:0]
```

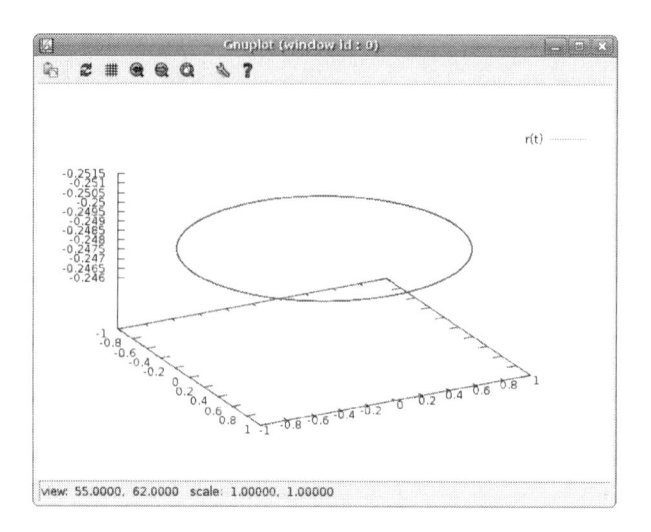

Figure 2.14: The plot of the particle trajectory $\vec{r}(t)$ of the program ConicalPendulum.cpp for $\omega = 6.28$, $l = 1.0$. We can click and drag with the mouse on the window and rotate the curve and see it from a different angle. At the bottom left of the window, we see the viewing direction, given by the angles $\theta = 55.0$ degrees (angle with the z axis) and $\phi = 62$ degrees (angle with the x axis).

```
gnuplot> t0=0;tf=10;dt=0.1
gnuplot> load "animate3D.gnu"
```

The result can be seen in figure 2.15. The program animate3D.gnu can be used on the data file of any program that prints t x y z as the first words on each of its lines. All we have to do is to change the value of the file variable in gnuplot.

Next, we will study the trajectory of a charged particle in a homogeneous magnetic field $\vec{B} = B\hat{z}$. At time t_0, the particle is at $\vec{r}_0 = x_0\hat{x}$ and its velocity is $\vec{v}_0 = v_{0y}\hat{y} + v_{0z}\hat{z}$, see figure 2.16. The magnetic force on the particle is $\vec{F} = q(\vec{v} \times \vec{B}) = qBv_y\hat{x} - qBv_x\hat{y}$ and the equations of motion are

$$\begin{aligned} a_x &= \frac{dv_x}{dt} = \omega v_y \qquad \omega \equiv \frac{qB}{m} \\ a_y &= \frac{dv_y}{dt} = -\omega v_x \\ a_z &= 0. \end{aligned} \qquad (2.18)$$

By integrating the above equations with the given initial conditions we

2.2. MOTION IN SPACE

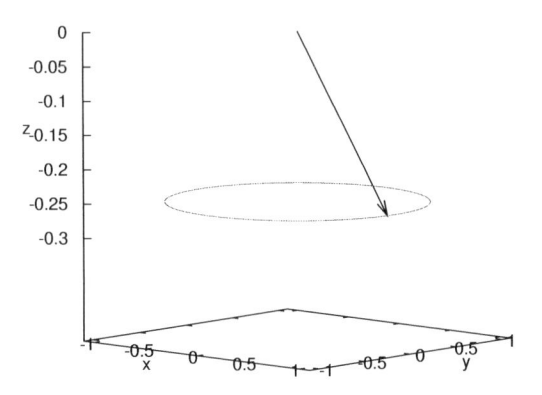

Figure 2.15: The particle trajectory $\vec{r}(t)$ computed by the program ConicalPendulum.cpp for $\omega = 6.28$, $l = 1.0$ and plotted by the gnuplot script animate3D.gnu. The title of the plot shows the current time and the particles coordinates.

obtain

$$\begin{aligned} v_x(t) &= v_{0y} \sin \omega t \\ v_y(t) &= v_{0y} \cos \omega t \\ v_z(t) &= v_{0z}\,. \end{aligned} \qquad (2.19)$$

Integrating once more, we obtain the position of the particle as a function of time

$$\begin{aligned} x(t) &= \left(x_0 + \frac{v_{0y}}{\omega}\right) - \frac{v_{0y}}{\omega} \cos \omega t = x_0 \cos \omega t \\ y(t) &= \frac{v_{0y}}{\omega} \sin \omega t = -x_0 \sin \omega t \quad \text{με} \quad x_0 = -\frac{v_{0y}}{\omega} \\ z(t) &= v_{0z} t\,, \end{aligned} \qquad (2.20)$$

where we have chosen $x_0 = -v_{0y}/\omega$. This choice places the center of the circle, which is the projection of the trajectory on the xy plane, to be at

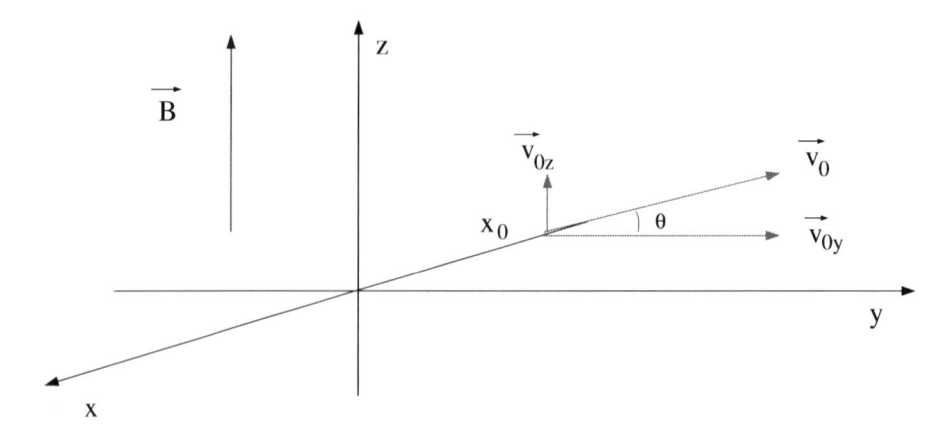

Figure 2.16: A particle at time $t_0 = 0$ is at the position $\vec{r}_0 = x_0\hat{x}$ with velocity $\vec{v}_0 = v_{0y}\hat{y} + v_{0z}\hat{z}$ in a homogeneous magnetic field $\vec{B} = B\hat{z}$.

the origin of the coordinate system. The trajectory is a helix with radius $R = -x_0$ and pitch $v_{0z}T = 2\pi v_{0z}/\omega$.

We are now ready to write a program that calculates the trajectory given by (2.20). The user enters the parameters v_0 and θ, shown in figure 2.16, as well as the angular frequency ω (Larmor frequency). The components of the initial velocity are $v_{0y} = v_0\cos\theta$ and $v_{0z} = v_0\sin\theta$. The initial position is calculated from the equation $x_0 = -v_{0y}/\omega$. The program can be found in the file ChargeInB.cpp:

```
//===========================================
// File ChargeInB.cpp
//A charged particle of mass m and charge q enters a magnetic
// field B in +z direction. It enters with velocity
//v0x=0,v0y=v0 cos(theta),v0z=v0 sin(theta), 0<=theta<pi/2
// at the position x0=-v0y/omega, omega=q B/m
//
// Enter v0 and theta and see trajectory from
//t0=0 to tf at step dt
//-------------------------------------------
#include <iostream>
#include <fstream>
#include <cstdlib>
#include <string>
#include <cmath>
```

2.2. MOTION IN SPACE

```cpp
using namespace std;

#define PI 3.1415926535897932

int main(){
//------------------------------------------------------------
//Declaration of variables
  double x,y,z,vx,vy,vz,t,tf,dt;
  double x0,y0,z0,v0x,v0y,v0z,v0;
  double theta,omega;
  string buf;
//------------------------------------------------------------
//Ask user for input:
  cout << "# Enter omega:\n";
  cin  >> omega;           getline(cin,buf);
  cout << "# Enter v0, theta (degrees):\n";
  cin  >> v0 >> theta;     getline(cin,buf);
  cout << "# Enter tf,dt:\n";
  cin  >> tf >> dt;        getline(cin,buf);
  cout << "# omega= " << omega
       << " T= "      << 2.0*PI/omega << endl;
  cout << "# v0=    " << v0
       << " theta= " << theta
       << "o(degrees)"<< endl;
  cout <<"# t0= "    << 0.0 << " tf= "    << tf
       << " dt= "    << dt  << endl;
//------------------------------------------------------------
//Initialize
  if(theta<0.0 || theta>=90.0) exit(1);
  theta = (PI/180.0)*theta; //convert to radians
  v0y   = v0*cos(theta);
  v0z   = v0*sin(theta);
  cout << "# v0x= " << 0.0
       << "  v0y= " << v0y
       << "  v0z= " << v0z << endl;
  x0    = - v0y/omega;
  cout << "# x0= " << x0
       << "  y0= " << y0
       << "  z0= " << z0   << endl;
  cout << "# xy plane: Circle with center (0,0) and R= "
       << abs(x0)          << endl;
  cout << "# step of helix: s=v0z*T= "
       << v0z*2.0*PI/omega << endl;
  ofstream myfile("ChargeInB.dat");
  myfile.precision(17);
```

```
//-----------------------------------------
//Compute:
  t    =  0.0;
  vz   =  v0z;
  while( t <= tf ){
    x  =  x0*cos(omega*t);
    y  = -x0*sin(omega*t);
    z  =  v0z*t;
    vx =  v0y*sin(omega*t);
    vy =  v0y*cos(omega*t);
    myfile << t     << " "
           << x  << " " << y << " " << z << " "
           << vx << " " << vy << " " << vz << " "
           << endl;
    t  =  t + dt;
  }
} //main()
```

A typical session in which we calculate the trajectories shown in figures 2.17 and 2.18 is shown below:

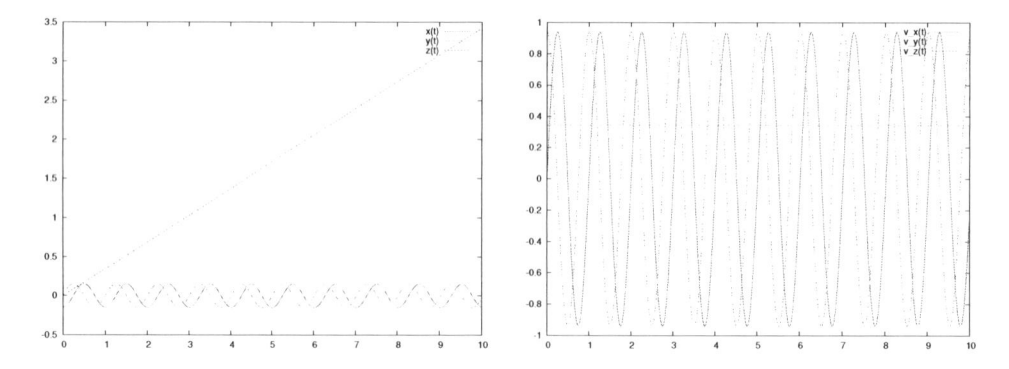

Figure 2.17: The plots of the $x(t), y(t), z(t), v_x(t), v_y(t), v_z(t)$ functions calculated by the program in ChargeInB.cpp for $\omega = 6.28$, $x_0 = 1.0$, $\theta = 20$ degrees.

```
> g++  ChargeInB.cpp -o chg
> ./chg
# Enter omega:
6.28
# Enter v0, theta (degrees):
1.0 20
```

```
# Enter tf,dt:
10 0.01
# omega= 6.28 T= 1.00051
# v0=    1 theta=   20o(degrees)
# t0= 0  tf= 10  dt= 0.01
# v0x= 0  v0y= 0.939693  v0z= 0.34202
# x0= -0.149633  y0= 0  z0= 3.11248e-317
# xy plane: Circle with center (0,0) and R= 0.149633
# step of helix: s=v0z*T= 0.342194
> gnuplot
gnuplot> plot    "ChargeInB.dat" u 1:2   w l title "x(t)"
gnuplot> replot "ChargeInB.dat" u 1:3   w l title "y(t)"
gnuplot> replot "ChargeInB.dat" u 1:4   w l title "z(t)"
gnuplot> plot    "ChargeInB.dat" u 1:5   w l title "v_x(t)"
gnuplot> replot "ChargeInB.dat" u 1:6   w l title "v_y(t)"
gnuplot> replot "ChargeInB.dat" u 1:7   w l title "v_z(t)"
gnuplot> splot  "ChargeInB.dat" u 2:3:4 w l title "r(t)"
gnuplot> file = "ChargeInB.dat"
gnuplot> set xrange [-0.65:0.65];set yrange [-0.65:0.65]
gnuplot> set zrange [0:1.3]
gnuplot> t0=0;tf=3.5;dt=0.1
gnuplot> load "animate3D.gnu"
```

2.3 Trapped in a Box

In this section we will study the motion of a particle that is free, except when bouncing elastically on a wall or on certain obstacles. This motion is calculated by *approximate* algorithms that introduce systematic errors. These types of errors[16] are also encountered in the study of more complicated dynamics, but the simplicity of the problem will allow us to control them in a systematic and easy to understand way.

2.3.1 The One Dimensional Box

The simplest example of such a motion is that of a particle in a "one dimensional box". The particle moves freely on the x axis for $0 < x < L$,

---
[16]In the previous sections, our calculations had a small systematic error due to the approximate nature of numerical floating point operations which approximate exact real number calculations. But the algorithms used were not introducing systematic errors like in the cases discussed in this section.

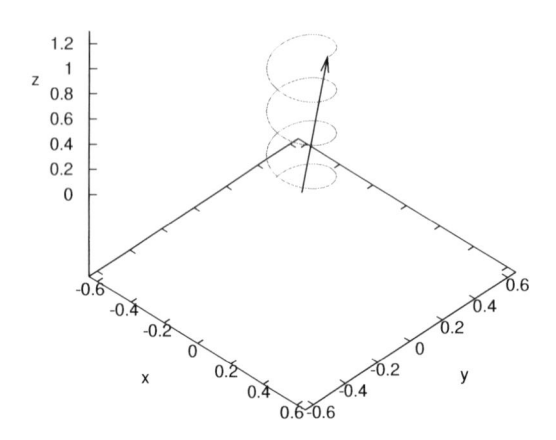

Figure 2.18: The trajectory $\vec{r}(t)$ calculated by the program in ChargeInB.cpp for $\omega = 6.28$, $v_0 = 1.0$, $\theta = 20$ degrees as shown by the program animate3D.gnu. The current time and the coordinates of the particle are printed on the title of the plot.

as can be seen in figure 2.19. When it reaches the boundaries $x = 0$ and $x = L$ it bounces and its velocity instantly reversed. Its potential energy is

$$V(x) = \begin{cases} 0 & 0 < x < L \\ +\infty & \text{elsewhere} \end{cases}, \quad (2.21)$$

which has the shape of an infinitely deep well. The force $F = -dV(x)/dx = 0$ within the box and $F = \pm\infty$ at the position of the walls.

Initially we have to know the position of the particle x_0 as well as its velocity v_0 (the sign of v_0 depends on the direction of the particle's motion) at time t_0. As long as the particle moves within the box, its motion is free and

$$\begin{aligned} x(t) &= x_0 + v_0(t - t_0) \\ v(t) &= v_0. \end{aligned} \quad (2.22)$$

For a small enough change in time δt, so that there is no bouncing on

2.3. TRAPPED IN A BOX

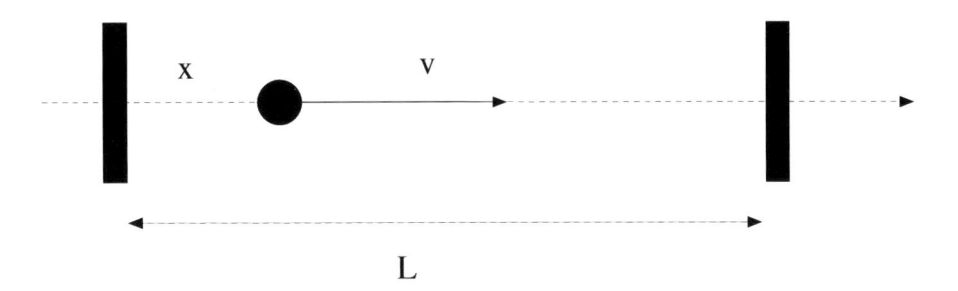

Figure 2.19: A particle in a one dimensional box with its walls located at $x = 0$ and $x = L$.

the wall in the time interval $(t, t + \delta t)$, we have that

$$\begin{aligned} x(t + \delta t) &= x(t) + v(t)\delta t \\ v(t + \delta t) &= v(t) \,. \end{aligned} \quad (2.23)$$

Therefore we could use the above relations in our program and when the particle bounces off a wall we could simple reverse its velocity $v(t) \to -v(t)$. The devil is hiding in the word "when". Since the time interval δt is finite in our program, there is no way to know the instant of the collision with accuracy better than $\sim \delta t$. However, our algorithm will change the direction of the velocity at time $t + \delta t$, when the particle will have already crossed the wall. This will introduce a systematic error, which is expected to decrease with decreasing δt. One way to implement the above idea is by constructing the loop

```
while(t <= tf){
    x += v*dt;
    t +=   dt;
    if(x < 0.0 || x > L) v = -v;
}
```

where the last line gives the testing condition for the wall collision and the subsequent change of the velocity.

The full program that realizes the proposed algorithm is listed below and can be found in the file box1D_1.cpp. The user can set the size of the box L, the initial conditions x0 and v0 at time t0, the final time tf and the time step dt:

```cpp
//=============================================
//File box1D_1.cpp
//Motion of a free particle in a box  0<x<L
//Use integration with time step dt: x = x + v*dt
//---------------------------------------------
#include <iostream>
#include <iomanip>
#include <fstream>
#include <cstdlib>
#include <string>
#include <cmath>
using namespace std;

int main(){
//---------------------------------------------
//Declaration of variables
  float L,x0,v0,t0,tf,dt,t,x,v;
  string buf;
//---------------------------------------------
//Ask user for input:
  cout << "# Enter L:\n";
  cin  >> L;            getline(cin,buf);
  cout << "# L = " << L << endl;
  cout << "# Enter x0,v0:\n";
  cin  >> x0 >> v0;     getline(cin,buf);
  cout << "# x0= " << x0 << " v0= " << v0 << endl;
  cout << "# Enter t0,tf,dt:\n";
  cin  >> t0 >> tf >> dt; getline(cin,buf);
  cout << "# t0= " << t0 << " tf= " << tf
       << " dt= " << dt << endl;
  if( L <= 0.0f){cerr << "L <=0\n"; exit(1);}
  if( x0<  0.0f){cerr << "x0<=0\n"; exit(1);}
  if( x0>  L   ){cerr << "x0> L\n"; exit(1);}
  if( v0== 0.0f){cerr << "v0 =0\n"; exit(1);}
//---------------------------------------------
//Initialize
  t = t0;
  x = x0;
  v = v0;
  ofstream myfile("box1D_1.dat");
  myfile.precision(9); // float precision (and a bit more...)
//---------------------------------------------
//Compute:
  while(t <= tf){
    myfile << setw(17) << t << " "    // set width of field
```

2.3. TRAPPED IN A BOX

```
           << setw(17) << x << " "      // to 17 characters
           << setw(17) << v << '\n';    // using setw(17)
    x += v*dt;
    t +=   dt;
    if(x < 0.0f || x > L) v = -v;
  }
  myfile.close();
} //main()
```

In this section we will study the effects of roundoff errors in numerical computations. Computers store numbers in memory, which is finite. Therefore, real numbers are represented in some approximation that depends on the amount of memory that is used for their storage. This approximation corresponds to what is termed as floating point numbers. C++ is supposed to provide at least three basic types of floating point numbers, float, double and long double. In most implementations[17], float uses 4 bytes of memory and double 8. In this case, float has an accuracy to, approximately, 7 significant digits and double 17. See Chapter 1 of [8] and [14] for details. Moreover, float represent numbers with magnitude in the, approximate, range $(10^{-38}, 10^{38})$ while double in $(10^{-308}, 10^{308})$. Note that variables of the integer type (int, long, ...) are exact representations of integers, whereas floating point numbers are approximations to reals.

In the program shown above, we used numbers of the float type instead of double in order to exaggerate roundoff errors. This way we can study the dependence of this type of errors on the accuracy of the floating point numbers used in a program[18]. In order to do that, we declared the floating point variables as float:

---

[17]Notice that the C++ standard states that the value representation of floating-point types is implementation-defined. The C standard requires that the type double provides at least as much precision as float, and the type long double provides at least as much precision as double. The gcc 5.4 version that we are using in this book represents float using 4 bytes and double with 8, but you should check whether this is true with the compiler that you are using.

[18]The use of float can be the preferred choice of a programmer for some applications. First, in order to save memory, because float occupies half the memory of a double. Second, it is not always true that increasing the accuracy of floating point numbers will increase the accuracy of a computation, although in most of the cases it will. The wisdom of the field is to always to use as much accuracy as you need and no more!

```
float L,x0,v0,t0,tf,dt,t,x,v;
```

We also used *numerical constants* of type `float`. This is indicated by the letter `f` at the end of their names: `2.0` is a constant of type `double` (the C++ default), whereas `2.0f` is a constant of type `float`. Determining the accuracy of floating point constants is a thorny issue that can be the cause on introducing subtle bugs in a program and the programmer should be very careful about doing it carefully.

Finally we changed the form of the output. Since a `float` represents a real number with at most 7 significant digits, there is no point of printing more. That is why we used the statements

```
myfile.precision(9);
myfile.setw(17);
```

For purposes of studying the numerical accuracy of our results, we used 9 digits of output, which is, of course, slightly redundant. `setw(17)` prints the numbers of the next output of the stream `myfile` using *at least* 17 character spaces. This improves the legibility of the results when inspecting the output files. The use of `setw` requires the header `iomanip`.

The computed data is recorded in the file `box1D_1.dat` in three columns. Compiling, running and plotting the trajectory using gnuplot can be done as follows:

```
> g++ box1D_1.cpp -o box1
> ./box1
# Enter L:
10
# L = 10
# Enter x0,v0:
0 1.0
# x0= 0 v0= 1
# Enter t0,tf,dt:
0 100 0.01
# t0= 0 tf= 100 dt= 0.01
> gnuplot
gnuplot> plot "box1D_1.dat" using  1:2 w l title "x(t)",\
                                 0 notitle,10 notitle
gnuplot> plot [:][-1.2:1.2] "box1D_1.dat" \
                           using  1:3 w l title "v(t)"
```

2.3. TRAPPED IN A BOX

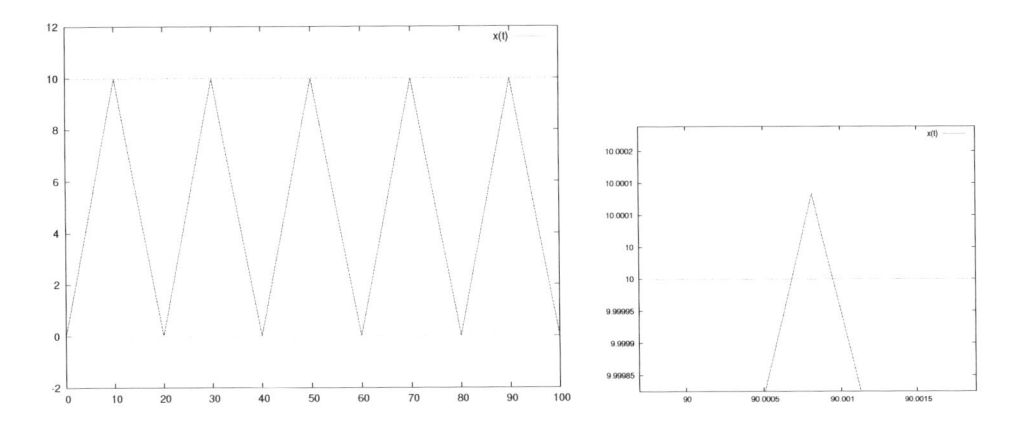

Figure 2.20: The trajectory $x(t)$ of a particle in a box with $L=10$, $x_0=0.0$, $v_0=1.0$, $t_0=0$, $\delta t = 0.01$. The plot to the right magnifies a detail when $t\approx 90$ which exposes the systematic errors in determining the exact moment of the collision of the particle with the wall at $t_k = 90$ and the corresponding maximum value of $x(t)$, $x_m = L = 10.0$.

The trajectory $x(t)$ is shown in figure 2.20. The effects of the systematic errors can be easily seen by noting that the expected collisions occur every $T/2 = L/v = 10$ units of time. Therefore, on the plot to the right of figure 2.20, the reversal of the particle's motion should have occurred at $t=90$, $x=L=10$.

The reader should have already realized that the above mentioned error can be made to vanish by taking arbitrarily small δt. Therefore, we naively expect that as long as we have the necessary computer power to take δt as small as possible and the corresponding time intervals as many as possible, we can achieve any precision that we want. Well, that is true only up to a point. The problem is that the next position is determined by the addition operation x+v*dt and the next moment in time by t+dt. Floating point numbers of the float type have a maximum accuracy of approximately 7 significant decimal digits. Therefore, if the operands x and v*dt are real numbers differing by more than 7 orders of magnitude (v*dt$\lesssim 10^{-7}$ x), the effect of the addition x+v*dt=x, which is null! The reason is that the floating point unit of the processor has to convert both numbers x and v*dt into a representation having the same exponent and in doing so, the corresponding significant digits of the smaller number v*dt are lost. The result is less catastrophic when v*dt$\lesssim 10^{-a}$ x with $0 < a < 7$, but some degree of accuracy is also lost at each addition operation. And since we have accumulation of such errors over many intervals t→t+dt, the error can become significant and destroy our calculation for large enough times. A similar error accumulates in the determination of the next instant of time t+dt, but we will discuss below how to make this contribution to the total error negligible. The above mentioned errors can become less detrimental by using floating point numbers of greater accuracy than the float type. For example double numbers have approximately 17 significant decimal digits. But again, the precision is finite and the same type of errors are there only to be revealed by a more demanding and complicated calculation.

The remedy to such a problem can only be a change in the algorithm. This is not always possible, but in the case at hand this is easy to do. For example, consider the equation that gives the position of a particle in free motion

$$x(t) = x_0 + v_0(t - t_0). \tag{2.24}$$

Let's use the above relation for the parts of the motion between two collisions. Then, all we have to do is to reverse the direction of the motion and reset the initial position and time to be the position and time of the collision. This can be done by using the loop:

```
while(t <= tf){
  x = x0 + v0*(t-t0);
  if(x < 0.0f || x > L) {
    x0= x;
    t0= t;
    v0= -v0;
  }
  t += dt;
}
```

In the above algorithm, the error in the time of the collision is not vanishing but we don't have the "instability" problem of the dt→ 0 limit[19]. Therefore we can isolate and study the effect of each type of error. The full program that implements the above algorithm is given below and can be found in the file box1D_2.cpp:

---

[19]We still have this problem in the t=t+dt operation. See discussion in the next section.

2.3. TRAPPED IN A BOX

```cpp
//================================================
//File box1D_2.cpp
//Motion of a free particle in a box   0<x<L
//Use constant velocity equation: x = x0 + v0*(t-t0)
//Reverse velocity and redefine x0,t0 on boundaries
//------------------------------------------------
#include <iostream>
#include <iomanip>
#include <fstream>
#include <cstdlib>
#include <string>
#include <cmath>
using namespace std;

int main(){
//------------------------------------------------
//Declaration of variables
  float L,x0,v0,t0,tf,dt,t,x,v;
  string buf;
//------------------------------------------------
//Ask user for input:
  cout << "# Enter L:\n";
  cin >> L;            getline(cin,buf);
  cout << "# L = " << L << endl;
  cout << "# Enter x0,v0:\n";
  cin >> x0 >> v0;     getline(cin,buf);
  cout << "# x0= " << x0 << " v0= " << v0 << endl;
  cout << "# Enter t0,tf,dt:\n";
  cin >> t0 >> tf >> dt; getline(cin,buf);
  cout << "# t0= " << t0 << " tf= " << tf
       << " dt= " << dt << endl;
  if( L <= 0.0f){cerr << "L <=0\n"; exit(1);}
  if( x0<  0.0f){cerr << "x0<=0\n"; exit(1);}
  if( x0>  L   ){cerr << "x0> L\n"; exit(1);}
  if( v0== 0.0f){cerr << "v0 =0\n"; exit(1);}
//------------------------------------------------
//Initialize
  t = t0;
  ofstream myfile("box1D_2.dat");
  myfile.precision(9); // float precision (and a bit more...)
//------------------------------------------------
//Compute:
  while(t <= tf){
    x = x0 + v0*(t-t0);
    myfile << setw(17) << t << " "
```

```
         << setw(17) << x  << " "
         << setw(17) << v0 << '\n';
    if(x < 0.0f || x > L) {
      x0= x;
      t0= t;
      v0= -v0;
    }
    t  += dt;
  }
  myfile.close();
} //main()
```

Compiling and running the above program is done as before and the results are stored in the file box1D_2.dat.

2.3.2 Errors

In this section we will study the effect of the systematic errors that we encountered in the previous section in more detail. We considered two types of errors: First, the systematic error of determining the instant of the collision of the particle with the wall. This error is *reduced* by taking a smaller time step δt. Then, the systematic error that accumulates with each addition of two numbers with increasing difference in their orders of magnitude. This error is *increased* with decreasing δt. The competition of the two effects makes the optimal choice of δt the result of a careful analysis. Such a situation is found in many interesting problems, therefore it is quite instructive to study it in more detail.

When the exact solution of the problem is not known, the systematic errors are controlled by studying the behavior of the solution as a function of δt. If the solutions are converging in a region of values of δt, one gains confidence that the true solution has been determined up to the accuracy of the convergence.

In the previous sections, we studied two different algorithms, programmed in the files box1D_1.cpp and box1D_2.cpp. We will refer to them as "method 1" and "method 2" respectively. We will study the convergence of the results as $\delta t \to 0$ by fixing all the parameters except δt and then study the dependence of the results on δt. We will take $L = 10$, $v_0 = 1.0$, $x_0 = 0.0$, $t_0 = 0.0$, $t_f = 95.0$, so that the particle will collide with the wall every 10 units of time. We will measure the position of

2.3. TRAPPED IN A BOX

the particle $x(t \approx 95)$[20] as a function of δt and study its convergence to a limit[21] as $\delta t \to 0$.

The analysis requires a lot of repetitive work: Compiling, setting the parameter values, running the program and calculating the value of $x(t \approx 95)$ for many values of δt. We write the values of the parameters read by the program in a file box1D_anal.in:

```
10          L
0  1.0      x0 v0
0  95  0.05 t0 tf dt
```

Then we compile the program

```
> g++ box1D_1.cpp -o box
```

and run it with the command:

```
> cat box1D_anal.in | ./box
```

By using the pipe |, we send the contents of box1D_anal.in to the stdin of the command ./box by using the command cat. The result $x(t \approx 95)$ can be found in the last line of the file box1D_1.dat:

```
> tail -n 1 box1D_1.dat
 94.9511948    5.45000267   -1.
```

The third number in the above line is the value of the velocity. In a file box1D_anal.dat we write δt and the first two numbers coming out from the command tail. Then we decrease the value $\delta t \to \delta t/2$ in the file box1D_anal.in and run again. We repeat for 12 more times until δt reaches the value[22] 0.000012. We do the same[23] using method 2 and we place the results for $x(t \approx 95)$ in two new columns in the file box1D_anal.dat. The result is

---

[20]Note the $\approx$!

[21]Of course we know the answer: $x(95) = 5$.

[22]Try the command sed 's/0.05/0.025/' box1D_anal.in | ./box by changing 0.025 with the desired value of δt.

[23]See the shell script box1D_anal.csh as a suggestion on how to automate this boring process.

```
# ------------------------------------------------
#   dt        t1_95      x1(95)     x2(95)
# ------------------------------------------------
0.050000  94.95119   5.450003   5.550126
0.025000  94.97849   5.275011   5.174837
0.012500  94.99519   5.124993   5.099736
0.006250  94.99850   4.987460   5.063134
0.003125  94.99734   5.021894   5.035365
0.001563  94.99923   5.034538   5.017764
0.000781  94.99939   4.919035   5.011735
0.000391  94.99979   4.695203   5.005493
0.000195  95.00000   5.434725   5.001935
0.000098  94.99991   5.528124   5.000745
0.000049  94.99998   3.358000   5.000330
0.000024  94.99998   2.724212   5.000232
0.000012  94.99999   9.240705   5.000158
```

Convergence is studied in figure 2.21. The 1st method maximizes its accuracy for $\delta t \approx 0.01$, whereas for $\delta t < 0.0001$ the error becomes $> 10\%$ and the method becomes useless. The 2nd method has much better behavior that the 1st one.

We observe that as δt decreases, the final value of t approaches the expected $t_f = 95$. Why don't we obtain $t = 95$, especially when $t/\delta t$ is an integer? How many steps does it really take to reach $t \approx 95$, when the expected number of those is $\approx 95/\delta t$? Each time you take a measurement, issue the command

```
> wc -l box1D_1.dat
```

which measures the number of lines in the file box1D_1.dat and compare this number with the expected one. The result is interesting:

```
# ------------------------------------
#   dt       N        N0
# ------------------------------------
0.050000  1900     1900
0.025000  3800     3800
0.012500  7601     7600
0.006250  15203    15200
0.003125  30394    30400
0.001563  60760    60780
```

2.3. TRAPPED IN A BOX

```
0.000781  121751   121638
0.000391  243753   242966
0.000195  485144   487179
0.000098  962662   969387
0.000049  1972589  1938775
0.000024  4067548  3958333
0.000012  7540956  7916666
```

where the second column has the number of steps computed by the program and the third one has the expected number of steps. We observe that the accuracy decreases with decreasing δt and in the end the difference is about 5%! Notice that the last line should have given $t_f = 0.000012 \times 7540956 \approx 90.5$, an error comparable to the period of the particle's motion.

We conclude that one important source of accumulation of systematic errors is the calculation of time. This type of errors become more significant with decreasing δt. We can improve the accuracy of the calculation significantly if we use the multiplication t=t0+i*dt instead of the addition t=t+dt, where i is a step counter:

```
//t = t + dt;    // Not accurate,    avoid
  t = t0 + i*dt; // Better accuracy, prefer
```

The main loop in the program box1D_1.cpp becomes:

```
t = t0;
x = x0;
v = v0;
i = 0;
while(t <= (tf+1.0e-5f)){
   i +=    1;
   x += v*dt;
   t  = t0 + i*dt;
   if(x < 0.0f || x > L) v = -v;
}
```

The full program can be found in the file box1D_4.cpp of the accompanying software. We call this "method 3". We perform the same change in the file box1D_2.cpp, which we store in the file box1D_5.cpp. We call this "method 4". We repeat the same analysis using methods 3 and 4 and we find that the problem of calculating time accurately practically vanishes. The result of the analysis can be found on the right plot of figure 2.21. Methods 2 and 4 have no significant difference in their results, whereas methods 1 and 3 do have a dramatic difference, with method 3 decreasing the error more than tenfold. The problem of the increase of systematic errors with decreasing δt does not vanish completely due to the operation x=x+v*dt. This type of error is harder to deal with and one has to invent more elaborate algorithms in order to reduce it significantly. This will be discussed further in chapter 4.

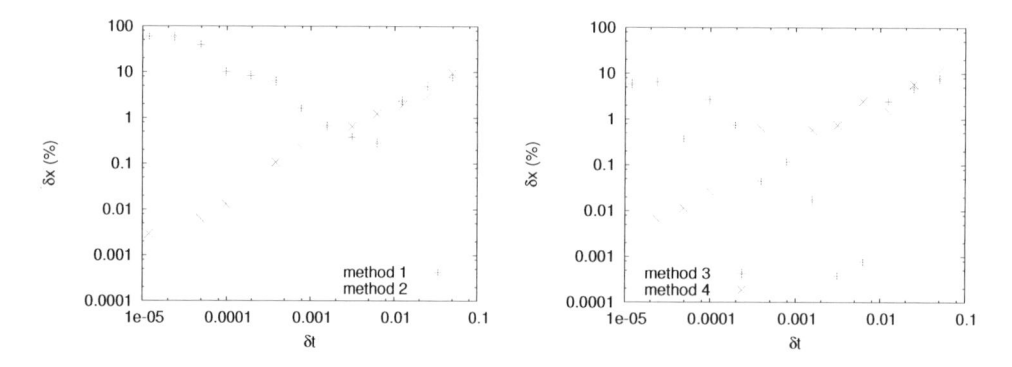

Figure 2.21: The error $\delta x = 2|x_i(95) - x(95)|/|x_i(95) + x(95)| \times 100$ where $x_i(95)$ is the value calculated by method $i = 1, 2, 3, 4$ and $x(95)$ the exact value according to the text.

2.3.3 The Two Dimensional Box

A particle is confined to move on the plane in the area $0 < x < L_x$ and $0 < y < L_y$. When it reaches the boundaries of this two dimensional box, it bounces elastically off its walls. The particle is found in an infinite depth orthogonal potential well. The particle starts moving at time t_0 from (x_0, y_0) and our program will calculate its trajectory until time t_f with time step δt. Such a trajectory can be seen in figure 2.23.

If the particle's position and velocity are known at time t, then at time

2.3. TRAPPED IN A BOX

$t + \delta t$ they will be given by the relations

$$\begin{aligned}
x(t + \delta t) &= x(t) + v_x(t)\delta t \\
y(t + \delta t) &= y(t) + v_y(t)\delta t \\
v_x(t + \delta t) &= v_x(t) \\
v_y(t + \delta t) &= v_y(t) \,.
\end{aligned} \quad (2.25)$$

The collision of the particle off the walls is modeled by reflection of the normal component of the velocity when the respective coordinate of the particle crosses the wall. This is a source of the systematic errors that we discussed in the previous section. The central loop of the program is:

```
i ++;
t  = t0 + i*dt;
x += vx*dt;
y += vy*dt;
if(x < 0.0 || x > Lx){
    vx = -vx;
    nx++;
}
if(y < 0.0 || y > Ly){
    vy = -vy;
    ny++;
}
```

The full program can be found in the file box2D_1.cpp. Notice that we introduced two counters nx and ny of the particle's collisions with the walls:

```
//===========================================================
//File box2D_1.cpp
//Motion of a free particle in a box  0<x<Lx 0<y<Ly
//Use integration with time step dt: x = x + vx*dt y=y+vy*dt
//-----------------------------------------------------------
#include <iostream>
#include <iomanip>
#include <fstream>
#include <cstdlib>
#include <string>
#include <cmath>
using namespace std;
```

```cpp
int main(){
//-----------------------------------------------------------
//Declaration of variables
  double Lx,Ly,x0,y0,v0x,v0y,t0,tf,dt,t,x,y,vx,vy;
  int    i,nx,ny;
  string buf;
//-----------------------------------------------------------
//Ask user for input:
  cout << "# Enter Lx,Ly:\n";
  cin  >> Lx >> Ly;                     getline(cin,buf);
  cout << "# Lx = "<< Lx  << " Ly= "  << Ly  << endl;
  cout << "# Enter x0,y0,v0x,v0y:\n";
  cin  >> x0 >> y0 >> v0x >> v0y;       getline(cin,buf);
  cout << "# x0= " << x0  << " y0= "  << y0
       << " v0x= " << v0x << " v0y= " << v0y << endl;
  cout << "# Enter t0,tf,dt:\n";
  cin  >> t0 >> tf >> dt;               getline(cin,buf);
  cout << "# t0= " << t0  << " tf= "  << tf
       << " dt= " << dt  << endl;
  if(Lx<= 0.0){cerr << "Lx<=0 \n"; exit(1);}
  if(Ly<= 0.0){cerr << "Ly<=0 \n"; exit(1);}
  if(x0<  0.0){cerr << "x0<=0 \n"; exit(1);}
  if(x0>  Lx ){cerr << "x0> Lx\n"; exit(1);}
  if(y0<  0.0){cerr << "x0<=0 \n"; exit(1);}
  if(y0>  Ly ){cerr << "y0> Ly\n"; exit(1);}
  if(v0x*v0x+v0y*v0y == 0.0 ){cerr << "v0 =0\n"; exit(1);}
//-----------------------------------------------------------
//Initialize
  i  = 0 ;
  nx = 0 ;   ny = 0 ;
  t  = t0 ;
  x  = x0 ;  y  = y0 ;
  vx = v0x;  vy = v0y;
  ofstream myfile("box2D_1.dat");
  myfile.precision(17);
//-----------------------------------------------------------
//Compute:
  while(t <= tf){
    myfile << setw(28) << t  << " "
           << setw(28) << x  << " "
           << setw(28) << y  << " "
           << setw(28) << vx << " "
           << setw(28) << vy << '\n';
    i +=     1;
```

2.3. TRAPPED IN A BOX

```
      t  = t0 + i*dt;
      x += vx*dt;
      y += vy*dt;
      if(x < 0.0 || x > Lx){
        vx = -vx;
        nx++;
      }
      if(y < 0.0 || y > Ly){
        vy = -vy;
        ny++;
      }
    }
    myfile.close();
    cout << "# Number of collisions:\n";
    cout << "# nx= " << nx << " ny= " << ny << endl;
} //main()
```

A typical session for the study of a particle's trajectory could be:

```
> g++ box2D_1.cpp -o box
> ./box
# Enter Lx,Ly:
10.0 5.0
# Lx = 10 Ly= 5
# Enter x0,y0,v0x,v0y:
5.0 0.0 1.27 1.33
# x0= 5 y0= 0 v0x= 1.27 v0y= 1.33
# Enter t0,tf,dt:
0 50 0.01
# t0= 0 tf= 50 dt= 0.01
# Number of collisions:
# nx= 6 ny= 13
> gnuplot
gnuplot> plot   "box2D_1.dat" using 1:2 w l title "x (t)"
gnuplot> replot "box2D_1.dat" using 1:3 w l title "y (t)"
gnuplot> plot   "box2D_1.dat" using 1:4 w l title "vx(t)"
gnuplot> replot "box2D_1.dat" using 1:5 w l title "vy(t)"
gnuplot> plot   "box2D_1.dat" using 2:3 w l title "x-y"
```

Notice the last line of output from the program: The particle bounces off the vertical walls 6 times (nx=6) and from the horizontal ones 13 (ny=13). The gnuplot commands construct the diagrams displayed in figures 2.22 and 2.23.

In order to animate the particle's trajectory, we can copy the file

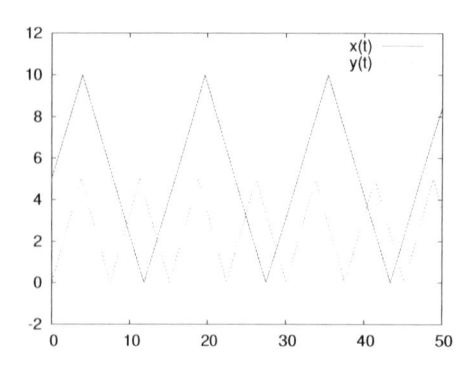

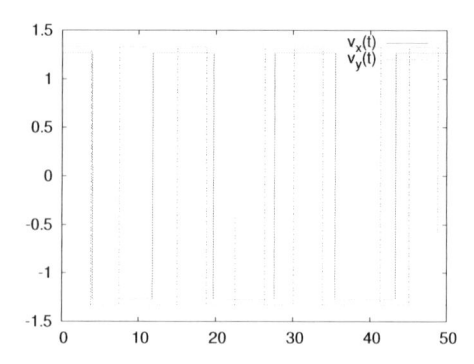

Figure 2.22: The results for the trajectory of a particle in a two dimensional box given by the program box2D_1.cpp. The parameters are $L_x = 10$, $L_y = 5$, $x_0 = 5$, $y_0 = 0$, $v_{0x} = 1.27$, $v_{0y} = 1.33$, $t_0 = 0$, $t_f = 50$, $\delta t = 0.01$.

box2D_animate.gnu of the accompanying software to the current directory and give the gnuplot commands:

```
gnuplot> file = "box2D_1.dat"
gnuplot> Lx = 10 ; Ly = 5
gnuplot> t0 = 0  ; tf = 50; dt = 1
gnuplot> load "box2D_animate.gnu"
gnuplot> t0 = 0  ; dt = 0.5; load "box2D_animate.gnu"
```

The last line repeats the same animation at half speed. You can also use the file animate2D.gnu discussed in section 2.1.1. We add new commands in the file box2D_animate.gnu so that the plot limits are calculated automatically and the box is drawn on the plot. The arrow drawn is not the position vector with respect to the origin of the coordinate axes, but the one connecting the initial with the current position of the particle.

The next step should be to test the accuracy of your results. This can be done by generalizing the discussion of the previous section and it is left as an exercise for the reader.

2.4 Applications

In this section we will study simple examples of motion in a box with different types of obstacles. We will start with a game of ... mini golf.

2.4. APPLICATIONS

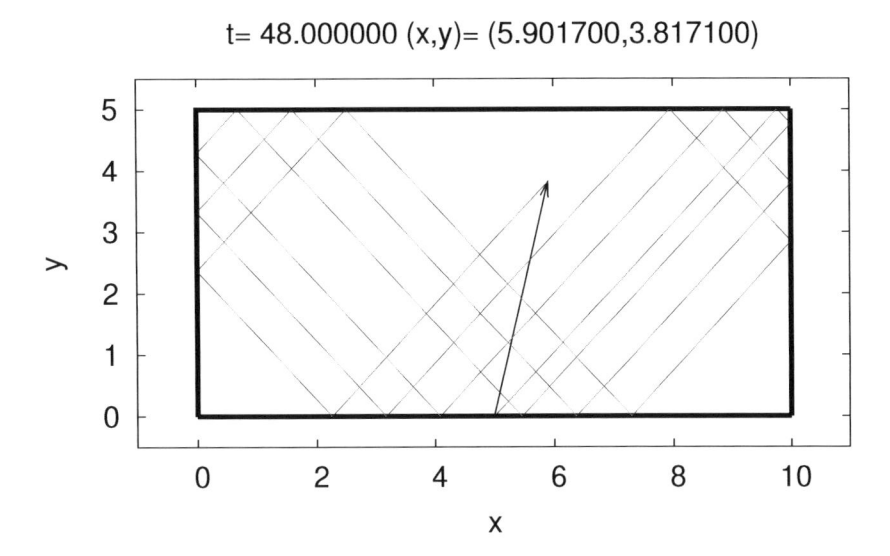

t= 48.000000 (x,y)= (5.901700,3.817100)

Figure 2.23: The trajectory of the particle of figure 2.22 until $t = 48$. The origin of the arrow is at the initial position of the particle and its end is at its current position. The bold lines mark the boundaries of the box.

The player shoots a (point) "ball" which moves in an orthogonal box of linear dimensions L_x and L_y and which is open on the $x = 0$ side. In the box there is a circular "hole" with center at (x_c, y_c) and radius R. If the "ball" falls in the "hole", the player wins. If the ball leaves out of the box through its open side, the player loses. In order to check if the ball is in the hole when it is at position (x, y), all we have to do is to check whether $(x - x_c)^2 + (y - y_c)^2 \leq R^2$.

Initially we place the ball at the position $(0, L_y/2)$ at time $t_0 = 0$. The player hits the ball which leaves with initial velocity of magnitude v_0 at an angle θ degrees with the x axis. The program is found in the file MiniGolf.cpp and is listed below:

```
//===========================================================
// File MiniGolf.cpp
// Motion of a free particle in a box  0<x<Lx 0<y<Ly
// The box is open at x=0 and has a hole at (xc,yc) of radius R
```

t= 45.300000 (x,y)= (7.854117,2.982556)

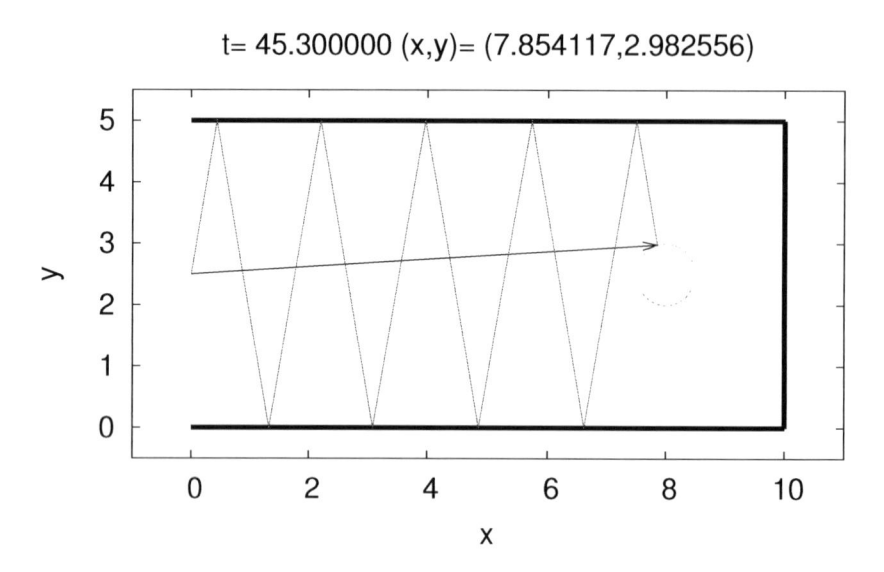

Figure 2.24: The trajectory of the particle calculated by the program MiniGolf.cpp using the parameters chosen in the text. The moment of ... success is shown. At time $t = 45.3$ the particle enters the hole's region which has its center at $(8, 2.5)$ and its radius is 0.5.

```
//Ball is shot at (0,Ly/2) with speed v0, angle theta (degrees)
//Use integration with time step dt: x = x + vx*dt y=y+vy*dt
//Ball stops in hole (success) or at x=0 (failure)
//------------------------------------------------------------
#include <iostream>
#include <iomanip>
#include <fstream>
#include <cstdlib>
#include <string>
#include <cmath>
using namespace std;

#define PI 3.14159265358979324

int main(){
//------------------------------------------------------------
//Declaration of variables
  double Lx,Ly,x0,y0,v0x,v0y,t0,tf,dt,t,x,y,vx,vy;
```

2.4. APPLICATIONS

```cpp
  double v0,theta,xc,yc,R,R2;
  int    i,nx,ny;
  string result;
  string buf;
//----------------------------------------------------------
//Ask user for input:
  cout << "# Enter Lx,Ly:\n";
  cin  >> Lx >> Ly;          getline(cin,buf);
  cout << "# Lx = " << Lx << " Ly= " << Ly  << endl;
  cout << "# Enter hole position and radius: (xc,yc), R:\n";
  cin  >> xc >> yc >> R;     getline(cin,buf);
  cout << "# (xc,yc)= ( " << xc << " , " << yc << ") "
       << " R= "      << R  << endl;
  cout << "# Enter v0, theta(degrees):\n";
  cin  >> v0 >> theta;       getline(cin,buf);
  cout << "# v0= " << v0 << " theta= "   << theta
       << " degrees "     << endl;
  cout << "# Enter dt:\n";
  cin  >> dt;                getline(cin,buf);
  if(Lx<=         0.0){cerr << "Lx<=0     \n"; exit(1);}
  if(Ly<=         0.0){cerr << "Ly<=0     \n"; exit(1);}
  if(v0<=         0.0){cerr << "v0<=0     \n"; exit(1);}
  if(abs(theta) > 90.0){cerr << "theta > 90\n"; exit(1);}
//----------------------------------------------------------
//Initialize
  t0    = 0.0;
  x0    = 0.00001; // small but non-zero
  y0    = Ly/2.0;
  R2    = R*R;
  theta = (PI/180.0)*theta;
  v0x   = v0*cos(theta);
  v0y   = v0*sin(theta);
  cout << "# x0= " << x0 << " y0= " << y0
       << " v0x= " << v0x << " v0y= " << v0y << endl;
  i  = 0 ;
  nx = 0 ;  ny = 0 ;
  t  = t0 ;
  x  = x0 ;  y = y0 ;
  vx = v0x; vy = v0y;
  ofstream myfile("MiniGolf.dat");
  myfile.precision(17);
//----------------------------------------------------------
//Compute:
  while(true){
    myfile << setw(28) << t << " "
```

```
          << setw(28) << x  << " "
          << setw(28) << y  << " "
          << setw(28) << vx << " "
          << setw(28) << vy << '\n';
    i ++;
    t  = t0 + i*dt;
    x += vx*dt;
    y += vy*dt;
    if(x > Lx ){vx = -vx; nx++;}
    if(y < 0.0){vy = -vy; ny++;}
    if(y > Ly ){vy = -vy; ny++;}
    if(x <=0.0)
      {result="Failure";break;} // exit loop
    if(((x-xc)*(x-xc)+(y-yc)*(y-yc)) <=  R2)
      {result="Success";break;} // exit loop
  }
  myfile.close();
  cout << "# Number of collisions:\n";
  cout << "# Result= "   << result
       << " nx= " << nx << " ny= " << ny << endl;
} //main()
```

In order to run it, we can use the commands:

```
> g++ MiniGolf.cpp -o mg
> ./mg
# Enter Lx,Ly:
10 5
# Lx = 10 Ly= 5
# Enter hole position and radius: (xc,yc), R:
8 2.5 0.5
# (xc,yc)= ( 8 , 2.5) R= 0.5
# Enter v0, theta(degrees):
1 80
# v0= 1 theta= 80 degrees
# Enter dt:
0.01
# x0= 1e-05  y0= 2.5 v0x= 0.173648 v0y= 0.984808
# Number of collisions:
# Result= Success nx= 0 ny= 9
```

You should construct the plots of the position and the velocity of the particle. You can also use the animation program found in the file MiniGolf_animate.gnu for fun. Copy it from the accompanying software

2.4. APPLICATIONS

to the current directory and give the gnuplot commands:

```
gnuplot> file = "MiniGolf.dat"
gnuplot> Lx = 10;Ly = 5
gnuplot> xc = 8; yc = 2.5 ; R = 0.5
gnuplot> t0 = 0; dt = 0.1
gnuplot> load "MiniGolf_animate.gnu"
```

The results are shown in figure 2.24.

The next example with be three dimensional. We will study the motion of a particle confined within a cylinder of radius R and height L. The collisions of the particle with the cylinder are elastic. We take the axis of the cylinder to be the z axis and the two bases of the cylinder to be located at $z = 0$ and $z = L$. This is shown in figure 2.26.

The collisions of the particle with the bases of the cylinder are easy to program: we follow the same steps as in the case of the simple box. For the collision with the cylinder's side, we consider the projection of the motion on the $x - y$ plane. The projection of the particle moves within a circle of radius R and center at the intersection of the z axis with the plane. This is shown in figure 2.25. At the collision, the r component of the velocity is reflected $v_r \to -v_r$, whereas v_θ remains the same. The velocity of the particle before the collision is

$$\begin{aligned} \vec{v} &= v_x \hat{x} + v_y \hat{y} \\ &= v_r \hat{r} + v_\theta \hat{\theta} \end{aligned} \tag{2.26}$$

and after the collision is

$$\begin{aligned} \vec{v}' &= v'_x \hat{x} + v'_y \hat{y} \\ &= -v_r \hat{r} + v_\theta \hat{\theta} \end{aligned} \tag{2.27}$$

From the relations

$$\begin{aligned} \hat{r} &= \cos\theta \hat{x} + \sin\theta \hat{y} \\ \hat{\theta} &= -\sin\theta \hat{x} + \cos\theta \hat{y}, \end{aligned} \tag{2.28}$$

and $v_r = \vec{v} \cdot \hat{r}$, $v_\theta = \vec{v} \cdot \hat{\theta}$, we have that

$$\begin{aligned} v_r &= v_x \cos\theta + v_y \sin\theta \\ v_\theta &= -v_x \sin\theta + v_y \cos\theta. \end{aligned} \tag{2.29}$$

The inverse relations are

$$\begin{aligned} v_x &= v_r \cos\theta - v_\theta \sin\theta \\ v_y &= v_r \sin\theta + v_\theta \cos\theta \, . \end{aligned} \quad (2.30)$$

With the transformation $v_r \to -v_r$, the new velocity in Cartesian coordinates will be

$$\begin{aligned} v'_x &= -v_r \cos\theta - v_\theta \sin\theta \\ v'_y &= -v_r \sin\theta + v_\theta \cos\theta \, . \end{aligned} \quad (2.31)$$

The transformation $v_x \to v'_x$, $v_y \to v'_y$ will be performed in the function

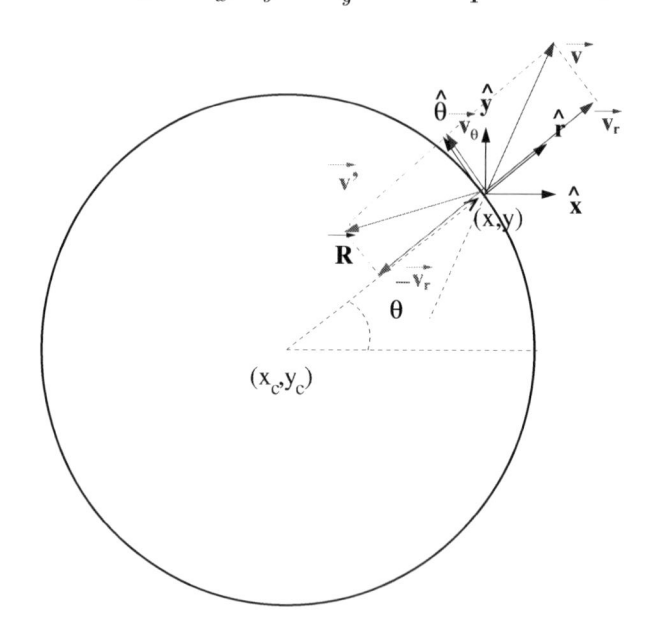

Figure 2.25: The elastic collision of the particle moving within the circle of radius $R = |\vec{R}|$ and center $\vec{r}_c = x_c\hat{x} + y_c\hat{y}$ at the point $\vec{r} = x\hat{x} + y\hat{y}$. We have that $\vec{R} = (x - x_c)\hat{x} + (y - y_c)\hat{y}$. The initial velocity is $\vec{v} = v_r\hat{r} + v_\theta\hat{\theta}$ where $\hat{r} \equiv \vec{R}/R$. After reflecting $v_r \to -v_r$ the new velocity of the particle is $\vec{v}' = -v_r\hat{r} + v_\theta\hat{\theta}$.

reflectVonCircle(vx,vy,x,y,xc,yc,R). Upon entry to the function, we provide the initial velocity (vx,vy), the collision point (x,y), the center of the circle (xc,yc) and the radius of the circle[24] R. Upon exit from the

---
[24] Of course one expects $R^2 = (x - x_c)^2 + (y - y_c)^2$, but because of systematic errors, we require R to be given.

2.4. APPLICATIONS

function, (vx,vy) have been replaced with the new values[25] (v'_x, v'_y).

The program can be found in the file Cylinder3D.cpp and is listed below:

```cpp
//===============================================================
// File Cylinder3D.cpp
// Motion of a free particle in a cylinder with axis the z-axis,
// radius R and 0<z<L
// Use integration with time step dt: x = x + vx*dt
//                                    y = y + vy*dt
//                                    z = z + vz*dt
// Use function reflectVonCircle for colisions at r=R
//---------------------------------------------------------------
#include <iostream>
#include <iomanip>
#include <fstream>
#include <cstdlib>
#include <string>
#include <cmath>
using namespace std;

void reflectVonCircle(double& vx,double& vy,
                      double& x ,double& y ,
                      const    double& xc,
                      const    double& yc,
                      const    double& R );

int main(){
//---------------------------------------------------------------
// Declaration of variables
  double x0,y0,z0,v0x,v0y,v0z,t0,tf,dt,t,x,y,z,vx,vy,vz;
  double L,R,R2,vxy,rxy,r2xy,xc,yc;
  int    i,nr,nz;
  string buf;
//---------------------------------------------------------------
// Ask user for input:
  cout << "# Enter R,L:\n";
  cin  >> R >> L;                          getline(cin,buf);
  cout << "# R= " << R << " L= " << L << endl;
  cout << "# Enter x0,y0,z0,v0x,v0y,v0z:\n";
  cin  >> x0>>y0>>z0>>v0x>>v0y>>v0z;       getline(cin,buf);
  rxy  = sqrt(x0*x0+y0*y0);
```

---

[25]Notice that upon exit, the particle is also placed exactly on the circle.

```cpp
  cout << "# x0 = " << x0
       << "  y0 = " << y0
       << "  z0 = " << z0
       << "  rxy= " << rxy << endl;
  cout << "# v0x= " << v0x
       << "  v0y= " << v0y
       << "  v0z= " << v0z << endl;
  cout << "# Enter t0,tf,dt:\n";
  cin  >> t0 >> tf >> dt;            getline(cin,buf);
  cout << "# t0= " << t0 << " tf= " << tf
       << " dt= " << dt << endl;
  if(R   <= 0.0){cerr << "R<=0  \n"; exit(1);}
  if(L   <= 0.0){cerr << "L<=0  \n"; exit(1);}
  if(z0  <  0.0){cerr << "z0<0  \n"; exit(1);}
  if(z0  >    L){cerr << "z0>L  \n"; exit(1);}
  if(rxy >    R){cerr << "rxy>R \n"; exit(1);}
  if(v0x*v0x+v0y*v0y+v0z*v0z == 0.0)
                {cerr << "v0=0  \n"; exit(1);}
//-----------------------------------------
//Initialize
  i  = 0 ;
  nr = 0 ;  nz = 0 ;
  t  = t0 ;
  x  = x0 ;  y  = y0 ;  z  = z0 ;
  vx = v0x;  vy = v0y;  vz = v0z;
  R2 = R*R;
  xc = 0.0; //center of circle which is the projection
  yc = 0.0; //of the cylinder on the xy plane
  ofstream myfile("Cylinder3D.dat");
  myfile.precision(17);
//-----------------------------------------
//Compute:
  while(t <= tf){
    myfile << setw(28) << t  << " "
           << setw(28) << x  << " "
           << setw(28) << y  << " "
           << setw(28) << z  << " "
           << setw(28) << vx << " "
           << setw(28) << vy << " "
           << setw(28) << vz << '\n';
    i ++;
    t  = t0 + i*dt;
    x += vx*dt;
    y += vy*dt;
    z += vz*dt;
```

2.4. APPLICATIONS

```
      if( z <= 0.0 || z > L){vz = -vz; nz++;}
      r2xy = x*x+y*y;
      if( r2xy > R2){
        reflectVonCircle(vx,vy,x,y,xc,yc,R);
        nr++;
      }
    }
    myfile.close();
    cout << "# Number of collisions:\n";
    cout << "# nr= " << nr << " nz= " << nz << endl;
} //main()
//-----------------------------------------------------------
//===========================================================
//-----------------------------------------------------------
void reflectVonCircle(double& vx,double& vy,
                      double& x ,double& y ,
                      const     double& xc,
                      const     double& yc,
                      const     double& R  ){
    double theta,cth,sth,vr,vth;

    theta = atan2(y-yc,x-xc);
    cth   = cos(theta);
    sth   = sin(theta);

    vr    =  vx*cth + vy *sth;
    vth   = -vx*sth + vy *cth;

    vx    = -vr*cth - vth*sth; //reflect vr -> -vr
    vy    = -vr*sth + vth*cth;

    x     = xc     + R*cth;   //put x,y on the circle
    y     = yc     + R*sth;
} //reflectVonCircle()
```

Note that the function atan2 is used for computing the angle theta. This function, when called with two arguments atan2(y,x), returns the angle $\theta = \tan^{-1}(y/x)$ in radians. The correct quadrant of the circle where (x,y) lies is chosen. The angle that we want to compute is given by atan2(y-yc,x-xc). Then we apply equations (2.29) and (2.31) and in the last two lines we enforce the particle to be at the point $(x_c + R\cos\theta, y_c + R\sin\theta)$, exactly on the circle.

A typical session is shown below:

t= 500.000000 (x,y,z)= (2.227212,0.469828,7.088600)

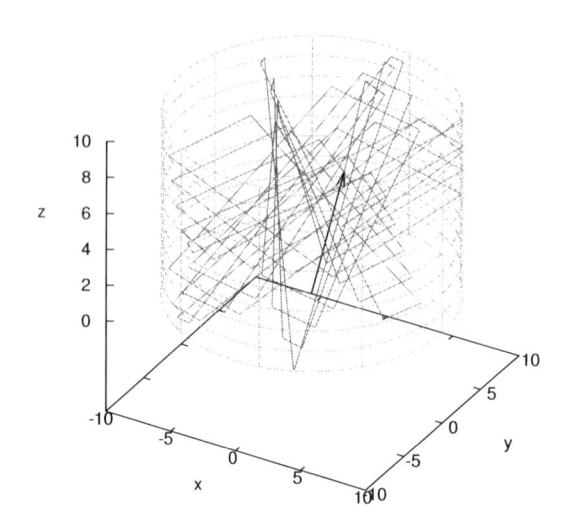

Figure 2.26: The trajectory of a particle moving inside a cylinder with $R = 10$, $L = 10$, computed by the program `Cylinder3D.cpp`. We have chosen $\vec{r}_0 = 1.0\hat{x} + 2.2\hat{y} + 3.1\hat{z}$, $\vec{v}_0 = 0.93\hat{x} - 0.89\hat{y} + 0.74\hat{z}$, $t_0 = 0$, $t_f = 500.0$, $\delta t = 0.01$.

```
> g++ Cylinder3D.cpp -o cl
> ./cl
# Enter R,L:
10.0 10.0
# R= 10 L= 10
# Enter x0,y0,z0,v0x,v0y,v0z:
1.0 2.2 3.1   0.93 -0.89 0.74
# x0 = 1   y0 = 2.2   z0 = 3.1   rxy= 2.41661
# v0x= 0.93   v0y= -0.89   v0z= 0.74
# Enter t0,tf,dt:
0.0 500.0 0.01
# t0= 0 tf= 500 dt= 0.01
# Number of collisions:
# nr= 33 nz= 37
```

In order to plot the position and the velocity as a function of time, we use the following `gnuplot` commands:

2.4. APPLICATIONS 133

```
gnuplot> file="Cylinder3D.dat"
gnuplot> plot file using 1:2 with lines title " x(t)",\
             file using 1:3 with lines title " y(t)",\
             file using 1:4 with lines title " z(t)"
gnuplot> plot file using 1:5 with lines title "v_x(t)",\
             file using 1:6 with lines title "v_y(t)",\
             file using 1:7 with lines title "v_z(t)"
```

We can also compute the distance of the particle from the cylinder's axis $r(t) = \sqrt{x(t)^2 + y(t)^2}$ as a function of time using the command:

```
gnuplot> plot file using 1:(sqrt($2**2+$3**2)) w l t "r(t)"
```

In order to plot the trajectory, together with the cylinder, we give the commands:

```
gnuplot> file="Cylinder3D.dat"
gnuplot> L = 10 ; R = 10
gnuplot> set urange [0:2.0*pi]
gnuplot> set vrange [0:L]
gnuplot> set parametric
gnuplot> splot file using 2:3:4 with lines notitle,\
             R*cos(u),R*sin(u),v notitle
```

The command set parametric is necessary if one wants to make a parametric plot of a surface $\vec{r}(u,v) = x(u,v)\,\hat{x} + y(u,v)\,\hat{y} + z(u,v)\,\hat{z}$. The cylinder (without the bases) is given by the parametric equations $\vec{r}(u,v) = R\cos u\,\hat{x} + R\sin u\,\hat{y} + v\,\hat{z}$ with $u \in [0, 2\pi)$, $v \in [0, L]$.

We can also animate the trajectory with the help of the gnuplot script file Cylinder3D_animate.gnu. Copy the file from the accompanying software to the current directory and give the gnuplot commands:

```
gnuplot> file="Cylinder3D.dat"
gnuplot> R=10;L=10;t0=0;tf=500;dt=10
gnuplot> load "Cylinder3D_animate.gnu"
```

The result is shown in figure 2.26.

The last example will be that of a simple model of a spacetime wormhole. This is a simple spacetime geometry which, in the framework of the theory of general relativity, describes the connection of two distant

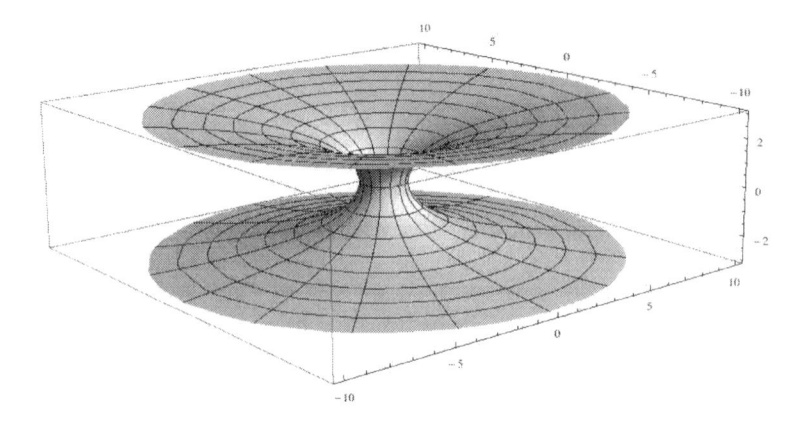

Figure 2.27: A typical geometry of space near a wormhole. Two asymptotically flat regions of space are connected through a "neck" which can be arranged to be of small length compared to the distance of the wormhole mouths when traveled from the outside space.

areas in space which are asymptotically flat. This means, that far enough from the wormhole's mouths, space is almost flat - free of gravity. Such a geometry is depicted in figure 2.27. The distance traveled by someone through the mouths could be much smaller than the distance traveled outside the wormhole and, at least theoretically, traversable wormholes could be used for interstellar/intergalactic traveling and/or communications between otherwise distant areas in the universe. Of course we should note that such macroscopic and stable wormholes are not known to be possible to exist in the framework of general relativity. One needs an exotic type of matter with negative energy density which has never been observed. Such exotic geometries may realize microscopically as quantum fluctuations of spacetime and make the small scale structure of the geometry[26] a "spacetime foam".

We will study a very simple model of the above geometry on the plane

---

[26]See K.S. Thorne "Black Holes and Time Wraps: Einstein's Outrageous Legacy", W.W. Norton, New York for a popular review of these concepts.

2.4. APPLICATIONS

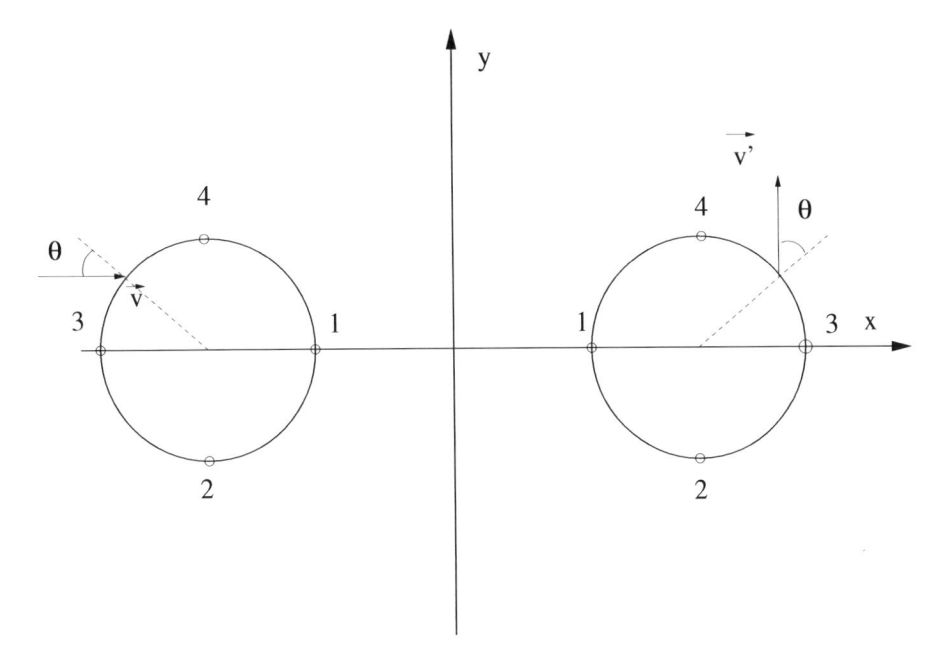

Figure 2.28: A simple model of the spacetime geometry of figure 2.27. The particle moves on the whole plane except withing the two disks that have been removed. The neck of the wormhole is modeled by the two circles $x(\theta) = \pm d/2 \pm R\cos\theta$, $y(\theta) = R\sin\theta$, $-\pi < \theta \leq \pi$ and has zero length since their points have been identified. There is a given direction in this identification, so that points with the same θ are the same (you can imagine how this happens by folding the plane across the y axis and then glue the two circles together). The entrance of the particle through one mouth and exit through the other is done as shown for the velocity vector $\vec{v} \to \vec{v}'$.

with a particle moving freely in it[27]. We take the two dimensional plane and cut two equal disks of radius R with centers at distance d like in figure 2.28. We identify the points on the two circles such that the point 1 of the left circle is the same as the point 1 on the right circle, the point 2 on the left with the point 2 on the right etc. The two circles are given by the parametric equations $x(\theta) = d/2 + R\cos\theta$, $y(\theta) = R\sin\theta$, $-\pi < \theta \leq \pi$ for the right circle and $x(\theta) = -d/2 - R\cos\theta$, $y(\theta) = R\sin\theta$, $-\pi < \theta \leq \pi$ for the left. Points on the two circles with the same θ are identified. A particle entering the wormhole from the left circle with velocity v is immediately exiting from the right with velocity v' as shown in figure

---

[27]This idea can be found as an exercise in the excellent introductory general relativity textbook J. B. Hartle, "Gravity: An Introduction to Einstein's General Relativity", Addison Wesley 2003, Ch. 7, Ex. 25.

2.28.
 Then we will do the following:

 1. Write a program that computes the trajectory of a particle moving in the geometry of figure 2.28. We set the limits of motion to be $-L/2 \leq x \leq L/2$ and $-L/2 \leq y \leq L/2$. We will use periodic boundary conditions in order to define what happens when the particle attempts to move outside these limits. This means that we identify the $x = -L/2$ line with the $x = +L/2$ line as well as the $y = -L/2$ line with the $y = +L/2$ line. The user enters the parameters R, d and L as well as the initial conditions (x_0, y_0), (v_0, ϕ) where $\vec{v}_0 = v_0(\cos\phi \hat{x} + \sin\phi \hat{y})$. The user will also provide the time parameters t_f and dt for motion in the time interval $t \in [t_0 = 0, t_f]$ with step dt.

 2. Plot the particle's trajectory with $(x_0, y_0) = (0, -1)$, $(v_0, \phi) = (1, 10°)$ με $t_f = 40$, $dt = 0.05$ in the geometry with $L = 20, d = 5, R = 1$.

 3. Find a closed trajectory which does not cross the boundaries $|x| = L/2$, $|y| = L/2$ and determine whether it is stable under small perturbations of the initial conditions.

 4. Find other closed trajectories that go through the mouths of the wormhole and study their stability under small perturbations of the initial conditions.

 5. Add to the program the option to calculate the distance traveled by the particle. If the particle starts from $(-x_0, 0)$ and moves in the $+x$ direction to the $(x_0, 0)$, $x_0 > R + d/2$ position, draw the trajectory and calculate the distance traveled on paper. Then confirm your calculation from the numerical result coming from your program.

 6. Change the boundary conditions, so that the particle bounces off elastically at $|x| = L/2$, $|y| = L/2$ and replot all the trajectories mentioned above.

Define the right circle c_1 by the parametric equations

$$x(\theta) = \frac{d}{2} + R\cos\theta, \qquad y(\theta) = R\sin\theta, \qquad -\pi < \theta \leq \pi, \qquad (2.32)$$

2.4. APPLICATIONS

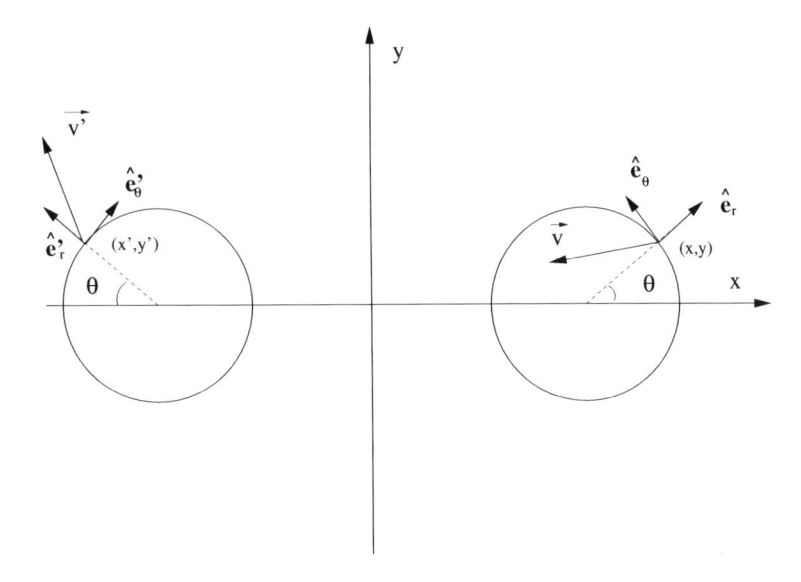

Figure 2.29: The particle crossing the wormhole through the right circle c_1 with velocity $\vec{v}$. It emerges from c_2 with velocity $\vec{v}'$. The unit vectors $(\hat{e}_r, \hat{e}_\theta)$, $(\hat{e}'_r, \hat{e}'_\theta)$ are computed from the parametric equations of the two circles c_1 and c_2.

and the left circle c_2 by the parametric equations

$$x(\theta) = -\frac{d}{2} - R\cos\theta, \qquad y(\theta) = R\sin\theta, \qquad -\pi < \theta \leq \pi. \qquad (2.33)$$

The particle's position changes at time dt by

$$\begin{aligned} t_i &= i\,dt \\ x_i &= x_{i-1} + v_x dt \\ y_i &= y_{i-1} + v_y dt \end{aligned} \qquad (2.34)$$

for $i = 1, 2, \ldots$ for given (x_0, y_0), $t_0 = 0$ and as long as $t_i \leq t_f$. If the point (x_i, y_i) is outside the boundaries $|x| = L/2$, $|y| = L/2$, we redefine $x_i \to x_i \pm L$, $y_i \to y_i \pm L$ in each case respectively. Points defined by the same value of θ are identified, i.e. they represent the *same* points of space. If the point (x_i, y_i) crosses either one of the circles c_1 or c_2, then we take the particle out from the other circle.

Crossing the circle c_1 is determined by the relation

$$\left(x_i - \frac{d}{2}\right)^2 + y_i^2 \leq R^2. \qquad (2.35)$$

The angle θ is calculated from the equation

$$\theta = \tan^{-1}\left(\frac{y_i}{x_i - \frac{d}{2}}\right), \tag{2.36}$$

and the point (x_i, y_i) is mapped to the point (x'_i, y'_i) where

$$x'_i = -\frac{d}{2} - R\cos\theta, \qquad y'_i = y_i, \tag{2.37}$$

as can be seen in figure 2.29. For mapping $\vec{v} \to \vec{v}'$, we first calculate the vectors

$$\left.\begin{array}{rl} \hat{e}_r = & \cos\theta\,\hat{x} + \sin\theta\,\hat{y} \\ \hat{e}_\theta = & -\sin\theta\,\hat{x} + \cos\theta\,\hat{y} \end{array}\right\} \to \left\{\begin{array}{rl} \hat{e}'_r = & -\cos\theta\,\hat{x} + \sin\theta\,\hat{y} \\ \hat{e}'_\theta = & \sin\theta\,\hat{x} + \cos\theta\,\hat{y} \end{array}\right., \tag{2.38}$$

so that the velocity

$$\vec{v} = v_r\,\hat{e}_r + v_\theta\,\hat{e}_\theta \quad \to \quad \vec{v}' = -v_r\,\hat{e}'_r + v_\theta\,\hat{e}'_\theta, \tag{2.39}$$

where the radial components are $v_r = \vec{v}\cdot\hat{e}_r$ and $v_\theta = \vec{v}\cdot\hat{e}_\theta$. Therefore, the relations that give the "emerging" velocity $\vec{v}'$ are:

$$\begin{array}{rl} v_r = & v_x\cos\theta + v_y\sin\theta \\ v_\theta = & -v_x\sin\theta + v_y\cos\theta \\ v'_x = & v_r\cos\theta + v_\theta\sin\theta \\ v'_y = & -v_r\sin\theta + v_\theta\cos\theta \end{array}. \tag{2.40}$$

Similarly we calculate the case of entering from c_2 and emerging from c_1. The condition now is:

$$\left(x_i + \frac{d}{2}\right)^2 + y_i^2 \leq R^2. \tag{2.41}$$

The angle θ is given by

$$\theta = \pi - \tan^{-1}\left(\frac{y_i}{x_i + \frac{d}{2}}\right), \tag{2.42}$$

and the point (x_i, y_i) is mapped to the point (x'_i, y'_i) where

$$x'_i = \frac{d}{2} + R\cos\theta, \qquad y'_i = y_i. \tag{2.43}$$

2.4. APPLICATIONS

For mapping $\vec{v} \to \vec{v}'$, we calculate the vectors

$$\left.\begin{array}{rl}\hat{e}_r = & -\cos\theta\,\hat{x} + \sin\theta\,\hat{y} \\ \hat{e}_\theta = & \sin\theta\,\hat{x} + \cos\theta\,\hat{y}\end{array}\right\} \to \left\{\begin{array}{rl}\hat{e}'_r = & \cos\theta\,\hat{x} + \sin\theta\,\hat{y} \\ \hat{e}'_\theta = & -\sin\theta\,\hat{x} + \cos\theta\,\hat{y}\end{array}\right., \quad (2.44)$$

so that the velocity

$$\vec{v} = v_r\,\hat{e}_r + v_\theta\,\hat{e}_\theta \quad \to \quad \vec{v}' = -v_r\,\hat{e}'_r + v_\theta\,\hat{e}'_\theta. \quad (2.45)$$

The emerging velocity $\vec{v}'$ is:

$$\begin{array}{rl}v_r = & -v_x\cos\theta + v_y\sin\theta \\ v_\theta = & v_x\sin\theta + v_y\cos\theta \\ v'_x = & -v_r\cos\theta - v_\theta\sin\theta \\ v'_y = & -v_r\sin\theta + v_\theta\cos\theta\end{array}. \quad (2.46)$$

Systematic errors are now coming from crossing the two mouths of the wormhole. There are no systematic errors from crossing the boundaries $|x| = L/2$, $|y| = L/2$ (why?). Try to think of ways to control those errors and study them.

The closed trajectories that we are looking for come from the initial conditions

$$(x_0, y_0, v_0, \phi) = (0, 0, 1, 0) \quad (2.47)$$

and they connect points 1 of figure 2.28. They are unstable, as can be seen by taking $\phi \to \phi + \epsilon$.

The closed trajectories that cross the wormhole and "wind" through space can come from the initial conditions

$$\begin{aligned}(x_0, y_0, v_0, \phi) &= (-9, 0, 1, 0) \\ (x_0, y_0, v_0, \phi) &= (2.5, -3, 1, 90^\circ)\end{aligned}$$

and cross the points $3 \to 3$ and $2 \to 2 \to 4 \to 4$ respectively. They are also unstable, as can be easily verified by using the program that you will write. The full program is listed below:

```
//==============================================
// File Wormhole.cpp
//----------------------------------------------
#include <iostream>
```

```cpp
#include <fstream>
#include <cstdlib>
#include <string>
#include <cmath>
using namespace std;

#define PI 3.1415926535897932

void crossC1(       double& x,        double& y,
                    double& vx,       double& vy,
              const double& dt, const double& R,
              const double& d);
void crossC2(       double& x,        double& y,
                    double& vx,       double& vy,
              const double& dt, const double& R,
              const double& d);

int main(){
//--------------------------------------------------
//Declaration of variables
  double Lx,Ly,L,R,d;
  double x0,y0,v0,theta;
  double t0,tf,dt;
  double t,x,y,vx,vy;
  double xc1,yc1,xc2,yc2,r1,r2;
  int    i;
  string buf;
//--------------------------------------------------
//Ask user for input:
  cout << "# Enter L,d,R:\n";
  cin  >> L >> d >> R;              getline(cin,buf);
  cout << "# Enter (x0,y0), v0, theta(degrees):\n";
  cin  >> x0 >> y0 >> v0 >> theta;  getline(cin,buf);
  cout << "# Enter tf,dt:\n";
  cin  >> tf >> dt; getline(cin,buf);
  cout << "# L= "  << L  << " d=     " << d
       << " R= "   << R  << endl;
  cout << "# x0= " << x0 << " y0=    " << y0    << endl;
  cout << "# v0= " << v0 << " theta= "
       << theta    << " degrees"                << endl;
  cout << "# tf= " << tf << " dt=    " << dt    << endl;
  if(L <=  d+2.0*R){cerr <<"L <= d+2*R \n";exit(1);}
  if(d <=    2.0*R){cerr <<"d <=   2*R \n";exit(1);}
  if(v0<=    0.0  ){cerr <<"v0<=     0 \n";exit(1);}
//--------------------------------------------------
```

2.4. APPLICATIONS

```
//Initialize
  theta = (PI/180.0)*theta;
  i     = 0;
  t     = 0.0;
  x     = x0              ; y    = y0;
  vx    = v0*cos(theta); vy    = v0*sin(theta);
  cout << "# x0= " << x0 << " y0= " << y0
       << " v0x= " << vx << " v0y= " << vy << endl;
//Wormhole's centers:
  xc1   =  0.5*d; yc1 =  0.0;
  xc2   = -0.5*d; yc2 =  0.0;
//Box limits coordinates:
  Lx    =  0.5*L; Ly  =  0.5*L;
//Test if already inside cut region:
  r1    = sqrt((x-xc1)*(x-xc1)+(y-yc1)*(y-yc1));
  r2    = sqrt((x-xc2)*(x-xc2)+(y-yc2)*(y-yc2));
  if(r1<=        R){cerr <<"r1 <=    R \n";exit(1);}
  if(r1<=        R){cerr <<"r2 <=    R \n";exit(1);}
//Test if outside box limits:
  if(abs(x) >= Lx){cerr <<"|x|>=   Lx \n";exit(1);}
  if(abs(y) >= Ly){cerr <<"|y|>=   Ly \n";exit(1);}
  ofstream myfile("Wormhole.dat");
  myfile.precision(17);
//----------------------------------------------------
//Compute:
  while( t < tf ){
    myfile << t   << " "
           << x   << " " << y << " "
           << vx  << " " << vy << endl;
    i++;
    t = i*dt;
    x += vx*dt; y += vy*dt;
// Toroidal boundary conditions:
    if( x >  Lx) x = x - L;
    if( x < -Lx) x = x + L;
    if( y >  Ly) y = y - L;
    if( y < -Ly) y = y + L;
    r1    = sqrt((x-xc1)*(x-xc1)+(y-yc1)*(y-yc1));
    r2    = sqrt((x-xc2)*(x-xc2)+(y-yc2)*(y-yc2));
// Notice: we pass r1 as radius of circle, not R
    if     (r1 < R)
      crossC1(x,y,vx,vy,dt,r1,d);
    else if(r2 < R)
      crossC2(x,y,vx,vy,dt,r2,d);
// small chance here that still in C1 or C2, but OK since
```

```cpp
// another dt-advance given at the beginning of for-loop
    }// while( t <= tf )
}  // main ()
//--------------------------------------------------------------
void crossC1(        double& x,        double& y,
                     double& vx,       double& vy,
              const double& dt, const double& R,
              const double& d){

  double vr,v0,theta,xc,yc;
  cout << "# Inside C1: (x,y,vx,vy,R)= "
       << x  << " " << y  << " "
       << vx << " " << vy << " " <<R << endl;
  xc    =  0.5*d;              //center of C1
  yc    =  0.0;
  theta =  atan2(y-yc,x-xc);
  x     = -xc - R*cos(theta); //new x-value, y invariant
//Velocity transformation:
  vr    =  vx*cos(theta)+vy*sin(theta);
  v0    = -vx*sin(theta)+vy*cos(theta);
  vx    =  vr*cos(theta)+v0*sin(theta);
  vy    = -vr*sin(theta)+v0*cos(theta);
//advance x,y, hopefully outside C2:
  x     = x + vx*dt;
  y     = y + vy*dt;
  cout << "# Exit    C2: (x,y,vx,vy  )= "
       << x  << " " << y  << " "
       << vx << " " << vy << endl;
}//void crossC1( )
//--------------------------------------------------------------
void crossC2(        double& x,        double& y,
                     double& vx,       double& vy,
              const double& dt, const double& R,
              const double& d){

  double vr,v0,theta,xc,yc;
  cout << "# Inside C2: (x,y,vx,vy,R)= "
       << x  << " " << y  << " "
       << vx << " " << vy << " " <<R << endl;
  xc    = -0.5*d;              //center of C2
  yc    =  0.0;
  theta =  PI-atan2(y-yc,x-xc);
  x     = -xc + R*cos(theta); //new x-value, y invariant
//Velocity transformation:
  vr    = -vx*cos(theta)+vy*sin(theta);
```

2.4. APPLICATIONS

```
    v0    =  vx*sin(theta)+vy*cos(theta);
    vx    = -vr*cos(theta)-v0*sin(theta);
    vy    = -vr*sin(theta)+v0*cos(theta);
//advance x,y, hopefully outside C1:
    x     = x + vx*dt;
    y     = y + vy*dt;
    cout << "# Exit    C1: (x,y,vx,vy  )= "
         << x  << " " << y  << " "
         << vx << " " << vy << endl;
}//void    crossC2( )
```

It is easy to compile and run the program. See also the files Wormhole.csh and Wormhole_animate.gnu of the accompanying software and run the gnuplot commands:

```
gnuplot> file  = "Wormhole.dat"
gnuplot> R=1;d=5;L=20;
gnuplot> ! ./Wormhole.csh
gnuplot> t0=0;dt=0.2;load "Wormhole_animate.gnu"
```

You are now ready to answer the rest of the questions that we asked in our list.

2.5 Problems

2.1 Change the program Circle.cpp so that it prints the number of full circles traversed by the particle.

2.2 Add all the necessary tests on the parameters entered by the user in the program Circle.cpp, so that the program is certain to run without problems. Do the same for the rest of the programs given in the same section.

2.3 A particle moves with constant angular velocity ω on a circle that has the origin of the coordinate system at its center. At time $t_0 = 0$, the particle is at (x_0, y_0). Write the program CircularMotion.cpp that will calculate the particle's trajectory. The user should enter the parameters $\omega, x_0, y_0, t_0, t_f, \delta t$. The program should print the results like the program Circle.cpp does.

2.4 Change the program SimplePendulum.cpp so that the user could enter a non zero initial velocity.

2.5 Study the $k \to 0$ limit in the projectile motion given by equations (2.10). Expand $e^{-kt} = 1 - kt + \frac{1}{2!}(kt)^2 + \ldots$ and keep the non vanishing terms as $k \to 0$. Then keep the next order leading terms which have a smaller power of k. Program these relations in a file ProjectileSmallAirResistance.cpp. Consider the initial conditions $\vec{v}_0 = \hat{x} + \hat{y}$ and calculate the range of the trajectory numerically by using the two programs
ProjectileSmallAirResistance.cpp, ProjectileAirResistance.cpp. Determine the range of values of k for which the two results agree within 5% accuracy.

2.6 Write a program for a projectile which moves through a fluid with fluid resistance proportional to the square of the velocity. Compare the range of the trajectory with the one calculated by the program ProjectileAirResistance.cpp for the parameters shown in figure 2.10.

2.7 Change the program Lissajous.cpp so that the user can enter a different amplitude and initial phase in each direction. Study the case where the amplitudes are the same and the phase difference

2.5. PROBLEMS

in the two directions are $\pi/4, \pi/2, \pi, -\pi$. Repeat by taking the amplitude in the y direction to be twice as much the amplitude in the x direction.

2.8 Change the program ProjectileAirResistance.cpp, so that it can calculate also the $k = 0$ case.

2.9 Change the program ProjectileAirResistance.cpp so that it can calculate the trajectory of the particle in three dimensional space. Plot the position coordinates and the velocity components as a function of time. Plot the three dimensional trajectory using splot in gnuplot and animate the trajectory using the gnuplot script animate3D.gnu.

2.10 Change the program ChargeInB.cpp so that it can calculate the number of full revolutions that the projected particle's position on the $x - y$ plane makes during its motion.

2.11 Change the program box1D_1.cpp so that it prints the number of the particle's collisions on the left wall, on the right wall and the total number of collisions to the stdout.

2.12 Do the same for the program box1D_2.cpp. Fill the table on page 115 the number of calculated collisions and comment on the results.

2.13 Run the program box1D_1.cpp and choose L= 10, v0=1. Decrease the step dt up to the point that the particle stops to move. For which value of dt this happens? Increase v0=10,100. Until which value of dt the particle moves now? Why?

2.14 Change the float declarations to double in the program box1D_1.cpp. Make sure that all the constants that you use become double precision (e.g. 1.0f changes to 1.0). Compare your results to those obtained in section 2.3.2. Repeat problem 2.13. What do you observe?

2.15 Change the program box1D_1.cpp so that you can study non elastic collisions $v' = -ev$, $0 < e \leq 1$ with the walls.

2.16 Change the program box2D_1.cpp so that you can study inelastic collisions with the walls, such that $v'_x = -ev_x$, $v'_y = -ev_y$, $0 < e \leq 1$.

2.17 Use the method of calculating time in the programs box1D_4.cpp and box1D_5.cpp in order to produce the results in figure 2.21.

2.18 Particle falls freely moving in the vertical direction. It starts with zero velocity at height h. Upon reaching the ground, it bounces inelastically such that $v'_y = -ev_y$ with $0 < e \leq 1$ a parameter. Write the necessary program in order to study numerically the particle's motion and study the cases $e = 0.1, 0.5, 0.9, 1.0$.

2.19 Generalize the program of the previous problem so that you can study the case $\vec{v}_0 = v_{0x}\hat{x}$. Animate the calculated trajectories.

2.20 Study the motion of a particle moving inside the box of figure 2.30. Count the number of collisions of the particle with the walls before it leaves the box.

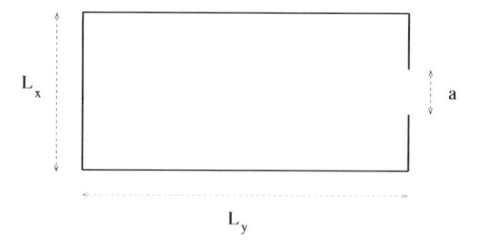

Figure 2.30: Problem 2.20.

2.21 Study the motion of the point particle on the "billiard table" of figure 2.31. Count the number of collisions with the walls before the particle enters into a hole. The program should print from which hole the particle left the table.

2.22 Write a program in order to study the motion of a particle in the box of figure 2.32. At the center of the box there is a disk on which the particle bounces off elastically (Hint: use the routine reflectVonCircle of the program Cylinder3D.cpp).

2.23 In the box of the previous problem, put four disks on which the particle bounces of elastically like in figure 2.33.

2.5. PROBLEMS

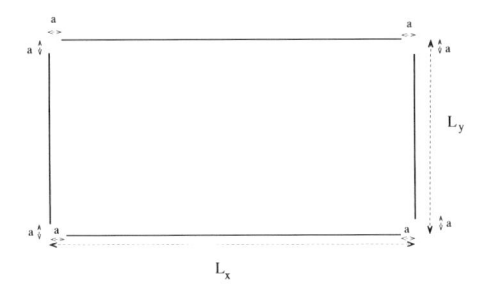

Figure 2.31: Problem 2.21.

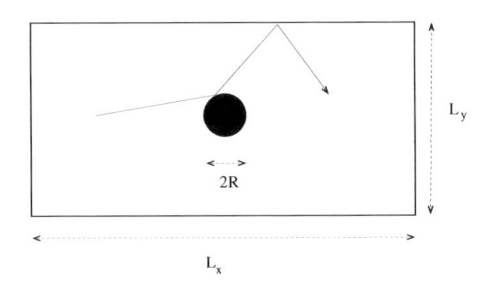

Figure 2.32: Problem 2.22.

2.24 Consider the arrangement of figure 2.34. Each time the particle bounces elastically off a circle, the circle disappears. The game is over successfully if all the circles vanish. Each time the particle bounces off on the wall to the left, you lose a point. Try to find trajectories that minimize the number of lost points.

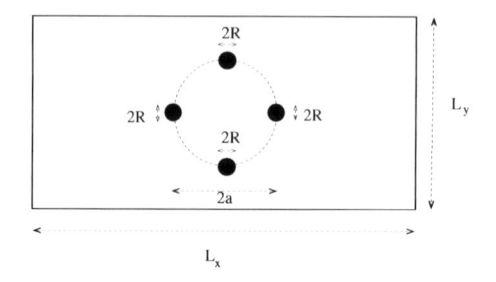

Figure 2.33: Problem 2.23.

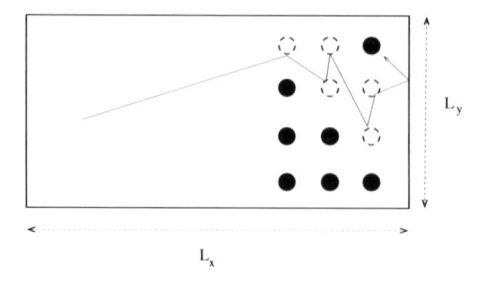

Figure 2.34: Problem 2.24.

Chapter 3

Logistic Map

Nonlinear differential equations model interesting dynamical systems in physics, biology and other branches of science. In this chapter we perform a numerical study of the discrete logistic map as a "simple mathematical model with complex dynamical properties" [23] similar to the ones encountered in more complicated and interesting dynamical systems. For certain values of the parameter of the map, one finds chaotic behavior giving us an opportunity to touch on this very interesting topic with important consequences in physical phenomena. Chaotic evolution restricts out ability for *useful* predictions in an otherwise fully deterministic dynamical system: measurements using slightly different initial conditions result in a distribution which is indistinguishable from the distribution coming from sampling a random process. This scientific field is huge and active and we refer the reader to the bibliography for a more complete introduction [23, 24, 25, 26, 27, 28, 29, 40].

3.1 Introduction

The most celebrated application of the logistic map comes from the study of population growth in biology. One considers populations which reproduce at fixed time intervals and whose generations do not overlap.

The simplest (and most naive) model is the one that makes the reasonable assumption that the rate of population growth $dP(t)/dt$ of a population $P(t)$ is proportional to the current population:

$$\frac{dP(t)}{dt} = kP(t). \tag{3.1}$$

The general solution of the above equation is $P(t) = P(0)e^{kt}$ showing an *exponential* population growth for $k > 0$ an decline for $k < 0$. It is obvious that this model is reasonable as long as the population is small enough so that the interaction with its environment (adequate food, diseases, predators etc) can be neglected. The simplest model that takes into account some of the factors of the interaction with the environment (e.g. starvation) is obtained by the introduction of a simple non linear term in the equation so that

$$\frac{dP(t)}{dt} = kP(t)(1 - bP(t)). \tag{3.2}$$

The parameter k gives the maximum growth rate of the population and b controls the ability of the species to maintain a certain population level. The equation (3.2) can be discretized in time by assuming that each generation reproduces every δt and that the n-th generation has population $P_n = P(t_n)$ where $t_n = t_0 + (n-1)\delta t$. Then $P(t_{n+1}) \approx P(t_n) + \delta t P'(t_n)$ and equation (3.1) becomes

$$P_{n+1} = rP_n, \tag{3.3}$$

where $r = 1 + k\delta t$. The solutions of the above equation are well approximated by $P_n \sim P_0 e^{kt_n} \propto e^{(r-1)n}$ so that we have population growth when $r > 1$ and decline when $r < 1$. Equation (3.2) can be discretized as follows:

$$P_{n+1} = P_n(r - bP_n). \tag{3.4}$$

Defining $x_n = (b/r)P_n$ we obtain the logistic map

$$x_{n+1} = rx_n(1 - x_n). \tag{3.5}$$

We define the functions

$$f(x) = rx(1-x), \qquad F(x, r) = rx(1-x) \tag{3.6}$$

(their only difference is that, in the first one, r is considered as a given parameter), so that

$$x_{n+1} = f(x_n) = f^{(2)}(x_{n-1}) = \ldots = f^{(n)}(x_1) = f^{(n+1)}(x_0), \tag{3.7}$$

3.1. INTRODUCTION

where we use the notation $f^{(1)}(x) = f(x)$, $f^{(2)}(x) = f(f(x))$, $f^{(3)}(x) = f(f(f(x)))$, ... for function composition. In what follows, the derivative of f will be useful:

$$f'(x) = \frac{\partial F(x,r)}{\partial x} = r(1-2x). \tag{3.8}$$

Since we interpret x_n to be the fraction of the population with respect to its maximum value, we should have $0 \leq x_n \leq 1$ for each[1] n. The function $f(x)$ has one global maximum for $x = 1/2$ which is equal to $f(1/2) = r/4$. Therefore, if $r > 4$, then $f(1/2) > 1$, which for an appropriate choice of x_0 will lead to $x_{n+1} = f(x_n) > 1$ for some value of n. Therefore, the interval of values of r which is of interest for our model is

$$0 < r \leq 4. \tag{3.9}$$

The logistic map (3.5) may be viewed as a finite difference equation and it is a one step inductive relation. Given an initial value x_0, a sequence of values $\{x_0, x_1, \ldots, x_n, \ldots\}$ is produced. This will be referred[2] to as the *trajectory* of x_0. In the following sections we will study the properties of these trajectories as a function of the parameter r.

The solutions of the logistic map are not known except in special cases. For $r = 2$ we have

$$x_n = \frac{1}{2}\left(1 - (1 - x_0)^{2n}\right), \tag{3.10}$$

and for[3] $r = 4$

$$x_n = \sin^2(2^n \pi \theta), \qquad \theta = \frac{1}{\pi}\sin^{-1}\sqrt{x_0}. \tag{3.11}$$

For $r = 2$, $\lim_{n \to \infty} x_n = 1/2$ whereas for $r = 4$ we have periodic trajectories resulting in rational θ and non periodic resulting in irrational θ. For other values of r we have to resort to a numerical computation of the trajectories of the logistic map.

---

[1] Note that if $x_n > 1$ then $x_{n+1} < 0$, so that if we want $x_n \geq 0$ for each n, then we should have $x_n \leq 1$ for each n.

[2] In the bibliography, the term "splinter of x_0" is frequently used.

[3] E. Schröder, "Über iterierte Funktionen", Math. Ann. **3** (1870) 296; E. Lorenz, "The problem of deducing the climate from the governing equations", Tellus **16** (1964) 1

3.2 Fixed Points and 2^n Cycles

It is obvious that if the point x^* is a solution of the equation $x = f(x)$, then $x_n = x^* \Rightarrow x_{n+k} = x^*$ for every $k \geq 0$. For the function $f(x) = rx(1-x)$ we have two solutions

$$x_1^* = 0 \quad \text{and} \quad x_2^* = 1 - 1/r. \tag{3.12}$$

We will see that for appropriate values of r, these solutions are *attractors* of most of the trajectories. This means that for a range of values for the initial point $0 \leq x_0 \leq 1$, the sequence $\{x_n\}$ approaches asymptotically one of these points as $n \to \infty$. Obviously the (measure zero) sets of initial values $\{x_0\} = \{x_1^*\}$ and $\{x_0\} = \{x_2^*\}$ result in trajectories attracted by x_1^* and x_2^* respectively. In order to determine which one of the two values is preferred, we need to study the *stability* of the fixed points x_1^* and x_2^*. For this, assume that for some value of n, x_n is infinitesimally close to the fixed point x^* so that

$$\begin{aligned} x_n &= x^* + \epsilon_n \\ x_{n+1} &= x^* + \epsilon_{n+1}. \end{aligned} \tag{3.13}$$

Since

$$x_{n+1} = f(x_n) = f(x^* + \epsilon_n) \approx f(x^*) + \epsilon_n f'(x^*) = x^* + \epsilon_n f'(x^*), \tag{3.14}$$

where we used the Taylor expansion of the analytic function $f(x^* + \epsilon_n)$ about x^* and the relation $x^* = f(x^*)$, we have that $\epsilon_{n+1} = \epsilon_n f'(x^*)$. Then we obtain

$$\left| \frac{\epsilon_{n+1}}{\epsilon_n} \right| = |f'(x^*)|. \tag{3.15}$$

Therefore, if $|f'(x^*)| < 1$ we obtain $\lim_{n \to \infty} \epsilon_n = 0$ and the fixed point x^* is *stable*: the sequence $\{x_{n+k}\}$ approaches x^* asymptotically. If $|f'(x^*)| > 1$ then the sequence $\{x_{n+k}\}$ deviates away from x^* and the fixed point is *unstable*. The limiting case $|f'(x^*)| = 1$ should be studied separately and it indicates a change in the stability properties of the fixed point. In the following discussion, these points will be shown to be bifurcation points.

For the function $f(x) = rx(1-x)$ with $f'(x) = r(1-2x)$ we have that $f'(0) = r$ and $f'(1 - 1/r) = 2 - r$. Therefore, if $r < 1$ the point $x_1^* = 0$ is an attractor, whereas the point $x_2^* = 1 - 1/r < 0$ is irrelevant. When

3.2. FIXED POINTS AND 2^N CYCLES

$r > 1$, the point $x_1^* = 0$ results in $|f'(x_1^*)| = r > 1$, therefore x_1^* is unstable. Any initial value x_0 near x_1^* deviates from it. Since for $1 < r < 3$ we have that $0 \leq |f'(x_2^*)| = |2 - r| < 1$, the point x_2^* is an attractor. Any initial value $x_0 \in (0, 1)$ approaches $x_2^* = 1 - 1/r$. When $r = r_c^{(1)} = 1$ we have the limiting case $x_1^* = x_2^* = 0$ and we say that at the critical value $r_c^{(1)} = 1$ the fixed point x_1^* *bifurcates* to the two fixed points x_1^* and x_2^*.

As r increases, the fixed points continue to bifurcate. Indeed, when $r = r_c^{(2)} = 3$ we have that $f'(x_2^*) = 2 - r = -1$ and for $r > r_c^{(2)}$ the point x_2^* becomes unstable. Consider the solution of the equation $x = f^{(2)}(x)$. If $0 < x^* < 1$ is one of its solutions and for some n we have that $x_n = x^*$, then $x_{n+2} = x_{n+4} = \ldots = x_{n+2k} = \ldots = x^*$ and $x_{n+1} = x_{n+3} = \ldots = x_{n+2k+1} = \ldots = f(x^*)$ (therefore $f(x^*)$ is also a solution). If $0 < x_3^* < x_4^* < 1$ are two such *different* solutions with $x_3^* = f(x_4^*)$, $x_4^* = f(x_3^*)$, then the trajectory is periodic with period 2. The points x_3^*, x_4^* are such that they are *real* solutions of the equation

$$f^{(2)}(x) = r^2 x(1-x)(1 - rx(1-x)) = x, \qquad (3.16)$$

and at the same time they are *not* the solutions $x_1^* = 0$ $x_2^* = 1 - 1/r$ of the equation[4] $x = f^{(2)}(x)$, the polynomial above can be written in the form (see [24] for more details)

$$x\left(x - \left(1 - \frac{1}{r}\right)\right)(Ax^2 + Bx + C) = 0. \qquad (3.17)$$

By expanding the polynomials (3.16), (3.17) and comparing their coefficients we conclude that $A = -r^3$, $B = r^2(r+1)$ and $C = -r(r+1)$. The roots of the trinomial in (3.17) are determined by the discriminant $\Delta = r^2(r+1)(r-3)$. For the values of r of interest ($1 < r \leq 4$), the discriminant becomes positive when $r > r_c^{(2)} = 3$ and we have two different solutions

$$x_\alpha^* = ((r+1) \mp \sqrt{r^2 - 2r - 3})/(2r) \qquad \alpha = 3, 4. \qquad (3.18)$$

When $r = r_c^{(2)}$ we have one double root, therefore a unique fixed point.

The study of the stability of the solutions of $x = f^{(2)}(x)$ requires the same steps that led to the equation (3.15) and we determine if the

---

[4]Because, if $x^* = f(x^*) \Rightarrow f^{(2)}(x^*) = f(f(x^*)) = f(x^*) = x^*$ etc, the point x^* is also a solution of $x^* = f^{(n)}(x^*)$.

absolute value of $f^{(2)\prime}(x)$ is greater, less or equal to one. By noting that[5] $f^{(2)\prime}(x_3) = f^{(2)\prime}(x_4) = f'(x_3)f'(x_4) = -r^2 + 2r + 4$, we see that for $r = r_c^{(2)} = 3$, $f^{(2)\prime}(x_3^*) = f^{(2)\prime}(x_4^*) = 1$ and for $r = r_c^{(3)} = 1 + \sqrt{6} \approx 3.4495$, $f^{(2)\prime}(x_3) = f^{(2)\prime}(x_4) = -1$. For the intermediate values $3 < r < 1 + \sqrt{6}$ the derivatives $|f^{(2)\prime}(x_\alpha^*)| < 1$ for $\alpha = 3, 4$. Therefore, these points are stable solutions of $x = f^{(2)}(x)$ and the points x_1^*, x_2^* bifurcate to x_α^*, $\alpha = 1, 2, 3, 4$ for $r = r_c^{(2)} = 3$. Almost all trajectories with initial points in the interval $[0, 1]$ are attracted by the periodic trajectory with period 2, the "2-cycle" $\{x_3^*, x_4^*\}$.

Using similar arguments we find that the fixed points x_α^*, $\alpha = 1, 2, 3, 4$ bifurcate to the eight fixed points x_α^*, $\alpha = 1, \ldots, 8$ when $r = r_c^{(3)} = 1 + \sqrt{6}$. These are real solutions of the equation that gives the 4-cycle $x = f^{(4)}(x)$. For $r_c^{(3)} < r < r_c^{(4)} \approx 3.5441$, the points x_α^*, $\alpha = 5, \ldots, 8$ are a stable 4-cycle which is an attractor of almost all trajectories of the logistic map[6]. Similarly, for $r_c^{(4)} < r < r_c^{(5)}$ the 16 fixed points of the equation $x = f^{(8)}(x)$ give a stable 8-cycle, for $r_c^{(5)} < r < r_c^{(6)}$ a stable 16-cycle etc[7]. This is the phenomenon which is called *period doubling* which continues ad infinitum. The points $r_c^{(n)}$ are getting closer to each other as n increases so that $\lim_{n\to\infty} r_c^{(n)} = r_c \approx 3.56994567$. As we will see, r_c marks the onset of the non-periodic, chaotic behavior of the trajectories of the logistic map.

Computing the bifurcation points becomes quickly intractable and we have to resort to a numerical computation of their values. Initially we will write a program that computes trajectories of the logistic map for chosen values of r and x_0. The program can be found in the file `logistic.cpp` and is listed below:

```
#include <iostream>
#include <fstream>
```

---

[5]The chain rule $dh(g(x))/dx = h'(g(x))g'(x)$ gives that $f^{(2)\prime}(x_3^*) = df(f(x_3^*))/dx = f'(f(x_3^*))f'(x_3^*) = f'(x_4^*)f'(x_3^*)$ and similarly for $f^{(2)\prime}(x_4^*)$. We can prove by induction that for the n solutions $x_{n+1}^*, x_{n+2}^*, \ldots, x_{2n}^*$ that belong to the n-cycle of the equation $x = f^{(n)}(x)$ we have that $f^{(n)\prime}(x_{n+i}) = f'(x_{n+1}) \, f'(x_{n+2}) \ldots f'(x_{2n})$ for every $i = 1, \ldots, n$.

[6]The points x_α^*, $\alpha = 1, \ldots, 4$ are unstable fixed points and 2-cycle.

[7]Generally, for $r_c^{(n)} < r < r_c^{(n+1)} < r_c \approx 3.56994567$ we have 2^n fixed points of the equation $x = f^{(2^{n-1})}(x)$ and stable 2^{n-1}-cycles, which are attractors of almost all trajectories.

3.2. FIXED POINTS AND 2^N CYCLES

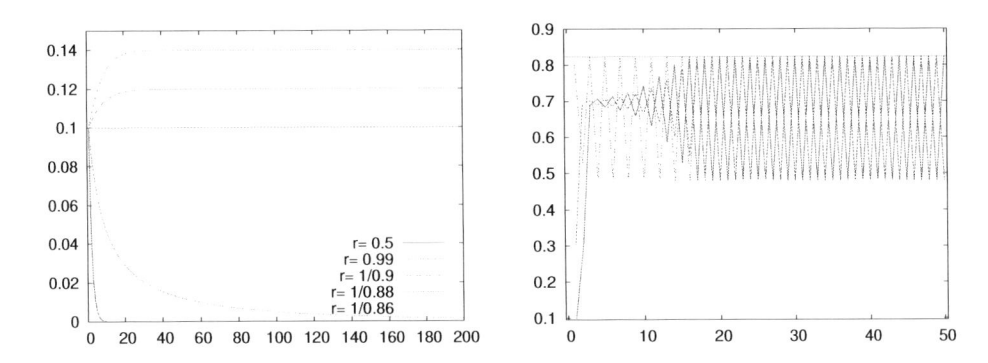

Figure 3.1: (Left) Some trajectories of the logistic map with $x_0 = 0.1$ and various values of r. We can see the first bifurcation for $r_c^{(1)} = 1$ from $x_1^* = 0$ to $x_2^* = 1 - 1/r$. (Right) Trajectories of the logistic map for $r_c^{(2)} < r = 3.5 < r_c^{(3)}$. The three curves start from three different initial points. After a transient period, depending on the initial point, one obtains a periodic trajectory which is a 2-cycle. The horizontal lines are the expected values $x_{3,4}^* = ((r+1) \mp \sqrt{r^2 - 2r - 3})/(2r)$ (see text).

```
#include <cstdlib>
#include <string>
#include <cmath>
using namespace std;

int main(){
  int    NSTEPS,i;
  double r,x0,x1;
  string buf;
  // ------ Input:
  cout << "# Enter NSTEPS, r, x0:\n";
  cin  >> NSTEPS  >> r >> x0;     getline(cin,buf);
  cout << "# NSTEPS = " << NSTEPS << endl;
  cout << "# r      = " << r      << endl;
  cout << "# x0     = " << x0     << endl;
  // ------ Initialize:
  ofstream myfile("log.dat");
  myfile.precision(17);
  // ------ Calculate:
  myfile << 0 << x0;
  for(i=1;i<=NSTEPS;i++){
    x1 = r * x0 * (1.0-x0);
    myfile << i << " " << x1 << "\n";
    x0 = x1;
```

```
    }
    myfile.close();
}//main()
```

The program is compiled and run using the commands:

```
> g++ logistic.cpp -o l
> echo "100 0.5 0.1" | ./l
```

The command echo prints to the stdout the values of the parameters NSTEPS=100, r=0.5 and x0=0.1. Its stdout is redirected to the stdin of the command ./l by using a pipe via the symbol |, from which the program reads their value and uses them in the calculation. The results can be found in two columns in the file log.dat and can be plotted using gnuplot. The plots are put in figure 3.1 and we can see the first two bifurcations when r goes past the values $r_c^{(1)}$ and $r_c^{(2)}$. Similarly, we can study trajectories which are 2^n-cycles when r crosses the values $r_c^{(n-1)}$.

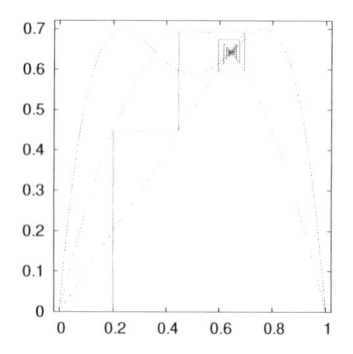

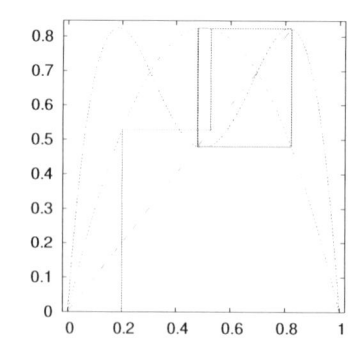

Figure 3.2: Cobweb plots of the logistic map for $r = 2.8$ and 3.3. (Left) The left plot is an example of a fixed point $x^* = f(x^*)$. The green line is $y = f(x)$ and the blue line is $y = f^{(2)}(x)$. The trajectory ends at the unique non zero intersection of the diagonal and $y = f(x)$ which is $x_2^* = 1 - 1/r$. The trajectory intersects the curve $y = f^{(2)}(x)$ at the same point. $y = f^{(2)}(x)$ does not intersect the diagonal anywhere else. (Right) The right plot shows an example of a 2-cycle. $y = f^{(2)}(x)$ intersects the diagonal at two additional points determined by x_3^* and x_4^*. The trajectory ends up on the orthogonal (x_3^*, x_3^*), (x_4^*, x_3^*), (x_4^*, x_4^*), (x_3^*, x_4^*).

Another way to depict the 2-cycles is by constructing the cobweb plots: We start from the point $(x_0, 0)$ and we calculate the point (x_0, x_1), where

3.2. FIXED POINTS AND 2^N CYCLES

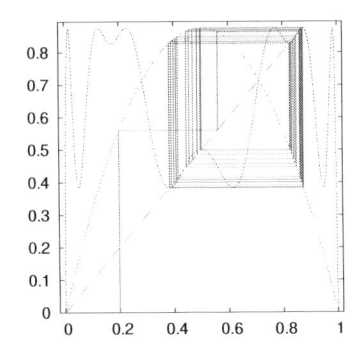

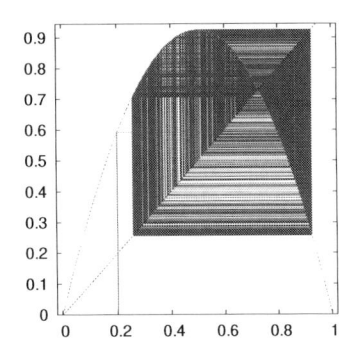

Figure 3.3: (Left) A 4-cycle for $r = 3.5$. The blue curve is $y = f^{(4)}(x)$ which intersects the diagonal at four points determined by x_α, $\alpha = 5, 6, 7, 8$. The four cycle passes through these points. (Right) a non periodic orbit for $r = 3.7$ when the system exhibits chaotic behavior.

$x_1 = f(x_0)$. This point belongs on the curve $y = f(x)$. The point (x_0, x_1) is then projected on the diagonal $y = x$ and we obtain the point (x_1, x_1). We repeat n times obtaining the points (x_n, x_{n+1}) and (x_{n+1}, x_{n+1}) on $y = f(x)$ and $y = x$ respectively. The fixed points $x^* = f(x^*)$ are at the intersections of these curves and, if they are attractors, the trajectories will converge on them. If we have a 2^n-cycle, we will observe a periodic trajectory going through points which are solutions to the equation $x = f^{(2^n)}(x)$. This exercise can be done by using the following program, which can be found in the file logistic1.cpp:

```
//===========================================================
// Discrete Logistic Map: Cobweb diagram
// Map the trajectory in 2d space (plane)
//===========================================================
#include <iostream>
#include <fstream>
#include <cstdlib>
#include <string>
#include <cmath>
using namespace std;

int main(){
  int    NSTEPS,i;
  double r,x0,x1;
```

```
    string buf;
    // ——— Input:
    cout << "# Enter NSTEPS, r, x0:\n";
    cin  >> NSTEPS  >> r  >> x0;           getline(cin,buf);
    cout << "# NSTEPS = "  << NSTEPS  << endl;
    cout << "# r      = "  << r       << endl;
    cout << "# x0     = "  << x0      << endl;
    // ——— Initialize:
    ofstream myfile("trj.dat");
    myfile.precision(17);
    // ——— Calculate:
    myfile   << 0      << " " << x0 << " " << 0  << '\n';
    for(i=1;i<=NSTEPS;i++){
      x1 = r * x0 * (1.0-x0);
      myfile << 2*i-3 << " " << x0 << " " << x1 << '\n';
      myfile << 2*i-2 << " " << x1 << " " << x1 << '\n';
      x0 = x1;
    }
    myfile.close();
}//main()
```

Compiling and running this program is done exactly as in the case of the program in logistic.cpp. We can plot the results using gnuplot. The plot in figure 3.2 can be constructed using the commands:

```
gnuplot> set size square
gnuplot> f(x) = r*x*(1.0-x)
gnuplot> r = 3.3
gnuplot> plot "<echo 50 3.3 0.2|./l;cat trj.dat" using 2:3 w l
gnuplot> replot f(x) ,f(f(x)),x
```

The plot command shown above, runs the program exactly as it is done on the command line. This is accomplished by using the symbol <, which reads the plot from the stdout of the command "echo 50 3.3 0.2|./l;cat trj.dat". Only the second command "echo trj.dat" writes to the stdout, therefore the plot is constructed from the contents of the file trj.dat. The following line adds the plots of the functions $f(x)$, $f^{(2)}(x) = f(f(x))$ and of the diagonal $y = x$. Figures 3.2 and 3.3 show examples of attractors which are fixed points, 2-cycles and 4-cycles. An example of a non periodic trajectory is also shown, which exhibits chaotic behavior which can happen when $r > r_c \approx 3.56994567$.

3.3 Bifurcation Diagrams

The bifurcations of the fixed points of the logistic map discussed in the previous section can be conveniently shown on the "bifurcation diagram". We remind to the reader that the first bifurcations happen at the critical values of r

$$r_c^{(1)} < r_c^{(2)} < r_c^{(3)} < \ldots < r_c^{(n)} < \ldots < r_c, \qquad (3.19)$$

where $r_c^{(1)} = 1$, $r_c^{(2)} = 3$, $r_c^{(3)} = 1 + \sqrt{6}$ and $r_c = \lim_{n\to\infty} r_c^{(n)} \approx 3.56994567$. For $r_c^{(n)} < r < r_c^{(n+1)}$ we have 2^n fixed points x_α^*, $\alpha = 1, 2, ..., 2^n$ of $x = f^{(2^n)}(x)$. By plotting these points $x_\alpha^*(r)$ as a function of r we construct the bifurcation diagram. These can be calculated numerically by using the program bifurcate.cpp. In this program, the user selects the values of r that she needs to study and for each one of them the program records the point of the 2^{n-1}-cycles[8] $x_\alpha^*(r)$, $\alpha = 2^{n-1} + 1, 2^{n-1} + 2, \ldots, 2^n$. This is easily done by computing the logistic map several times until we are sure that the trajectories reach the stable state. The parameter NTRANS in the program determines the number of points that we throw away, which should contain all the *transient* behavior. After NTRANS steps, the program records NSTEPS points, where NSTEPS should be large enough to cover all the points of the 2^{n-1}-cycles or depict a dense enough set of values of the non periodic orbits. The program is listed below:

```
//==========================================================
// Bifurcation Diagram of the Logistic Map
//==========================================================
#include <iostream>
#include <fstream>
#include <cstdlib>
#include <string>
#include <cmath>
using namespace std;

int main(){
  const double rmin   = 2.5;
  const double rmax   = 4.0;
  const double NTRANS = 500;    //Number of discarted steps
```

---

[8]If we want to be more precise, the bifurcation diagram contains also the unstable points. What we really construct is the orbit diagram which contains only the stable points.

```cpp
  const double NSTEPS = 100;    //Number of recorded steps
  const double RSTEPS = 2000;   //Number of values of r
  int i;
  double r,dr,x0,x1;

  // ——— Initialize:
  dr      = (rmax-rmin)/RSTEPS; //Increment in r
  ofstream myfile("bif.dat");
  myfile.precision(17);
  // ——— Calculate:
  r       = rmin;
  while( r <= rmax){
    x0    = 0.5;
    // ——— Transient steps: skip
    for(i=1;i<=NTRANS;i++){
      x1 = r * x0 * (1.0-x0);
      x0 = x1;
    }
    for(i=1;i<=NSTEPS;i++){
      x1 = r * x0 * (1.0-x0);
      myfile << r << " " << x1 << '\n';
      x0 = x1;
    }
    r += dr;
  }//while( r <= rmax)
  myfile.close();
}//main()
```

The program can be compiled and run using the commands:

```
> g++ bifurcate.cpp -o b
> ./b;
```

The left plot of figure 3.4 can be constructed by the gnuplot commands:

```
gnuplot> plot "bif.dat" with dots
```

We observe the fixed points and the 2^n-cycles for $r < r_c$. When r goes past r_c, the trajectories become non-periodic and exhibit chaotic behavior. Chaotic behavior will be discussed more extensively in the next section. For the time being, we note that if we measure the distance between the points $\Delta r^{(n)} = r_c^{(n+1)} - r_c^{(n)}$, we find that it decreases constantly with n so

3.3. BIFURCATION DIAGRAMS

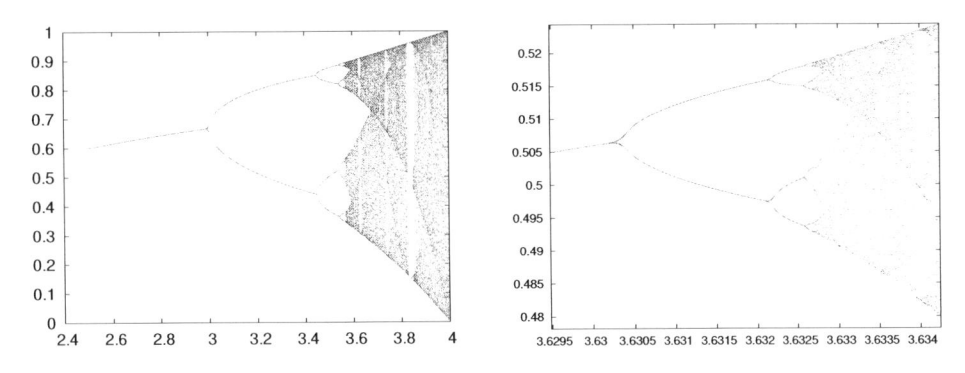

Figure 3.4: (Left) The bifurcation diagram computed by the program `bifurcate.cpp` for $2.5 < r < 4$. Notice the first bifurcation points followed by intervals of chaotic, non-periodic orbits interrupted by intermissions of stable periodic trajectories. The chaotic trajectories take values in subsets of the interval $(0, 1)$. For $r = 4$ they take values within the whole $(0, 1)$. One can see that for $r = 1 + \sqrt{8} \approx 3.8284$ we obtain a 3-cycle which subsequently bifurcates to $3 \cdot 2^n$-cycles. (Right) The diagram on the left is magnified in a range of r showing the self-similarity of the diagram at all scales.

that
$$\lim_{n \to \infty} \frac{\Delta r^{(n)}}{\Delta r^{(n+1)}} = \delta \approx 4.669\,201\,609\,, \qquad (3.20)$$

where δ is the Feigenbaum constant. An additional constant α, defined by the quotient of the separation of adjacent elements Δw_n of period doubled attractors from one double to the next Δw_{n+1}, is

$$\lim_{n \to \infty} \frac{\Delta w_n}{\Delta w_{n+1}} = \alpha \approx 2.502\,907\,875\,. \qquad (3.21)$$

It is also interesting to note the appearance of a 3-cycle right after $r = 1 + \sqrt{8} \approx 3.8284 > r_c$! By using the theorem of Sharkovskii, Li and Yorke[9] showed that any one dimensional system has 3-cycles, therefore it will have cycles of any length and chaotic trajectories. The stability of the 3-cycle can be studied from the solutions of $x = f^{(3)}(x)$ in exactly the same way that we did in equations (3.16) and (3.17) (see [24] for details). Figure 3.5 magnifies a branch of the 3-cycle. By magnifying different regions in the bifurcation plot, as shown in the right plot of figure 3.4, we find similar shapes to the branching of the 3-cycle. Figure 3.4 shows that

---

[9] T.Y. Li, J.A. Yorke, "Period Three Implies Chaos", American Mathematical Monthly 82 (1975) 985.

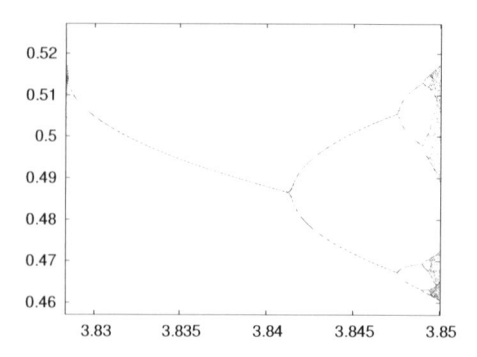

Figure 3.5: Magnification of one of the three branches of the 3-cycle for $r > 1 + \sqrt{8}$. To the left, we observe the temporary halt of the chaotic behavior of the trajectory, which comes back as shown in the plot to the right after an intermission of stable periodic trajectories.

between intervals of chaotic behavior we obtain "windows" of periodic trajectories. These are infinite but countable. It is also quite interesting to note that if we magnify a branch withing these windows, we obtain a diagram that is similar to the whole diagram! We say that the bifurcation diagram exhibits self similarity. There are more interesting properties of the bifurcation diagram and we refer the reader to the bibliography for a more complete exposition.

We close this section by mentioning that the qualitative properties of the bifurcation diagram are the same for a whole class of functions. Feigenbaum discovered that if one takes any function that is concave and has a unique global maximum, its bifurcation diagram behaves qualitatively the same way as that of the logistic map. Examples of such functions[10] studied in the literature are $g(x) = xe^{r(1-x)}$, $u(x) = r\sin(\pi x)$ and $w(x) = b - x^2$. The constants δ and α of equations (3.20) and (3.21) are the same of all these mappings. The functions that result in chaotic behavior are studied extensively in the literature and you can find a list of those in [30].

---

[10] The function $x \exp(r(1-x))$ has been used as a model for populations whose large density is restricted by epidemics. The populations are always positive independently of the (positive) initial conditions and the value of r.

3.4 The Newton-Raphson Method

In order to determine the bifurcation points, one has to solve the nonlinear, polynomial, algebraic equations $x = f^{(n)}(x)$ and $f^{(n)\prime}(x) = -1$. For this reason, one has to use an approximate numerical calculation of the roots, and the simple Newton-Raphson method will prove to be a good choice.

Newton-Raphson's method uses an initial guess x_0 for the solution of the equation $g(x) = 0$ and computes a sequence of points $x_1, x_2, \ldots, x_n, x_{n+1}, \ldots$ that presumably converges to one of the roots of the equation. The computation stops at a finite n, when we decide that the desired level of accuracy has been achieved. In order to understand how it works, we assume that $g(x)$ is an analytic function for all the values of x used in the computation. Then, by Taylor expanding around x_n we obtain

$$g(x_{n+1}) = g(x_n) + (x_{n+1} - x_n)g'(x) + \ldots. \qquad (3.22)$$

If we wish to have $g(x_{n+1}) \approx 0$, we choose

$$x_{n+1} = x_n - \frac{g(x_n)}{g'(x_n)}. \qquad (3.23)$$

The equation above gives the Newton-Raphson method for one equation $g(x) = 0$ of one variable x. Different choices for x_0 will possibly lead to different roots. When $g'(x)$, $g''(x)$ are non zero at the root and $g'''(x)$ is bounded, the convergence of the method is quadratic with the number of iterations. This means that there is a neighborhood of the root α such that the distance $\Delta x_{n+1} = x_{n+1} - \alpha$ is $\Delta x_{n+1} \propto (\Delta x_n)^2$. If the root α has multiplicity larger than 1, convergence is slower. The proofs of these statements are simple and can be found in [31].

The Newton-Raphson method is simple to program and, most of the times, sufficient for the solution of many problems. In the general case it works well only close enough to a root. We should also keep in mind that there are simple reasons for the method to fail. For example, when $g'(x_n) = 0$ for some n, the method stops. For functions that tend to 0 as $x \to \pm\infty$, it is easy to make a bad choice for x_0 that does not lead to convergence to a root. Sometimes it is a good idea to combine the Newton-Raphson method with the *bisection* method. When the derivative $g'(x)$ diverges at the root we might get into trouble. For example, the equation $|x|^\nu = 0$ with $0 < \nu < 1/2$, does not lead to a convergent sequence. In some cases, we might enter into non-convergent cycles [8]. For some functions the basin of attraction of a root (the values of x_0 that will converge to the root) can be tiny. See problem 13.

As a test case of our program, consider the equation

$$\epsilon \tan \epsilon = \sqrt{\rho^2 - \epsilon^2} \qquad (3.24)$$

which results from the solution of Schrödinger's equation for the energy spectrum of a quantum mechanical particle of mass m in a one dimensional potential well of depth V_0 and width L. The parameters $\epsilon = \sqrt{mL^2 E/(2\hbar)}$ and $\rho = \sqrt{mL^2 V_0/(2\hbar)}$. Given ρ, we solve for ϵ which gives the energy E. The function $g(x)$ and its derivative $g'(x)$ are

$$\begin{aligned} g(x) &= x \tan x - \sqrt{\rho^2 - x^2} \\ g'(x) &= \frac{x}{\sqrt{\rho^2 - x^2}} + \frac{x}{\cos^2 x} + \tan x \, . \end{aligned} \qquad (3.25)$$

The program of the Newton-Raphson method for solving the equation $g(x) = 0$ can be found in the file nr.cpp:

```
//===================================================
//Newton Raphson of function of one variable
//===================================================
#include <iostream>
#include <fstream>
#include <cstdlib>
#include <string>
#include <cmath>
using namespace std;

int main(){
  const double rho  = 15.0;
  const double eps  = 1.0e-6;
  const int    NMAX = 1000;
  double x0, x1, err, g, gp;
  int    i;
  string buf;
  // ——— Input:
  cout << "# Enter x0:\n";
  cin  >> x0;    getline(cin,buf);
  err = 1.0;
```

3.4. THE NEWTON-RAPHSON METHOD

```
    cout << "iter              x                        error        \n";
    cout << "_____\n";
    cout << 0    <<  " "   <<  x0  << " "  <<         err << '\n';

cout.precision(17);
for(i=1;i<=NMAX;i++){
    //value of function g(x):
    g    = x0*tan(x0)-sqrt(rho*rho-x0*x0);
    //value of the derivative g'(x):
    gp   = x0/sqrt(rho*rho-x0*x0)+x0/(cos(x0)*cos(x0))+tan(x0);
    x1   = x0 - g/gp;
    err  = abs(x1-x0);
    cout << i << " " << x1 << " " << err << '\n';
    if(err < eps) break;
    x0   = x1;
}
}//main()
```

In the program listed above, the user is asked to set the initial point x_0. We fix $\rho =$ rho $= 15$. It is instructive to make the plot of the left and right hand sides of (3.24) and make a graphical determination of the roots from their intersections. Then we can make appropriate choices of the initial point x_0. Using gnuplot, the plots are made with the commands:

```
gnuplot> g1(x) = x*tan(x)
gnuplot> g2(x) = sqrt(rho*rho-x*x)
gnuplot> plot [0:20][0:20]  g1(x), g2(x)
```

The compilation and running of the program can be done as follows:

```
> g++ nr.cpp -o n
> echo "1.4"|./n
# Enter x0:
iter              x                        error
_____
0 1.4                   1
1 1.5254292024457967    0.12542920244579681
2 1.5009739120496131    0.02445529039618366
3 1.48072070172022      0.02025321032939309
4 1.4731630533073483    0.0075576484128716537
5 1.4724779331237687    0.00068512018357957949
6 1.4724731072313519    4.8258924167932093e-06
7 1.4724731069952235    2.3612845012621619e-10
```

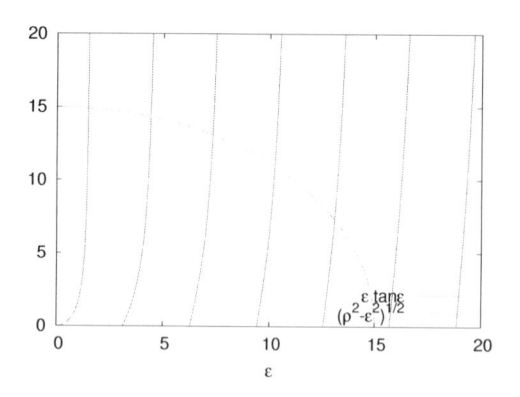

Figure 3.6: Plots of the right and left hand sides of equation (3.24). The intersections of the curves determine the solutions of the equation and their approximate graphical estimation can serve as initial points x_0 for the Newton-Raphson method.

We conclude that one of the roots of the equation is $\epsilon \approx 1.472473107$. The reader can compute more of these roots by following these steps by herself.

The method discussed above can be easily generalized to the case of two equations. Suppose that we need to solve simultaneously two algebraic equations $g_1(x_1, x_2) = 0$ and $g_2(x_1, x_2) = 0$. In order to compute a sequence (x_{10}, x_{20}), (x_{11}, x_{21}), ..., (x_{1n}, x_{2n}), $(x_{1(n+1)}, x_{2(n+1)})$, ... that may converge to a root of the above system of equations, we Taylor expand the two functions around (x_{1n}, x_{2n})

$$\begin{aligned}
g_1(x_{1(n+1)}, x_{2(n+1)}) &= g_1(x_{1n}, x_{2n}) + (x_{1(n+1)} - x_{1n})\frac{\partial g_1(x_{1n}, x_{2n})}{\partial x_1} \\
&\quad + (x_{2(n+1)} - x_{2n})\frac{\partial g_1(x_{1n}, x_{2n})}{\partial x_2} + \ldots \\
g_2(x_{1(n+1)}, x_{2(n+1)}) &= g_2(x_{1n}, x_{2n}) + (x_{1(n+1)} - x_{1n})\frac{\partial g_2(x_{1n}, x_{2n})}{\partial x_1} \\
&\quad + (x_{2(n+1)} - x_{2n})\frac{\partial g_2(x_{1n}, x_{2n})}{\partial x_2} + \ldots . \quad (3.26)
\end{aligned}$$

Defining $\delta x_1 = (x_{1(n+1)} - x_{1n})$ and $\delta x_2 = (x_{2(n+1)} - x_{2n})$ and setting

3.4. THE NEWTON-RAPHSON METHOD

$g_1(x_{1(n+1)}, x_{2(n+1)}) \approx 0$, $g_2(x_{1(n+1)}, x_{2(n+1)}) \approx 0$, we obtain

$$\delta x_1 \frac{\partial g_1}{\partial x_1} + \delta x_2 \frac{\partial g_1}{\partial x_2} = -g_1$$
$$\delta x_1 \frac{\partial g_2}{\partial x_1} + \delta x_2 \frac{\partial g_2}{\partial x_2} = -g_2. \quad (3.27)$$

This is a linear 2×2 system of equations

$$A_{11}\delta x_1 + A_{12}\delta x_2 = b_1$$
$$A_{21}\delta x_1 + A_{22}\delta x_2 = b_2, \quad (3.28)$$

where $A_{ij} = \partial g_i / \partial x_j$ and $b_i = -g_i$, with $i, j = 1, 2$. Solving for δx_i we obtain

$$x_{1(n+1)} = x_{1n} + \delta x_1$$
$$x_{2(n+1)} = x_{2n} + \delta x_2. \quad (3.29)$$

The iterations stop when δx_i become small enough.

As an example, consider the equations with $g_1(x) = 2x^2 - 3xy + y - 2$, $g_2(x) = 3x + xy + y - 1$. We have $A_{11} = 4x - 3y$, $A_{12} = 1 - 3x$, $A_{21} = 3 + y$, $A_{22} = 1 + x$. The program can be found in the file nr2.cpp:

```
//=================================================
//Newton Raphson of two functions of two variables
//=================================================
#include <iostream>
#include <fstream>
#include <cstdlib>
#include <string>
#include <cmath>
using namespace std;

void solve2x2(double A[2][2],double b[2],double dx[2]);

int main(){
  const double eps  = 1.0e-6;
  const int    NMAX = 1000;
  double A[2][2],b[2],dx[2];
  double x, y, err;
  int    i;
  string buf;
```

```cpp
//  ———— Input:
cout << "# Enter x0,y0:\n";
cin  >> x >> y;    getline(cin,buf);
err = 1.0;
cout << "iter          x                y             error     \n";
cout << "————————————————————————————————————————————————————————————\n";
cout << 0 << " " << x << " " << y << " " <<   err << '\n';

cout.precision(17);
for(i=1;i<=NMAX;i++){
   b[0] =  -(2.0*x*x-3.0*x*y + y - 2.0);  // -g1(x,y)
   b[1] =  -(3.0*x +    x*y + y - 1.0);   // -g2(x,y)
   // dg1/dx                     dg1/dy
   A[0][0] = 4.0*x-3.0*y; A[0][1] = 1.0-3.0*x;
   // dg2/dx                     dg2/dy
   A[1][0] = 3.0 +     y; A[1][1] = 1.0+      x;
   solve2x2(A,b,dx);
   x += dx[0];
   y += dx[1];
   err = 0.5*sqrt(dx[0]*dx[0]+dx[1]*dx[1]);
   cout << i << " " << x << " " << y << " " << err << endl;
   if(err < eps) break;
}
}//main()
void solve2x2(double A[2][2],double b[2],double dx[2]){
  double num0,num1,det;

  num0   = A[1][1] * b[0]    - A[0][1] * b[1];
  num1   = A[0][0] * b[1]    - A[1][0] * b[0];
  det    = A[0][0] * A[1][1] - A[0][1] * A[1][0];
  if(det == 0.0){cerr << "solve2x2: det=0\n";exit(1);}
  dx[0]  = num0/det;
  dx[1]  = num1/det;

}//solve2x2()
```

In order to guess the region where the real roots of the systems lie, we make a 3-dimensional plot using gnuplot:

```
gnuplot> set isosamples 20
gnuplot> set hidden3d
gnuplot> splot 2*x**2-3*x*y+y-2,3*x+y*x+y-1,0
```

We plot the functions $g_i(x,y)$ together with the plane $x=0$. The intersection of the three surfaces determine the roots we are looking for. Compiling and running the program can be done by using the commands:

```
> g++ nr2.cpp -o n
> echo 2.2 1.5 1./n
# Enter x0,y0:
iter    x            y            error
0    2.20000000   1.50000000    1.0000
1    0.76427104   0.26899383    0.9456
2    0.73939531  -0.68668275    0.4780
3    0.74744506  -0.71105605    1.2834e-2
4    0.74735933  -0.71083147    1.2019e-4
5    0.74735932  -0.71083145    1.2029e-8
> echo 0 1 1./n
................
5   -0.10899022   1.48928857    4.3461e-12
> echo -5 0 1./n
6   -6.13836909  -3.77845711    3.2165e-13
```

The computation above leads to the roots $(0.74735932, -0.71083145)$, $(-0.10899022, 1.48928857)$, $(-6.13836909, -3.77845711)$.

The Newton-Raphson method for many variables becomes hard quite soon: One needs to calculate the functions as well as their derivatives, which is prohibitively expensive for many problems. It is also hard to determine the roots, since the method converges satisfactorily only very close to the roots. We refer the reader to [8] for more information on how one can deal with these problems.

3.5 Calculation of the Bifurcation Points

In order to determine the bifurcation points for $r < r_c$ we will solve the algebraic equations $x = f^{(k)}(x)$ and $f^{(k)\prime}(x) = -1$. At these points, k-cycles become unstable and $2k$-cycles appear and are stable. This happens when $r = r_c^{(n)}$, where $k = 2^{n-2}$. We will look for solutions $(x_\alpha^*, r_c^{(n)})$ for $\alpha = k+1, k+2, \ldots, 2k$.

We define the functions $F(x, r) = f(x) = rx(1-x)$ and $F^{(k)}(x, r) =$

$f^{(k)}(x)$ as in equation (3.6). We will solve the algebraic equations:

$$\begin{aligned} g_1(x,r) &= x - F^{(k)}(x,r) = 0 \\ g_2(x,r) &= \frac{\partial F^{(k)}(x,r)}{\partial x} + 1 = 0. \end{aligned} \qquad (3.30)$$

According to the discussion of the previous section, in order to calculate the roots of these equations we have to solve the linear system (3.28), where the coefficients are

$$\begin{aligned} b_1 &= -g_1(x,r) = -x + F^{(k)}(x,r) \\ b_2 &= -g_2(x,r) = -\frac{\partial F^{(k)}(x,r)}{\partial x} - 1 \\ A_{11} &= \frac{\partial g_1(x,r)}{\partial x} = 1 - \frac{\partial F^{(k)}(x,r)}{\partial x} \\ A_{12} &= \frac{\partial g_1(x,r)}{\partial r} = -\frac{\partial F^{(k)}(x,r)}{\partial r} \\ A_{21} &= \frac{\partial g_2(x,r)}{\partial x} = \frac{\partial^2 F^{(k)}(x,r)}{\partial x^2} \\ A_{22} &= \frac{\partial g_2(x,r)}{\partial r} = \frac{\partial^2 F^{(k)}(x,r)}{\partial x \partial r}. \end{aligned} \qquad (3.31)$$

The derivatives will be calculated approximately using finite differences

$$\begin{aligned} \frac{\partial F^{(k)}(x,r)}{\partial x} &\approx \frac{F^{(k)}(x+\epsilon,r) - F^{(k)}(x-\epsilon,r)}{2\epsilon} \\ \frac{\partial F^{(k)}(x,r)}{\partial r} &\approx \frac{F^{(k)}(x,r+\epsilon) - F^{(k)}(x,r-\epsilon)}{2\epsilon}, \end{aligned} \qquad (3.32)$$

3.5. CALCULATION OF THE BIFURCATION POINTS

and similarly for the second derivatives

$$\frac{\partial^2 F^{(k)}(x,r)}{\partial x^2} \approx \frac{\frac{\partial F^{(k)}(x+\frac{\epsilon}{2},r)}{\partial x} - \frac{\partial F^{(k)}(x-\frac{\epsilon}{2},r)}{\partial x}}{2\frac{\epsilon}{2}}$$

$$= \frac{1}{\epsilon}\left\{\frac{F^{(k)}(x+\epsilon,r) - F^{(k)}(x,r)}{\epsilon} - \frac{F^{(k)}(x,r) - F^{(k)}(x-\epsilon,r)}{\epsilon}\right\}$$

$$= \frac{1}{\epsilon^2}\left\{F^{(k)}(x+\epsilon,r) - 2F^{(k)}(x,r) + F^{(k)}(x-\epsilon,r)\right\}$$

$$\frac{\partial^2 F^{(k)}(x,r)}{\partial x \partial r} \approx \frac{\frac{\partial F^{(k)}(x+\epsilon_x,r)}{\partial r} - \frac{\partial F^{(k)}(x-\epsilon_x,r)}{\partial r}}{2\epsilon_x}$$

$$= \frac{1}{2\epsilon_x}\left\{\frac{F^{(k)}(x+\epsilon_x,r+\epsilon_r) - F^{(k)}(x+\epsilon_x,r-\epsilon_r)}{2\epsilon_r}\right.$$
$$\left. -\frac{F^{(k)}(x-\epsilon_x,r+\epsilon_r) - F^{(k)}(x-\epsilon_x,r-\epsilon_r)}{2\epsilon_r}\right\}$$

$$= \frac{1}{4\epsilon_x\epsilon_r}\left\{F^{(k)}(x+\epsilon_x,r+\epsilon_r) - F^{(k)}(x+\epsilon_x,r-\epsilon_r)\right.$$
$$\left. -F^{(k)}(x-\epsilon_x,r+\epsilon_r) + F^{(k)}(x-\epsilon_x,r-\epsilon_r)\right\} \quad (3.33)$$

We are now ready to write the program for the Newton-Raphson method like in the previous section. The only difference is the approximate calculation of the derivatives using the relations above and the calculation of the function $F^{(k)}(x,r)$ by a routine that will compose the function $f(x)$ k-times. The program can be found in the file bifurcationPoints.cpp:

```
//=========================================================
//         bifurcationPoints.cpp
// Calculate bifurcation points of the discrete logistic map
// at period k by solving the condition
// g1(x,r) = x - F(k,x,r)    = 0
// g2(x,r) = dF(k,x,r)/dx+1 = 0
// determining when the Floquet multiplier bacomes 1
// F(k,x,r) iterates F(x,r) = r*x*(x-1) k times
// The equations are solved by using a Newton-Raphson method
//=========================================================
#include <iostream>
#include <fstream>
#include <cstdlib>
#include <string>
```

```cpp
#include <cmath>
using namespace std;
double F       (const int& k,const double& x,const double& r);
double dFdx    (const int& k,const double& x,const double& r);
double dFdr    (const int& k,const double& x,const double& r);
double d2Fdx2  (const int& k,const double& x,const double& r);
double d2Fdrdx (const int& k,const double& x,const double& r);
void solve2x2 (double A[2][2],double b[2],double dx[2]);

int main(){
  const double tol = 1.0e-10;
  double r0,x0;
  double A[2][2],B[2],dX[2];
  double error;
  int k,iter;
  string buf;

  // ------ Input:
  cout << "# Enter k,r0,x0:\n";
  cin  >> k >> r0 >> x0;           getline(cin,buf);
  cout << "# Period k= "                 << k << endl;
  cout << "# r0= " << r0 << " x0= " << x0 << endl;
  // ------ Initialize
  error = 1.0; //initial large value of error>tol
  iter  =   0;
  cout.precision(17);
  while(error > tol){
    // ---- Calculate jacobian matrix
    A[0][0] = 1.0  -dFdx(k,x0,r0);
    A[0][1] = -dFdr    (k,x0,r0);
    A[1][0] = d2Fdx2   (k,x0,r0);
    A[1][1] = d2Fdrdx  (k,x0,r0);
    B[0]    = -x0 +    F(k,x0,r0);
    B[1]    = -dFdx    (k,x0,r0) -1.0;
    // ---- Solve a 2x2 linear system:
    solve2x2(A,B,dX);
    x0    = x0 + dX[0];
    r0    = r0 + dX[1];
    error = 0.5*sqrt(dX[0]*dX[0]+dX[1]*dX[1]);
    iter++;
    cout <<iter
         << " x0= " << x0
         << " r0= " << r0
         << " err= " << error << '\n';
  }//while(error > tol)
```

3.5. CALCULATION OF THE BIFURCATION POINTS

```
}//main()
//=========================================================
//Function F(k,x,r) and its derivatives
double F       (const int& k,const double& x,const double& r){
  double x0;
  int    i;
  x0 = x;
  for(i=1;i<=k;i++) x0 = r*x0*(1.0-x0);
  return x0;
}
//----------------------------------------
double dFdx    (const int& k,const double& x,const double& r){
  double eps;
  eps    = 1.0e-6*x;
  return (F(k,x+eps,r)-F(k,x-eps,r))/(2.0*eps);
}
//----------------------------------------
double dFdr    (const int& k,const double& x,const double& r){
  double eps;
  eps    = 1.0e-6*r;
  return (F(k,x,r+eps)-F(k,x,r-eps))/(2.0*eps);
}
//----------------------------------------
double d2Fdx2  (const int& k,const double& x,const double& r){
  double eps;
  eps    = 1.0e-6*x;
  return (F(k,x+eps,r)-2.0*F(k,x,r)+F(k,x-eps,r))/(eps*eps);
}
//----------------------------------------
double d2Fdrdx(const int& k,const double& x,const double& r){
  double epsx,epsr;
  epsx   = 1.0e-6*x;
  epsr   = 1.0e-6*r;
  return ( F(k,x+epsx,r+epsr)-F(k,x+epsx,r-epsr)
          -F(k,x-epsx,r+epsr)+F(k,x-epsx,r-epsr))
         /(4.0*epsx*epsr);
}
//=========================================================
void solve2x2(double A[2][2],double b[2],double dx[2]){
  double num0,num1,det;

  num0   = A[1][1] * b[0]    - A[0][1] * b[1];
  num1   = A[0][0] * b[1]    - A[1][0] * b[0];
  det    = A[0][0] * A[1][1] - A[0][1] * A[1][0];
  if(det == 0.0){cerr << "solve2x2: det=0\n";exit(1);}
```

```
    dx[0]  =  num0/det;
    dx[1]  =  num1/det;

}//solve2x2()
```

Compiling and running the program can be done as follows:

```
> g++ bifurcationPoints.cpp -o b
> echo 2 3.5 0.5 |./b
# Enter k,r0,x0:
# Period k=               2
# r0= 3.5000000000000   x0=  0.50000000000
1  x0= 0.4455758353187  r0= 3.38523275827  err= 6.35088e-2
2  x0= 0.4396562547624  r0= 3.45290970406  err= 3.39676e-2
3  x0= 0.4399593001407  r0= 3.44949859951  err= 1.71226e-3
4  x0= 0.4399601690333  r0= 3.44948974267  err= 4.44967e-6
5  x0= 0.4399601689937  r0= 3.44948974281  err= 7.22160e-11
> echo 2 3.5 0.85 | ./b
.................
4  x0= 0.8499377795512  r0= 3.44948974275  err= 1.85082e-11
> echo 4 3.5 0.5 |./b
.................
5  x0= 0.5235947861540  r0= 3.54409035953  err= 1.86318e-11
> echo 4 3.5 0.35 | ./b
.................
5  x0= 0.3632903374118  r0= 3.54409035955  err= 5.91653e-13
```

The above listing shows the points of the 2-cycle and some of the points of the 4-cycle. It is also possible to compare the calculated value $r_c^{(3)} = 3.449490132$ with the expected one $r_c^{(3)} = 1+\sqrt{6} \approx 3.449489742$. Improving the accuracy of the calculation is left as an exercise for the reader who has to control the systematic errors of the calculations and achieve better accuracy in the computation of $r_c^{(4)}$.

3.6 Liapunov Exponents

We have seen that when $r > r_c \approx 3.56994567$, the trajectories of the logistic map become non periodic and exhibit chaotic behavior. Chaotic behavior mostly means sensitivity of the evolution of a dynamical system to the choice of initial conditions. More precisely, it means that two different trajectories constructed from infinitesimally close initial conditions,

3.6. LIAPUNOV EXPONENTS

diverge very fast from each other. This implies that there is a set of initial conditions that densely cover subintervals of $(0,1)$ whose trajectories do not approach arbitrarily close to any cycle of finite length.

Assume that two trajectories have x_0, $\tilde{x}_0$ as initial points and $\Delta x_0 = x_0 - \tilde{x}_0$. When the points x_n, $\tilde{x}_n$ have a distance $\Delta x_n = \tilde{x}_n - x_n$ that for small enough n increases exponentially with n (the "time"), i.e.

$$\Delta x_n \sim \Delta x_0 e^{\lambda n}, \qquad \lambda > 0, \qquad (3.34)$$

the system is most likely exhibiting chaotic behavior[11]. The exponent λ is called a *Liapunov exponent*. A useful equation for the calculation of λ is

$$\lambda = \lim_{n \to \infty} \frac{1}{n} \sum_{k=0}^{n-1} \ln |f'(x_k)|. \qquad (3.35)$$

This relation can be easily proved by considering infinitesimal $\epsilon \equiv |\Delta x_0|$ so that $\lambda = \lim_{n \to \infty} \lim_{\epsilon \to 0} \frac{1}{n} \ln |\Delta x_n|/\epsilon$. Then we obtain

$$\begin{aligned}
\tilde{x}_1 &= f(\tilde{x}_0) = f(x_0 + \epsilon) \approx f(x_0) + \epsilon f'(x_0) \\
&= x_1 + \epsilon f'(x_0) \Rightarrow \\
\frac{\Delta x_1}{\epsilon} &= \frac{\tilde{x}_1 - x_1}{\epsilon} \approx f'(x_0)
\end{aligned}$$

$$\begin{aligned}
\tilde{x}_2 &= f(\tilde{x}_1) = f(x_1 + \epsilon f'(x_0)) \approx f(x_1) + (\epsilon f'(x_0))f'(x_1) \\
&= x_2 + \epsilon f'(x_0)f'(x_1) \Rightarrow \\
\frac{\Delta x_2}{\epsilon} &= \frac{\tilde{x}_2 - x_2}{\epsilon} \approx f'(x_0)f'(x_1)
\end{aligned}$$

$$\begin{aligned}
\tilde{x}_3 &= f(\tilde{x}_2) = f(x_2 + \epsilon f'(x_0)f'(x_1)) \approx f(x_2) + (\epsilon f'(x_0)f'(x_1))f'(x_2) \\
&= x_3 + \epsilon f'(x_0)f'(x_1)f'(x_2) \Rightarrow \\
\frac{\Delta x_3}{\epsilon} &= \frac{\tilde{x}_3 - x_3}{\epsilon} \approx f'(x_0)f'(x_1)f'(x_2). \qquad (3.36)
\end{aligned}$$

We can show by induction that $|\Delta x_n|/\epsilon \approx f'(x_0)f'(x_1)f'(x_2)\ldots f'(x_{n-1})$ and by taking the logarithm and the limits we can prove (3.35).

---

[11]Sensitivity to the initial condition alone does not necessarily imply chaos. It is necessary to have topological mixing and dense periodic orbits. Topological mixing means that every open set in phase space will evolve to a set that for large enough time will have non zero intersection with any open set. Dense periodic orbits means that every point in phase space lies infinitesimally close to a periodic orbit.

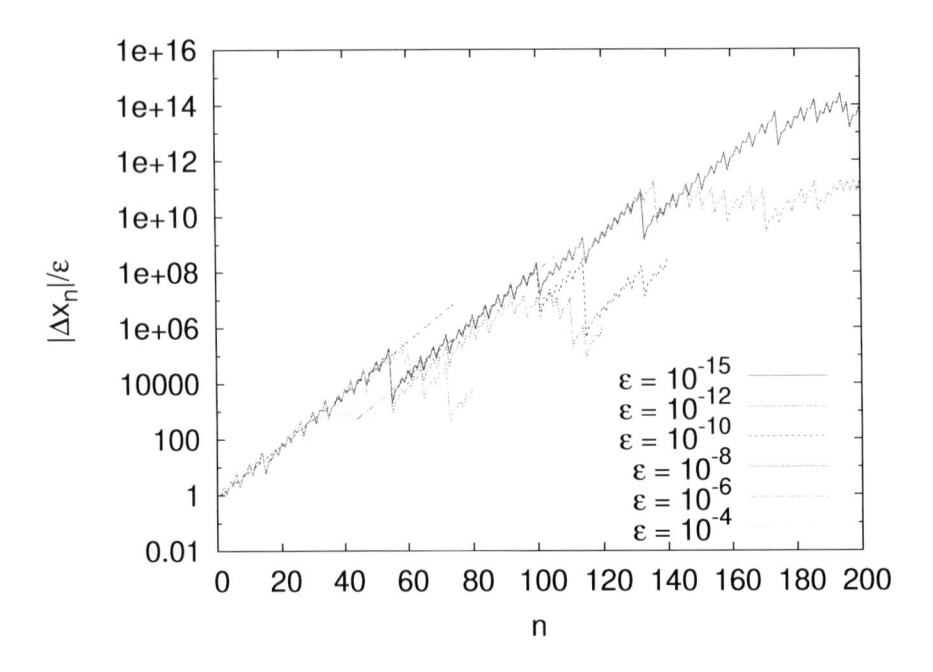

Figure 3.7: A plot of $|\Delta x_n|/\epsilon$ for the logistic map for $r = 3.6$, $x_0 = 0.2$. Note the convergence of the curves as $\epsilon \to 0$ and the approximate exponential behavior in this limit. The two lines are fits to the equation (3.34) and give $\lambda = 0.213(4)$ and $\lambda = 0.217(6)$ respectively.

A first attempt to calculate the Liapunov exponents could be made by using the definition (3.34). We modify the program `logistic.cpp` so that it calculates two trajectories whose initial distance is ϵ = epsilon:

```cpp
//============================================
//Discrete Logistic Map:
//Two trajectories with close initial conditions.
//============================================
#include <iostream>
#include <fstream>
#include <cstdlib>
#include <string>
#include <cmath>
using namespace std;

int main(){
```

3.6. LIAPUNOV EXPONENTS

```
   int    NSTEPS,i;
   double r,x0,x1,x0t,x1t,epsilon;
   string buf;
// ------- Input:
   cout << "# Enter NSTEPS, r, x0, epsilon:\n";
   cin  >> NSTEPS    >> r >> x0 >> epsilon;getline(cin,buf);
   cout << "# NSTEPS  = " << NSTEPS  << endl;
   cout << "# r       = " << r       << endl;
   cout << "# x0      = " << x0      << endl;
   cout << "# epsilon = " << epsilon << endl;

   x0t = x0+epsilon;
// ------- Initialize:
   ofstream myfile("lia.dat");
   myfile.precision(17);
// ------- Calculate:
   myfile   << 1    << " "    << x0 << " " << x0t << " "
            << abs(x0t-x0)/epsilon << "\n";
   for(i=2;i<=NSTEPS;i++){
     x1  = r * x0  * (1.0-x0 );
     x1t = r * x0t * (1.0-x0t);
     myfile << i    << " "    << x1 << " " << x1t << " "
            << abs(x1t-x1)/epsilon << "\n";
     x0  = x1; x0t  = x1t;
   }
   myfile.close();
}//main()
```

After running the program, the quantity $|\Delta x_n|/\epsilon$ is found at the fourth column of the file lia.dat. The curves of figure 3.7 can be constructed by using the commands:

```
> g++ liapunov1.cpp -o l
> gnuplot
gnuplot> set logscale y
gnuplot> plot \
  "<echo 200 3.6 0.2 1e-15 |./l;cat lia.dat" u 1:4 w l
```

The last line plots the stdout of the command "echo 200 3.6 0.2 1e-15 |./l;cat lia.dat", i.e. the contents of the file lia.dat produced after running our program using the parameters NSTEPS = 200, r = 3.6, x0 = 0.2 and epsilon = 10^{-15}. The gnuplot command set logscale y, puts the y axis in a logarithmic scale. Therefore an exponential function is shown as a straight line and this is what we see in figure 3.7: The points $|\Delta x_n|/\epsilon$ tend to lie on a straight line as ϵ decreases. The slopes of these lines are equal to the Liapunov exponent λ. Deviations from the straight line behavior indicates corrections and systematic errors, as we point out in figure 3.7. A different initial condition results in a slightly different value of λ, and the true value can be estimated as the average over several such choices. We estimate the error of our computation from the standard error of the mean. The reader should perform such a computation as an exercise.

One can perform a fit of the points $|\Delta x_n|/\epsilon$ to the exponential function in the following way: Since $|\Delta x_n|/\epsilon \sim C\exp(\lambda n) \Rightarrow \ln(|\Delta x_n|/\epsilon) = \lambda n + c$, we can make a fit to a straight line instead. Using gnuplot, the relevant commands are:

```
gnuplot> fit [5:53] a*x+b \
         "<echo 500 3.6 0.2 1e-15 |./l;cat lia.dat"\
         using 1:(log($4)) via a,b
gnuplot> replot exp(a*x+b)
```

The command shown above fits the data to the function a*x+b by taking the 1st column and the logarithm of the 4th column (using 1:(log($4))) of the stdout of the command that we used for creating the previous plot. We choose data for which $5 \leq n \leq 53$ ([5:53]) and the fitting parameters are a,b (via a,b). The second line, adds the fitted function to the plot.

Now we are going to use equation (3.35) for calculating λ. This equation is approximately correct when (a) we have already reached the steady state and (b) in the large n limit. For this reason we should study if we obtain a satisfactory convergence when we (a) "throw away" a number of NTRANS steps, (b) calculate the sum (3.35) for increasing NSTEPS= n (c) calculate the sum (3.35) for many values of the initial point x_0. This has to be carefully repeated for all values of r since each factor will contribute differently to the quality of the convergence: In regions that manifest chaotic behavior (large λ) convergence will be slower. The program can be found in the file liapunov2.cpp:

```
//=====================================================
//Discrete Logistic Map:
//Liapunov exponent from sum_i ln|f'(x_i)|
```

3.6. LIAPUNOV EXPONENTS

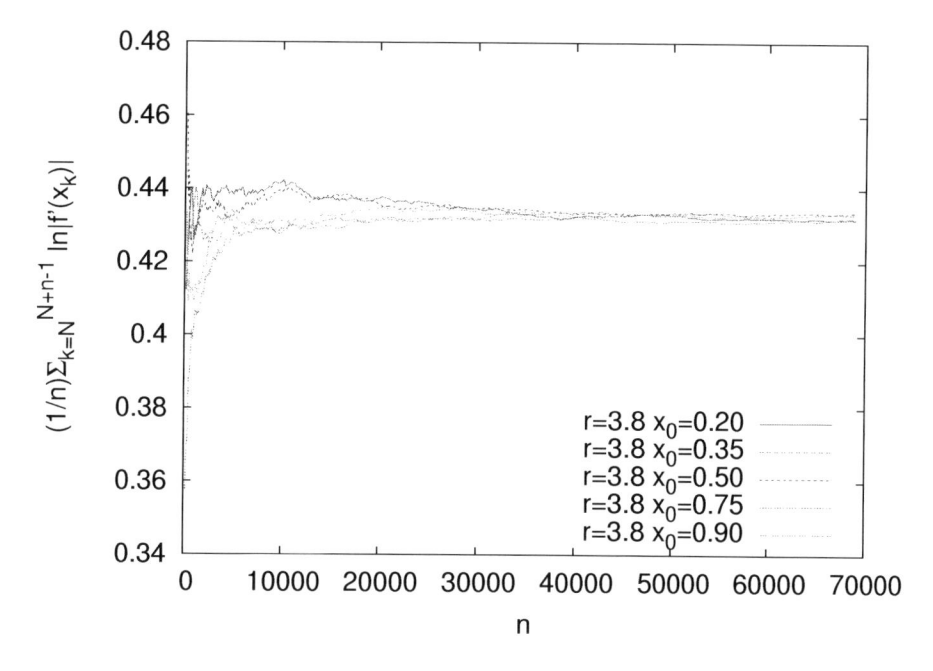

Figure 3.8: Plot of the sum $(1/n)\sum_{k=N}^{N+n-1} \ln|f'(x_k)|$ as a function of n for the logistic map with $r = 3.8$, $N = 2000$ for different initial conditions $x_0 = 0.20, 0.35, 0.50, 0.75, 0.90$. The different curves converge in the limit $n \to \infty$ to $\lambda = 0.4325(10)$.

```
// NTRANS: number of discarted iterations in order to discart
//         transient behaviour
// NSTEPS: number of terms in the sum
//==========================================================
#include <iostream>
#include <fstream>
#include <cstdlib>
#include <string>
#include <cmath>
using namespace std;

int main(){
  int     NTRANS,NSTEPS,i;
  double  r,x0,x1,sum;
  string  buf;
  // ——— Input:
  cout  << "# Enter NTRANS, NSTEPS, r, x0:\n";
  cin   >> NTRANS >> NSTEPS >> r   >>   x0; getline(cin,buf);
```

```cpp
cout    <<  "# NTRANS = "  <<  NTRANS  << endl;
cout    <<  "# NSTEPS = "  <<  NSTEPS  << endl;
cout    <<  "# r      = "  <<  r       << endl;
cout    <<  "# x0     = "  <<  x0      << endl;

for(i=1;i<=NTRANS;i++){
   x1  = r * x0  * (1.0 - x0);
   x0  = x1;
}
sum    = log(abs(r*(1.0 - 2.0*x0)));
// ——— Initialize:
ofstream myfile("lia.dat");
myfile.precision(17);
// ——— Calculate:
myfile   << 1   << " " << x0  << " " << sum    << "\n";
for(i=2;i<=NSTEPS;i++){
   x1   = r * x0  * (1.0-x0 );
   sum += log(abs(r*(1.0-2.0*x1)));
   myfile << i   << " " << x1  << " " << sum/i << "\n";
   x0   = x1;
}
myfile.close();
}//main()
```

After NTRANS steps, the program calculates NSTEPS times the sum of the terms $\ln|f'(x_k)| = \ln|r(1-2x_k)|$. At each step the sum divided by the number of steps i is printed to the file lia.dat. Figure 3.6 shows the results for $r = 3.8$. This is a point where the system exhibits strong chaotic behavior and convergence is achieved after we compute a large number of steps. Using NTRANS = 2 000 and NSTEPS $\approx 70\,000$ the achieved accuracy is about 0.2% with $\lambda = 0.4325 \pm 0.0010 \equiv 0.4325(10)$. The main contribution to the error comes from the different paths followed by each initial point chosen. The plot can be constructed with the gnuplot commands:

```
> g++ liapunov2.cpp -o l
> gnuplot
gnuplot> plot \
 "<echo 2000 70000 3.8 0.20 |./l;cat lia.dat" u 1:3 w l,\
 "<echo 2000 70000 3.8 0.35 |./l;cat lia.dat" u 1:3 w l,\
 "<echo 2000 70000 3.8 0.50 |./l;cat lia.dat" u 1:3 w l,\
 "<echo 2000 70000 3.8 0.75 |./l;cat lia.dat" u 1:3 w l,\
 "<echo 2000 70000 3.8 0.90 |./l;cat lia.dat" u 1:3 w l
```

3.6. LIAPUNOV EXPONENTS

The plot command runs the program using the parameters NTRANS = 2 000, NSTEPS = 70 000, r = 3.8 and x0 = 0.20, 0.35, 0.50, 0.75, 0.90 and plots the results from the contents of the file lia.dat.

In order to determine the regions of chaotic behavior we have to study the dependence of the Liapunov exponent λ on the value of r. Using our experience coming from the careful computation of λ before, we will run the program for several values of r using the parameters NTRANS = 2 000, NSTEPS = 60 000 from the initial point x0 = 0.2. This calculation gives accuracy of the order of 1%. If we wish to measure λ carefully and estimate the error of the results, we have to follow the steps described in figures 3.7 and 3.8. The program can be found in the file liapunov3.cpp and it is a simple modification of the previous program so that it can perform the calculation for many values of r.

```
//=============================================
//Discrete Logistic Map:
//Liapunov exponent from sum_i ln|f'(x_i)|
//Calculation for r in [rmin,rmax] with RSTEPS steps
// RSTEPS: values or r studied: r=rmin+(rmax-rmin)/RSTEPS
// NTRANS: number of discarted iterations in order to discart
//         transient behaviour
// NSTEPS: number of terms in the sum
// xstart: value of initial x0 for every r
//=============================================
#include <iostream>
#include <fstream>
#include <cstdlib>
#include <string>
#include <cmath>
using namespace std;

int main(){
  const double rmin    = 2.5;
  const double rmax    = 4.0;
  const double xstart  = 0.2;
  const int    RSTEPS  = 1000;
  const int    NSTEPS  = 60000;
  const int    NTRANS  = 2000;
  int    i,ir;
  double r,x0,x1,sum,dr;
```

```cpp
  string buf;

//  ―――― Initialize:
  ofstream myfile("lia.dat");
  myfile.precision(17);
//  ―――― Calculate:
  dr       = (rmax-rmin)/(RSTEPS-1);
  for(ir   = 0; ir < RSTEPS;ir++){
    r      = rmin+ir*dr;
    x0     = xstart;
    for(i = 1; i <= NTRANS; i++){
      x1   = r * x0 * (1.0-x0 );
      x0   = x1;
    }
    sum    = log(abs(r*(1.0-2.0*x0)));
//Calculate:
    for(i = 2; i <= NSTEPS; i++){
      x1   = r * x0 * (1.0-x0 );
      sum+= log(abs(r*(1.0-2.0*x1)));
      x0   = x1;
    }
    myfile << r << " " << sum/NSTEPS << '\n';
  }// for(ir=0;ir<RSTEPS;ir++)
  myfile.close();
}//main()
```

The program can be compiled and run using the commands:

```
> g++ liapunov3.cpp -o l
> ./l &
```

The character & makes the program ./l to run in the background. This is recommended for programs that run for a long time, so that the shell returns the prompt to the user and the program continues to run even after the shell is terminated.

The data are saved in the file lia.dat and we can make the plot shown in figure 3.7 using gnuplot:

```
gnuplot> plot "lia.dat" with lines notitle ,0 notitle
```

Now we can compare figure 3.9 with the bifurcation diagram shown in figure 3.4. The intervals with $\lambda < 0$ correspond to stable k-cycles. The

3.6. LIAPUNOV EXPONENTS

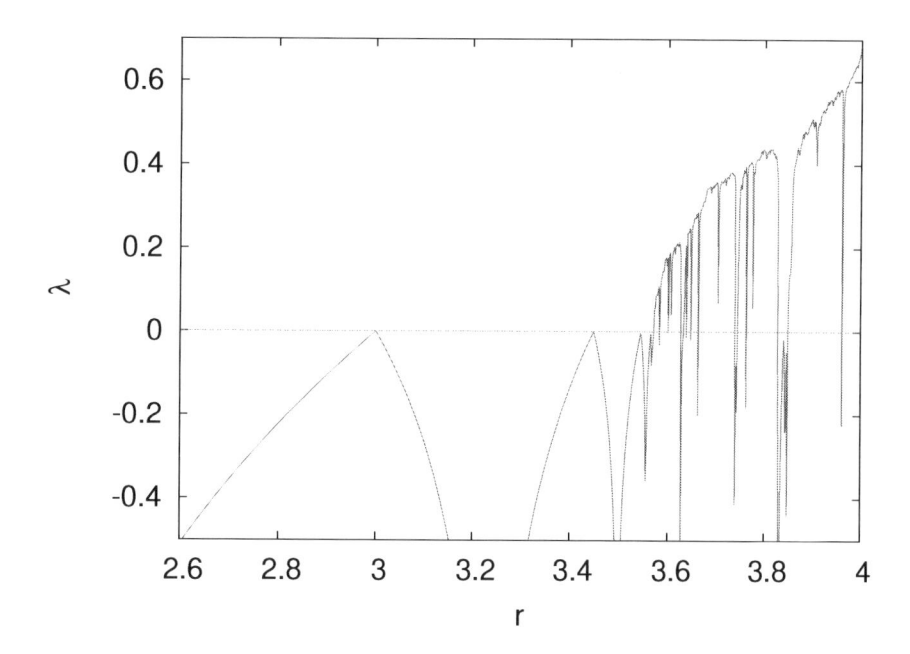

Figure 3.9: The Liapunov exponent λ of the logistic map calculated via equation (3.35). Note the chaotic behavior that manifests for the values of r where $\lambda > 0$ and the windows of stable k-cycles where $\lambda < 0$. Compare this plot with the bifurcation diagram of figure 3.4. At the points where $\lambda = 0$ we have onset of chaos (or "edge of chaos") with manifestation of weak chaos (i.e. $|\Delta x_n| \sim |\Delta x_0| n^\omega$). At these points we have transitions from stable k-cycles to strong chaos. We observe the onset of chaos for the first time when $r = r_c \approx 3.5699$, at which point $\lambda = 0$ (for smaller r the plot seems to touch the $\lambda = 0$ line, but in fact λ takes negative values with $|\lambda|$ very small).

intervals where $\lambda > 0$ correspond to manifestation of *strong chaos*. These intervals are separated by points with $\lambda = 0$ where the system exhibits *weak chaos*. This means that neighboring trajectories diverge from each other with a power law $|\Delta x_n| \sim |\Delta x_0| n^\omega$ instead of an exponential, where $\omega = 1/(1-q)$ is a positive exponent that needs to be determined. The parameter q is the one usually used in the literature. Strong chaos is obtained in the $q \to 1$ limit. For larger r, switching between chaotic and stable periodic trajectories is observed each time λ changes sign. The critical values of r can be computed with relatively high accuracy by restricting the calculation to a small enough neighborhood of the critical point. You can do this using the program listed above by setting the parameters `rmin` and `rmax`.

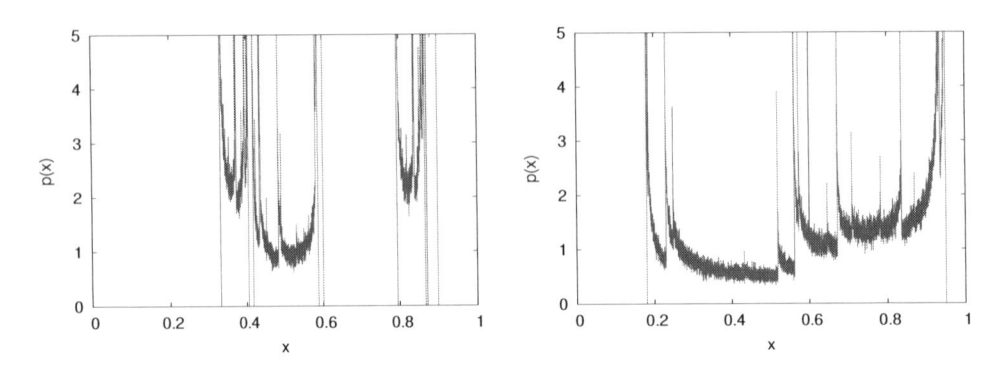

Figure 3.10: The distribution functions $p(x)$ of x of the logistic map for $r = 3.59$ (left) and 3.8 (right). The chaotic behavior appears to be weaker for $r = 3.59$, and this is reflected on the value of the entropy. One sees that there exist intervals of x with $p(x) = 0$ which become smaller and vanish as r gets close to 4. This distribution is very hard to be distinguished from a truly random distribution.

We can also study the chaotic properties of the trajectories of the logistic map by computing the distribution $p(x)$ of the values of x in the interval $(0, 1)$. After the transitional period, the distribution $p(x)$ for the k cycles will have support only at the points of the k cycles, whereas for the chaotic regimes it will have support on subintervals of $(0, 1)$. The distribution function $p(x)$ is independent for most of the initial points of the trajectories. If one obtains a large number of points from many trajectories of the logistic map, it will be practically impossible to understand that these are produced by a deterministic rule. For this reason, chaotic systems can be used for the production of *pseudorandom* numbers, as we will see in chapter ??. By measuring the *entropy*, which is a measure of disorder in a system, we can quantify the "randomness" of the distribution. As we will see in chapter ??, it is given by the equation

$$S = -\sum_k p_k \ln p_k, \qquad (3.37)$$

where p_k is the probability of observing the state k. In our case, we can make an approximate calculation of S by dividing the interval $(0, 1)$ to N subintervals of width Δx. For given r we obtain a large number M of values x_n of the logistic map and we compute the histogram h_k of

3.6. LIAPUNOV EXPONENTS

their distribution in the intervals $(x_k, x_k + \Delta x)$. The probability density is obtained from the limit of $p_k = h_k/(M\Delta x)$ as M becomes large and Δx small (large N). Indeed, $\sum_{k=1}^{N} p_k \Delta x = 1$ converges to $\int_0^1 p(x)\,dx = 1$. We will define $S = -\sum_{k=1}^{N} p_k \ln p_k \Delta x$.

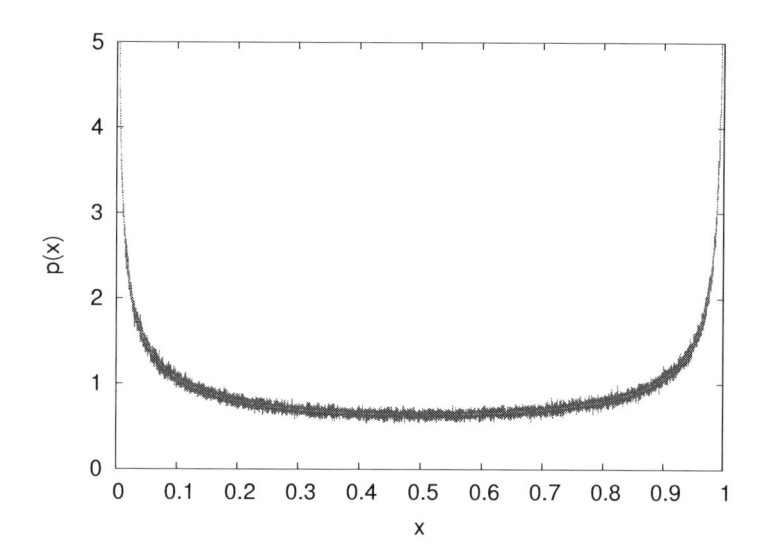

Figure 3.11: The distribution $p(x)$ of x for the logistic map for $r = 4$. We observe strong chaotic behavior, $p(x)$ has support over the whole interval $(0, 1)$ and the entropy is large. The solid line is the analytic form of the distribution $p(x) = \pi^{-1} x^{-1/2}(1-x)^{-1/2}$ which is known for $r = 4$ [32]. This is the beta distribution for $a = 1/2$, $b = 1/2$.

The program listed below calculates p_k for chosen values of r, and then the entropy S is calculated using (3.37). It is a simple modification of the program in liapunov3.cpp where we add the parameter NHIST counting the number of intervals N for the histograms. The probability density is calculated in the array p[NHIST]. The program can be found in the file entropy.cpp:

```
//=================================================================
// Discrete Logistic Map:
// Liapunov exponent from sum_i ln|f'(x_i)|
// Calculation for r in [rmin,rmax] with RSTEPS steps
//  RSTEPS: values or r studied: r=rmin+(rmax-rmin)/RSTEPS
//  NTRANS: number of discarted iterations in order to discart
//          transient behaviour
```

```cpp
// NSTEPS: number of terms in the sum
// xstart: value of initial x0 for every r
//==============================================
#include <iostream>
#include <fstream>
#include <cstdlib>
#include <string>
#include <cmath>
using namespace std;

int main(){
  const double rmin   = 2.5;
  const double rmax   = 4.0;
  const double xstart = 0.2;
  const int    RSTEPS = 1000;
  const int    NHIST  = 10000;
  const int    NTRANS = 2000;
  const int    NSTEPS = 5000000;
  const double xmin=0.0,xmax=1.0;
  int    i,ir,isum,n;
  double r,x0,x1,sum,dr,dx;
  double p[NHIST],S;
  string buf;

  // ------ Initialize:
  ofstream myfile("entropy.dat");
  myfile.precision(17);
  // ------ Calculate:
  for(i=0;i<NHIST;i++) p[i] = 0.0;
  dr = (rmax-rmin)/(RSTEPS-1);
  dx = (xmax-xmin)/(NHIST -1);
  for(ir=0;ir<RSTEPS;ir++){
    r = rmin+ir*dr;
    x0= xstart;
    for(i=1;i<=NTRANS;i++){
      x1  = r * x0  * (1.0-x0 );
      x0  = x1;
    }
    //make histogram:
    n=int(x0/dx); p[n]+=1.0;
    for(i=2;i<=NSTEPS;i++){
      x1    = r * x0  * (1.0-x0 );
      n     = int(x1/dx);
      p[n] += 1.0;
      x0    = x1;
```

3.6. LIAPUNOV EXPONENTS

```
    }
   //p[k] is now histogram of x-values.
   //Normalize so that sum_k p[k]*dx=1
   //to get probability distribution:
    for(i=0;i < NHIST;i++) p[i] /= (NSTEPS*dx);
   //sum all non zero terms: p[n]*log(p[n])*dx
    S = 0.0;
    for(i=0;i < NHIST;i++)
      if(p[i] > 0.0)
        S -= p[i]*log(p[i])*dx;
    myfile << r << " " << S << '\n';
  }//for(ir=0;ir<RSTEPS;ir++)
  myfile.close();

  myfile.open("entropy_hist.dat");
  myfile.precision(17);
  for(n=0;n<NHIST;n++){
    x0 = xmin + n*dx + 0.5*dx;
    myfile << r << " " << x0 << " " << p[n] << '\n';
  }
  myfile.close();
}//main()
```

For the calculation of the distribution functions and the entropy we have to choose the parameters which control the systematic error. The parameter NTRANS should be large enough so that the transitional behavior will not contaminate our results. Our measurements must be checked for being independent of its value. The same should be done for the initial point xstart. The parameter NHIST controls the partitioning of the interval $(0, 1)$ and the width Δx, so it should be large enough. The parameter NSTEPS is the number of "measurements" for each value of r and it should be large enough in order to reduce the "noise" in p_k. It is obvious that NSTEPS should be larger when Δx becomes smaller. Appropriate choices lead to the plots shown in figures 3.10 and 3.11 for $r = 3.59, 3.58$ and 4. We see that stronger chaotic behavior means a wider distribution of the values of x.

The entropy is shown in figure 3.12. The stable periodic trajectories lead to small entropy, whereas the chaotic ones lead to large entropy. There is a sudden increase in the value of the entropy at the beginning of chaos at $r = r_c$, which increases even further as the chaotic behavior becomes stronger. During the intermissions of the chaotic behavior there are sudden drops in the value of the entropy. It is quite instructive to

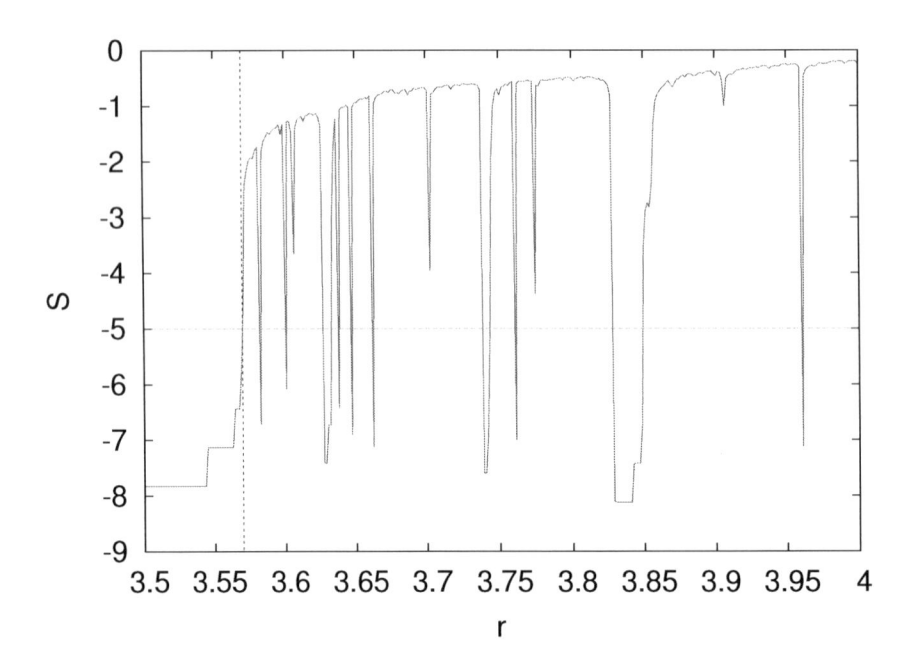

Figure 3.12: The entropy $S = -\sum_k p_k \ln p_k \, \Delta x$ for the logistic map as a function of r. The vertical line is $r_c \approx 3.56994567$ which marks the beginning of chaos and the horizontal is the corresponding entropy. The entropy is low for small values of r, where we have the stable 2^n cycles, and large in the chaotic regimes. S drops suddenly when we pass to a (temporary) periodic behavior interval. We clearly observe the 3-cycle for $r = 1 + \sqrt{8} \approx 3.8284$ and the subsequent bifurcations that we observed in the bifurcation diagram (figure 3.4) and the Liapunov exponent diagram (figure 3.9).

compare the entropy diagrams with the corresponding bifurcation diagrams (see figure 3.4) and the Liapunov exponent diagrams (see figure 3.9). The entropy is increasing until r reaches its maximum value 4, but this is not done smoothly. By magnifying the corresponding areas in the plot, we can see an infinite number of sudden drops in the entropy in intervals of r that become more and more narrow.

3.7 Problems

Several of the programs that you need to write for solving the problems of this chapter can be found in the Problems directory of the accompanying software of this chapter.

3.1 Confirm that the trajectories of the logistic map for $r < 1$ are falling off exponentially for large enough n.

(a) Choose $r = 0.5$ and plot the trajectories for $x_0 = 0.1 - 0.9$ with step 0.1 for $n = 1, \ldots, 1000$. Put the y axis in a logarithmic scale. From the resulting curves discuss whether you obtain an exponential falloff.

(b) Fit the points x_n for $n > 20$ to the function ce^{-ax} and determine the fitting parameters a and c. How do these parameters depend on the initial point x_0? You can use the following gnuplot commands for your calculation:

```
gnuplot> !g++ logistic.cpp -o l
gnuplot> a=0.7;c=0.4;
gnuplot> fit [10:] c*exp(-a*x) \
  "<echo 1000 0.5 0.5|./l;cat log.dat" via a,c
gnuplot> plot c*exp(-a*x),\
  "<echo 1000 0.5 0.5|./l;cat log.dat" w l
```

As you can see, we set NSTEPS = 1000, r = 0.5, x0 = 0.5. By setting the limits [10:] to the fit command, the fit includes only the points $x_n \geq 10$, therefore avoiding the transitional period and the deviation from the exponential falloff for small n.

(c) Repeat for $r = 0.3 - 0.9$ with step 0.1 and for $r = 0.99, 0.999$. As you will be approaching $r = 1$, you will need to discard more points from near the origin. You might also need to increase NSTEPS. You should always check graphically whether the fitted exponential function is a good fit to the points x_n for large n. Construct a table for the values of a as a function of r.

The solutions of the equation (3.3) is $e^{(r-1)x}$. How is this related to the values that you computed in your table?

3.2 Consider the logistic map for $r = 2$. Choose NSTEPS=100 and calculate the corresponding trajectories for x0=0.2, 0.3, 0.5, 0.7, 0.9. Plot them on the same graph. Calculate the fixed point x_2^* and compare your result to the known value $1 - 1/r$. Repeat for x0= $10^{-\alpha}$ for $\alpha = -1, -2, -5, -10, -20, -25$. What do you conclude about the point $x_1^* = 0$?

3.3 Consider the logistic map for $r = 2.9, 2.99, 2.999$. Calculate the stable point x_2^* and compare your result to the known value $1 - 1/r$. How large should NSTEPS be chosen each time? You may choose x0=0.3.

3.4 Consider the logistic map for $r = 3.2$. Take x0=0.3, 0.5, 0.9 and NSTEPS=300 and plot the resulting trajectories. Calculate the fixed points x_3^* and x_4^* by using the command tail log.dat. Increase NSTEPS and repeat so that you make sure that the trajectory has converged to the 2-cycle. Compare their values to the ones given by equation (3.18). Make the following plots:

```
gnuplot> plot \
"<echo 300 3.2 0.3|./l;awk 'NR%2==0' log.dat" w l
gnuplot> replot \
"<echo 300 3.2 0.3|./l;awk 'NR%2==1' log.dat" w l
```

What do you observe?

3.5 Repeat the previous problem for $r = 3.4494$. How big should NSTEPS be chosen so that you obtain x_3^* and x_4^* with an accuracy of 6 significant digits?

3.6 Repeat the previous problem for $r = 3.5$ and 3.55. Choose NSTEPS = 1000, x0 = 0.5. Show that the trajectories approach a 4-cycle and an 8-cycle respectively. Calculate the fixed points x_5^*-x_8^* and x_9^*-x_{16}^*.

3.7 Plot the functions $f(x)$, $f^{(2)}(x)$, $f^{(4)}(x)$, x for given r on the same graph. Use the commands:

```
gnuplot> set samples 1000
gnuplot> f(x) = r*x*(1-x)
gnuplot> r=1;plot [0:1] x,f(x),f(f(x)),f(f(f(f(x))))
```

3.7. PROBLEMS

The command r=1 sets the value of r. Take $r = 2.5, 3, 3.2, 1+\sqrt{6}, 3.5$. Determine the fixed points and the k-cycles from the intersections of the plots with the diagonal $y = x$.

3.8 Construct the cobweb plots of figures 3.2 and 3.4 for $r = 2.8, 3.3$ and 3.5. Repeat by dropping from the plot an increasing number of initial points, so that in the end only the k-cycles will remain. Do the same for $r = 3.55$.

3.9 Construct the bifurcation diagrams shown in figure 3.4.

3.10 Construct the bifurcation diagram of the logistic map for $3.840 < r < 3.851$ and for $0.458 < x < 0.523$. Compute the first four bifurcation points with an accuracy of 5 significant digits by magnifying the appropriate parts of the plots. Take NTRANS=15000.

3.11 Construct the bifurcation diagram of the logistic map for $2.9 < r < 3.57$. Compute graphically the bifurcation points $r_c^{(n)}$ for $n = 2, 3, 4, 5, 6, 7, 8$. Make sure that your results are stable against variations of the parameters NTRANS, NSTEPS as well as from the choice of branching point. From the known values of $r_c^{(n)}$ for $n = 2, 3$, and from the dependence of your results on the choices of NTRANS, NSTEPS, estimate the accuracy achieved by this graphical method. Compute the ratios $(r_c^{(n)} - r_c^{(n-1)})/(r_c^{(n+1)} - r_c^{(n)})$ and compare your results to equation (3.20).

3.12 Choose the values of ρ in equation (3.24) so that you obtain only one energy level. Compute the resulting value of the energy. When do we have three energy levels?

3.13 Consider the polynomial $g(x) = x^3 - 2x^2 - 11x + 12$. Find the roots obtained by the Newton-Raphson method when you choose $x_0 = 2.35287527, 2.35284172, 2.35283735, 2.352836327, 2.352836323$. What do you conclude concerning the basins of attraction of each root of the polynomial? Make a plot of the polynomial in a neighborhood of its roots and try other initial points that will converge to each one of the roots.

3.14 Use the Newton-Raphson method in order to compute the 4-cycle $x_5^*, \ldots, x_8^*$ of the logistic map. Use appropriate areas of the bifur-

n	$r_c^{(n)}$	n	$r_c^{(n)}$
2	3.0000000000	10	3.56994317604
3	3.4494897429	11	3.569945137342
4	3.544090360	12	3.5699455573912
5	3.564407266	13	3.569945647353
6	3.5687594195	14	3.5699456666199
7	3.5696916098	15	3.5699456707464
8	3.56989125938	16	3.56994567163008
9	3.56993401837	17	3.5699456718193
$r_c = 3.56994567\ldots$				

Table 3.1: The values of $r_c^{(n)}$ for the logistic map calculated for problem 17. $r_c^{(\infty)} \equiv r_c$ is taken from the bibliography.

cation diagram so that you can choose the initial points correctly. Check that your result for $r_c^{(4)}$ is the same for all x_α^*. Tune the parameters chosen in your calculation on order to improve the accuracy of your measurements.

3.15 Repeat the previous problem for the 8-cycle $x_9^*, \ldots, x_{16}^*$ and $r_c^{(5)}$.

3.16 Repeat the previous problem for the 16-cycle $x_{17}^*, \ldots, x_{32}^*$ and $r_c^{(6)}$.

3.17 Calculate the critical points $r_c^{(n)}$ for $n = 3, \ldots, 17$ of the logistic map using the Newton-Raphson method. In order to achieve that, you should determine the bifurcation points graphically in the bifurcation diagram first and then choose the initial points in the Newton-Raphson method appropriately. The program in bifurcationPoints.cpp should read the parameters eps, epsx, epsr from the stdin so that they can be tuned for increasing n. If these parameters are too small the convergence will be unstable and if they are too large you will have large systematic errors. Using this method, try to reproduce table 3.1

3.18 Calculate the ratios $\Delta r^{(n)}/\Delta r^{(n+1)}$ of equation (3.20) using the results of table 3.1. Calculate Feigenbaum's constant and comment on the accuracy achieved by your calculation.

3.7. PROBLEMS

3.19 Estimate Feigenbaum's constant δ and the critical value r_c by assuming that for large enough n, $r_c^{(n)} \approx r_c - C\delta^{-n}$. This behavior is a result of equation (3.20). Fit the results of table 3.1 to this function and calculate δ and r_c. This hypothesis is confirmed in figure 3.13 where we can observe the exponential convergence of $r_c^{(n)}$ to r_c. Construct the same plot using the parameters of your calculation.

Hint: You can use the following gnuplot commands:

```
gnuplot> nmin=2;nmax=17
gnuplot> r(x)= rc-c*d**(-x)
gnuplot> fit [nmin:nmax] r(x)  "rcrit" u 1:2 via rc,c,d
gnuplot> plot "rcrit", r(x)
gnuplot> print rc,d
```

The file rcrit contains the values of table 3.1. You should vary the parameters nmin, nmax and repeat until you obtain a stable fit.

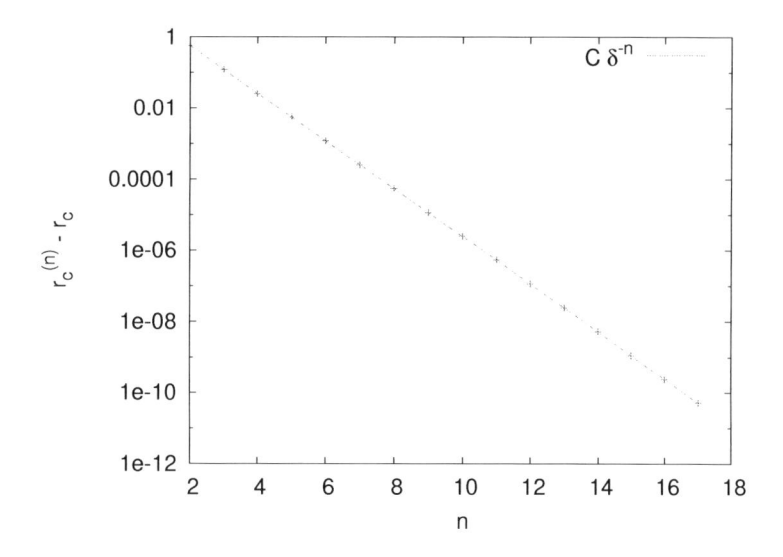

Figure 3.13: Test of the relation $r_c^{(n)} \approx r_c - C\delta^{-n}$ discussed in problem 17. The parameters used in the plot are approximately $r_c = 3.5699457$, $\delta = 4.669196$ and $C = 12.292$.

3.20 Use the Newton-Raphson method to calculate the first three bifurcation points after the appearance of the 3-cycle for $r = 1 + \sqrt{8}$. Choose one bifurcation point of the 3-cycle, one of the 6-cycle and one of the 12-cycle and magnify the bifurcation diagram in their neighborhood.

3.21 Consider the map describing the evolution of a population
$$x_{n+1} = p(x_n) = x_n e^{r(1-x_n)}. \tag{3.38}$$

(a) Plot the functions x, $p(x)$, $p^{(2)}(x)$, $p^{(4)}(x)$ for $r = 1.8, 2, 2.6, 2.67, 2.689$ for $0 < x < 8$. For which values of r do you expect to obtain stable k-cycles?

(b) For the same values of r plot the trajectories with initial points $x_0 = 0.2, 0.5, 0.7$. For each r make a separate plot.

(c) Use the Newton-Raphson method in order to determine the points $r_c^{(n)}$ for $n = 3, 4, 5$ as well as the first two bifurcation points of the 3-cycle.

(d) Construct the bifurcation diagram for $1.8 < r < 4$. Determine the point marking the onset of chaos as well as the point where the 3-cycle starts. Magnify the diagram around a branch that you will choose.

(e) Estimate Feigenbaum's constant δ as in problem 17. Is your result compatible with the expectation of universality for the value of δ? Is the value of r_c the same as that of the logistic map?

3.22 Consider the sine map:
$$x_{n+1} = s(x_n) = r \sin(\pi x_n). \tag{3.39}$$

(a) Plot the functions x, $s(x)$, $s^{(2)}(x)$, $s^{(4)}(x)$, $s^{(8)}(x)$ for $r = 0.65, 0.75, 0.84, 0.86, 0.88$. Which values of r are expected to lead to stable k-cycles?

(b) For the same values of r, plot the trajectories with initial points $x_0 = 0.2, 0.5, 0.7$. Make one plot for each r.

(c) Use the Newton-Raphson method in order to determine the points $r_c^{(n)}$ for $n = 3, 4, 5$ as well as the first two bifurcation points of the 3-cycle.

3.7. PROBLEMS

(d) Construct the bifurcation diagram for $0.6 < r < 1$. Within which limits do the values of x lie in? Repeat for $0.6 < r < 2$. What do you observe? Determine the point marking the onset of chaos as well as the point where the 3-cycle starts. Magnify the diagram around a branch that you will choose.

3.23 Consider the map:
$$x_{n+1} = 1 - rx_n^2. \tag{3.40}$$

(a) Construct the bifurcation diagram for $0 < r < 2$. Within which limits do the values of x lie in? Determine the point marking the onset of chaos as well as the point where the 3-cycle starts. Magnify the diagram around a branch that you will choose.

(b) Use the Newton-Raphson method in order to determine the points $r_c^{(n)}$ for $n = 3, 4, 5$ as well as the first two bifurcation points of the 3-cycle.

3.24 Consider the tent map:
$$x_{n+1} = r\min\{x_n, 1-x_n\} = \begin{cases} rx_n & 0 \le x_n \le \frac{1}{2} \\ r(1-x_n) & \frac{1}{2} < x_n \le 1 \end{cases}. \tag{3.41}$$

Construct the bifurcation diagram for $0 < r < 2$. Within which limits do the values of x lie in? On the same graph, plot the functions $r/2$, $r - r^2/2$.

Magnify the diagram in the area $1.407 < r < 1.416$ and $0.580 < x < 0.588$. At which point do the two disconnected intervals within which x_n take their values merge into one? Magnify the areas $1.0 < r < 1.1$, $0.4998 < x < 0.5004$ and $1.00 < r < 1.03$, $0.4999998 < x < 0.5000003$ and determine the merging points of two disconnected intervals within which x_n take their values.

3.25 Consider the Gauss map (or mouse map):
$$x_{n+1} = e^{-rx_n^2} + q. \tag{3.42}$$

Construct the bifurcation diagram for $-1 < q < 1$ and $r = 4.5, 4.9, 7.5$. Make your program to take as the initial point of the new trajectory to be the last one of the previous trajectory and choose

$x_0 = 0$ for $q = -1$. Repeat for $x_0 = 0.7, 0.5, -0.7$. What do you observe? Note that as q is increased, we obtain bifurcations and "anti-bifurcations".

3.26 Consider the circle map:

$$x_{n+1} = [x_n + r - q\sin(2\pi x_n)] \mod 1. \qquad (3.43)$$

(Make sure that your program keeps the values of x_n so that $0 \leq x_n < 1$). Construct the bifurcation diagram for $0 < q < 2$ and $r = 1/3$.

3.27 Use the program in liapunov.cpp in order to compute the distance between two trajectories of the logistic map for $r = 3.6$ that originally are at a distance $\Delta x_0 = 10^{-15}$. Choose $x_0 = 0.1, 0.2, 0.3, 0.4, 0.5, 0.6, 0.7, 0.8, 0.9, 0.99, 0.999$ and calculate the Liapunov exponent by fitting to a straight line appropriately. Compute the mean value and the standard error of the mean.

3.28 Calculate the Liapunov exponent for $r = 3.58, 3.60, 3.65, 3.70, 3.80$ for the logistic map. Use both ways mentioned in the text. Choose at least 5 different initial points and calculate the mean and the standard error of the mean of your results. Compare the values of λ that you obtain with each method and comment.

3.29 Compute the critical value r_c numerically as the limit $\lim_{n\to\infty} r_c^{(n)}$ for the logistic map with an accuracy of nine significant digits. Use the calculation of the Liapunov exponent λ given by equation (3.35).

3.30 Compute the values of r of the logistic map numerically for which we (a) enter a stable 3-cycle (b) reenter into the chaotic behavior. Do the calculation by computing the Liapunov exponent λ and compare your results with the ones obtained from the bifurcation diagram.

3.31 Calculate the Liapunov exponent using equation (3.35) for the fol-

3.7. PROBLEMS

lowing maps:

$$\begin{aligned}
x_{n+1} &= x_n e^{r(1-x_n)}, & 1.8 < r < 4 \\
x_{n+1} &= r\sin(\pi x_n), & 0.6 < r < 1 \\
x_{n+1} &= 1 - rx_n^2, & 0 < r < 2 \\
x_{n+1} &= e^{-rx_n^2} + q, & r = 7.5, -1 < q < 1 \\
x_{n+1} &= \left[x_n + \frac{1}{3} - q\sin(2\pi x_n)\right] \mod 1, & 0 < q < 2, \quad (3.44)
\end{aligned}$$

and construct the diagrams similar to the ones in figure 3.9. Compare your plots with the respective bifurcation diagrams (you may put both graphs on the same plot). Use two different initial points $x_0 = 0, 0.2$ for the Gauss map ($x_{n+1} = e^{-rx_n^2} + q$) and observe the differences. For the circle map ($x_{n+1} = [x_n + 1/3 - q\sin(2\pi x_n)] \mod 1$) study carefully the values $0 < q < 0.15$.

3.32 Reproduce the plots in figures 3.10, 3.11 and 3.12. Compute the function $p(x)$ for $r = 3.68, 3.80, 3.93$ and 3.98. Determine the points where you have stronger chaos by observing $p(x)$ and the corresponding values of the entropy. Compute the entropy for $r \in (3.95, 4.00)$ by taking RSTEPS=2000 and estimate the values of r where we enter to and exit from chaos. Compare your results with the computation of the Liapunov exponent.

3.33 Consider the Hénon map:

$$\begin{aligned}
x_{n+1} &= y_n + 1 - ax_n^2 \\
y_{n+1} &= bx_n
\end{aligned} \quad (3.45)$$

(a) Construct the two bifurcation diagrams for x_n and y_n for $b = 0.3$, $1.0 < a < 1.5$. Check if the values $a = 1.01, 1.4$ that we will use below correspond to stable periodic trajectories or chaotic behavior.

(b) Write a program in a file attractor.cpp which will take NINIT = NL × NL initial conditions $(x_0(i), y_0(i))$ $i = 1, \ldots,$ NL on a NL×NL lattice of the square $x_m \leq x_0 \leq x_M$, $y_m \leq y_0 \leq y_M$. Each of the points $(x_0(i), y_0(i))$ will evolve according to equation (3.45) for $n =$ NSTEPS steps. The program will print the points $(x_n(i), y_n(i))$ to the stdout. Choose $x_m = y_m = 0.6$, $x_M = y_M = 0.8$, NL= 200.

(c) Choose $a = 1.01$, $b = 0.3$ and plot the points $(x_n(i), y_n(i))$ for $n = 0, 1, 2, 3, 10, 20, 30, 40, 60, 1000$ on the same diagram.

(d) Choose $a = 1.4$, $b = 0.3$ and plot the points $(x_n(i), y_n(i))$ for $n = 0, \ldots, 7$ on the same diagram.

(e) Choose $a = 1.4$, $b = 0.3$ and plot the points $(x_n(i), y_n(i))$ for $n = 999$ on the same diagram. Observe the Hénon strange attractor and its fractal properties. It is characterized by a Hausdorff[12] dimension $d_H = 1.261 \pm 0.003$. Then magnify the regions

$$\begin{aligned}
\{(x,y)| & \quad -1.290 < x < -1.270, \quad 0.378 < y < 0.384\}, \\
\{(x,y)| & \quad 1.150 < x < -1.130, \quad 0.366 < y < 0.372\}, \\
\{(x,y)| & \quad 0.108 < x < 0.114, \quad 0.238 < y < 0.241\}, \\
\{(x,y)| & \quad 0.300 < x < 0.320, \quad 0.204 < y < 0.213\}, \\
\{(x,y)| & \quad 1.076 < x < 1.084, \quad 0.090 < y < 0.096\}, \\
\{(x,y)| & \quad 1.216 < x < 1.226, \quad 0.032 < y < 0.034\}.
\end{aligned}$$

3.34 Consider the Duffing map:

$$\begin{aligned}
x_{n+1} &= y_n \\
y_{n+1} &= -bx_n + ay_n - y_n^3.
\end{aligned} \quad (3.46)$$

(a) Construct the two bifurcation diagrams for x_n and y_n for $b = 0.3$, $0 < a < 2.78$. Choose four different initial conditions $(x_0, y_0) = (\pm 1/\sqrt{2}, \pm 1/\sqrt{2})$. What do you observe?

(b) Use the program attractor.cpp from problem 33 in order to study the attractor of the map for $b = 0.3$, $a = 2.75$.

3.35 Consider the Tinkerbell map:

$$\begin{aligned}
x_{n+1} &= x_n^2 - y_n^2 + ax_n + by_n \\
y_{n+1} &= 2x_n y_n + cx_n + dy_n.
\end{aligned} \quad (3.47)$$

---

[12]D.A. Russel, J.D. Hanson, and E. Ott, "Dimension of strange attractors", Phys. Rev. Lett. **45** (1980) 1175. See "Hausdorff dimension" in Wikipedia.

(a) Choose $a = 0.9$, $b = -0.6013$, $c = 2.0$, $d = 0.50$. Plot a trajectory on the plane by plotting the points (x_n, y_n) for $n = 0, \ldots, 10\,000$ with $(x_0, y_0) = (-0.72, -0.64)$.

(b) Use the program attractor.cpp from problem 33 in order to study the attractor of the map for the values of the parameters a, b, c, d given above. Choose $x_m = -0.68$, $x_M = -0.76$, $y_m = -0.60$, $y_M = -0.68$, $n = 10\,000$.

(c) Repeat the previous question by taking $d = 0.27$.

Chapter 4

Motion of a Particle

In this chapter we will study the numerical solution of classical equations of motion of one dimensional mechanical systems, e.g. a point particle moving on the line, the simple pendulum etc. We will make an introduction to the numerical integration of ordinary differential equations with initial conditions and in particular to the Euler and Runge-Kutta methods. We study in detail the examples of the damped harmonic oscillator and of the damped pendulum under the influence of an external periodic force. The latter system is nonlinear and exhibits interesting chaotic behavior.

4.1 Numerical Integration of Newton's Equations

Consider the problem of the solution of the dynamical equations of motion of one particle under the influence of a dynamical field given by Newton's law. The equations can be written in the form

$$\frac{d^2\vec{x}}{dt^2} = \vec{a}(t, \vec{x}, \vec{v}), \qquad (4.1)$$

where

$$\vec{a}(t, \vec{x}, \vec{v}) \equiv \frac{\vec{F}}{m} \qquad \vec{v} = \frac{d\vec{x}}{dt}. \qquad (4.2)$$

From the numerical analysis point of view, the problems that we will discuss are initial value problems for ordinary differential equations where the initial conditions

$$\vec{x}(t_0) = \vec{x}_0 \qquad \vec{v}(t_0) = \vec{v}_0, \qquad (4.3)$$

determine a unique solution $\vec{x}(t)$. The equations (4.1) are of second order with respect to time and it is convenient to write them as a system of twice as many first order equations:

$$\frac{d\vec{x}}{dt} = \vec{v} \qquad \frac{d\vec{v}}{dt} = \vec{a}(t, \vec{x}, \vec{v}). \qquad (4.4)$$

In particular, we will be interested in the study of the motion of a particle moving on a line (1 dimension), therefore the above equations become

$$\begin{aligned}\frac{dx}{dt} &= v & \frac{dv}{dt} &= a(t, x, v) \quad \text{1-dimension} \\ x(t_0) &= x_0 & v(t_0) &= v_0.\end{aligned} \qquad (4.5)$$

When the particle moves on the plane (2 dimensions) the equations of motion become

$$\begin{aligned}\frac{dx}{dt} &= v_x & \frac{dv_x}{dt} &= a_x(t, x, v_x, y, v_y) \quad \text{2-dimensions} \\ \frac{dy}{dt} &= v_y & \frac{dv_y}{dt} &= a_y(t, x, v_x, y, v_y) \\ x(t_0) &= x_0 & v_x(t_0) &= v_{0x} \\ y(t_0) &= y_0 & v_y(t_0) &= v_{0y},\end{aligned} \qquad (4.6)$$

4.2 Prelude: Euler Methods

As a first attempt to tackle the problem, we will study a simple pendulum of length l in a homogeneous gravitational field g (figure 4.1). The equations of motion are given by the differential equations

$$\begin{aligned}\frac{d^2\theta}{dt^2} &= -\frac{g}{l}\sin\theta \\ \frac{d\theta}{dt} &= \omega,\end{aligned} \qquad (4.7)$$

4.2. PRELUDE: EULER METHODS

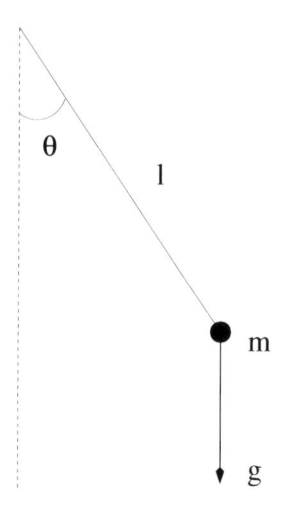

Figure 4.1: A simple pendulum of length l in a homogeneous gravitational field g.

which can be rewritten as a first order system of differential equations

$$\begin{aligned} \frac{d\theta}{dt} &= \omega \\ \frac{d\omega}{dt} &= -\frac{g}{l}\sin\theta \end{aligned} \quad , \qquad (4.8)$$

The equations above need to be written in a discrete form appropriate for a numerical solution with the aid of a computer. We split the interval of time of integration $[t_i, t_f]$ to $N-1$ equal intervals[1] of width $\Delta t \equiv h$, where $h = (t_f - t_i)/(N-1)$. The derivatives are approximated by the relations $(x_{n+1} - x_n)/\Delta t \approx x'_n$, so that

$$\begin{aligned} \omega_{n+1} &= \omega_n + \alpha_n \Delta t \\ \theta_{n+1} &= \theta_n + \omega_n \Delta t \, . \end{aligned} \qquad (4.9)$$

where $\alpha = -(g/l)\sin\theta$ is the angular acceleration. This is the so-called *Euler* method. The error at each step is estimated to be of order $(\Delta t)^2$. This is most easily seen by Taylor expanding around the point t_n and neglecting all terms starting from the second derivative and beyond[2].

---

[1] We have N discrete time points $t_i \equiv t_1, \ldots, t_{N-1}, t_N \equiv t_f$
[2] See appendix 4.7 for retails.

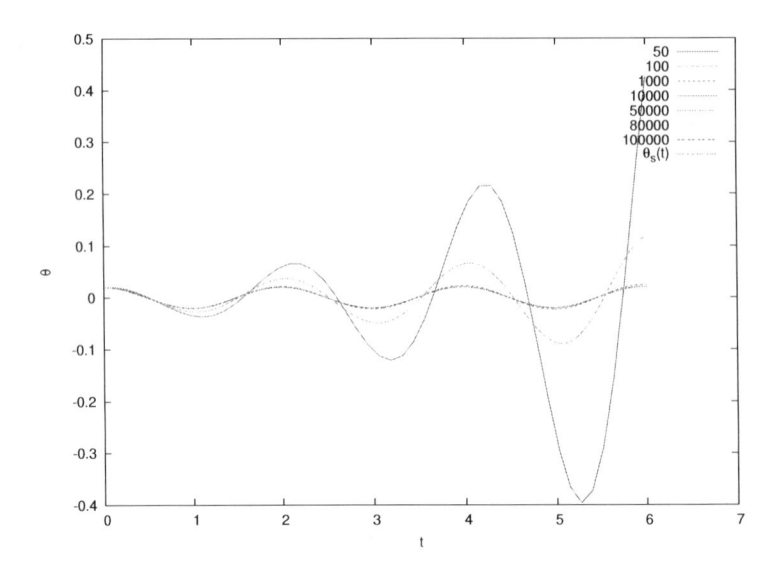

Figure 4.2: Convergence of Euler's method for a simple pendulum with period $T \approx 1.987(\omega^2 = 10.0)$ for several values of the time step Δt which is determined by the number of integration steps Nt= 50−100,000. The solution is given for $\theta_0 = 0.2$, $\omega_0 = 0.0$ and we compare it with the known solution for small angles with $\alpha(t) \approx -(g/l)\,\theta$.

What we are mostly interested in is in the *total* error of the estimate of the functions we integrate for at time t_f! We expect that errors accumulate in an additive way at each integration step, and since the number of steps is $N \propto 1/\Delta t$ the total error should be $\propto (\Delta t)^2 \times (1/\Delta t) = \Delta t$. This is indeed what happens, and we say that Euler's method is a first order method. Its range of applicability is limited and we only study it for academic reasons. Euler's method is *asymmetric* because it uses information only from the beginning of the integration interval $(t, t + \Delta t)$. It can be put in a more balanced form by using the velocity *at the end* of the interval $(t, t + \Delta t)$. This way we obtain the Euler-Cromer method with a slightly improved behavior, but which is still of first order

$$\begin{aligned} \omega_{n+1} &= \omega_n + \alpha_n \Delta t \\ \theta_{n+1} &= \theta_n + \omega_{n+1} \Delta t\,. \end{aligned} \qquad (4.10)$$

An improved algorithm is the Euler–Verlet method which is of second

4.2. PRELUDE: EULER METHODS

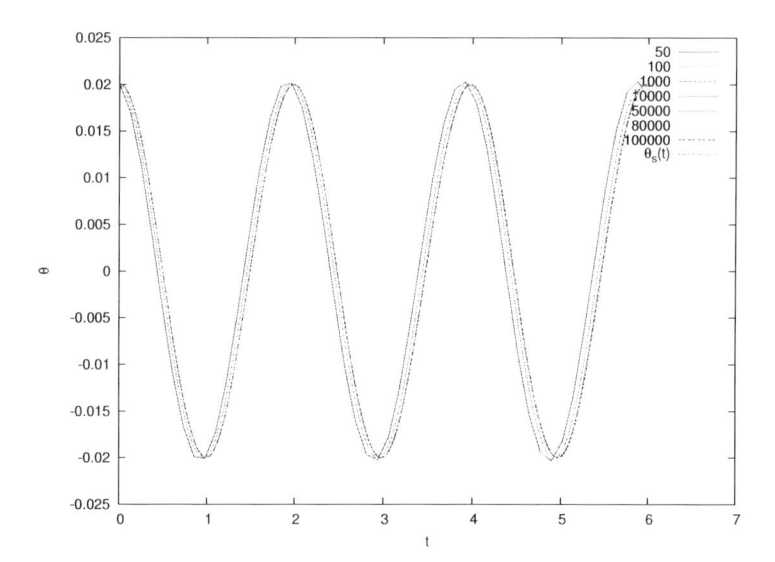

Figure 4.3: Convergence of the Euler-Cromer method, similarly to figure 4.2. We observe a faster convergence compared to Euler's method.

order and gives total error[3] $\sim (\Delta t)^2$. This is given by the equations

$$\begin{aligned} \theta_{n+1} &= 2\theta_n - \theta_{n-1} + \alpha_n (\Delta t)^2 \\ \omega_n &= \frac{\theta_{n+1} - \theta_{n-1}}{2\Delta t}. \end{aligned} \quad (4.11)$$

The price that we have to pay is that we have to use a two step relation in order to advance the solution to the next step. This implies that we have to carefully determine the initial conditions of the problem which are given only at one given time t_i. We make one Euler time step *backwards* in order to define the value of θ_0. If the initial conditions are $\theta_1 = \theta(t_i)$, $\omega_1 = \omega(t_i)$, then we *define*

$$\theta_0 = \theta_1 - \omega_1 \Delta t + \frac{1}{2}\alpha_1(\Delta t)^2. \quad (4.12)$$

It is important that at this step the error introduced is not larger than $\mathcal{O}(\Delta t^2)$, otherwise it will spoil and eventually dominate the $\mathcal{O}(\Delta t^2)$ total error of the method introduced by the intermediate steps. At the last

---
[3]See appendix 4.7 for details.

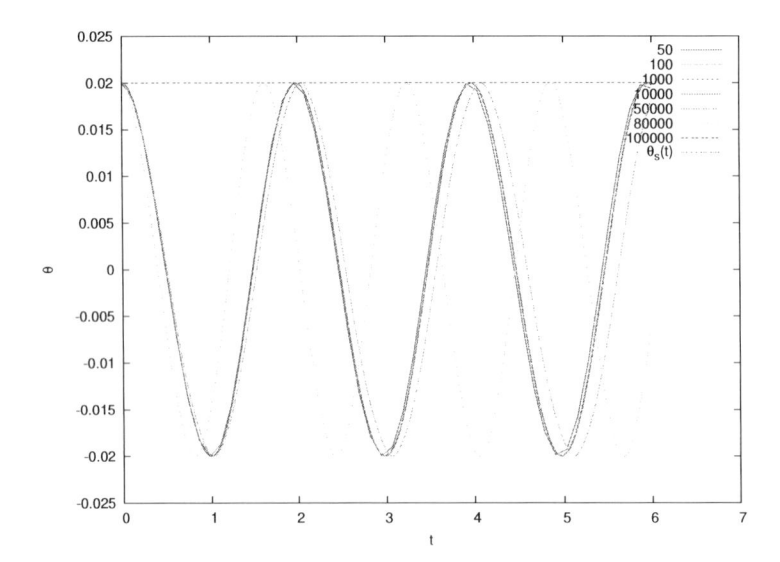

Figure 4.4: Convergence of the Euler-Verlet method, similarly to figure 4.2. We observe a faster convergence than Euler's method, but the roundoff errors make the results useless for Nt$\gtrsim$ 50,000 (note what happens when Nt= 100,000. Why?).

step we also have to take

$$\omega_N = \frac{\theta_N - \theta_{N-1}}{\Delta t}. \qquad (4.13)$$

Even though the method has smaller total error than the Euler method, it becomes unstable for small enough Δt due to roundoff errors. In particular, the second equation in (4.11) gives the angular velocity as the ratio of two small numbers. The problem is that the numerator is the result of the subtraction of two almost equal numbers. For small enough Δt, this difference has to be computed from the last digits of the finite representation of the numbers θ_{n+1} and θ_n in the computer memory. The accuracy in the determination of $(\theta_{n+1} - \theta_n)$ decreases until it eventually becomes *exactly* zero. For the first equation of (4.11), the term $\alpha_n \Delta t^2$ is smaller by a factor Δt compared to the term $\alpha_n \Delta t$ in Euler's method. At some point, by decreasing Δt, we obtain $\alpha_n \Delta t^2 \ll 2\theta_n - \theta_{n-1}$ and the accuracy of the method vanishes due to the finite representation of real number in the memory of the computer. When the numbers $\alpha_n \Delta t^2$ and $2\theta_n - \theta_{n-1}$ differ from each other by more that approximately sixteen

4.2. PRELUDE: EULER METHODS

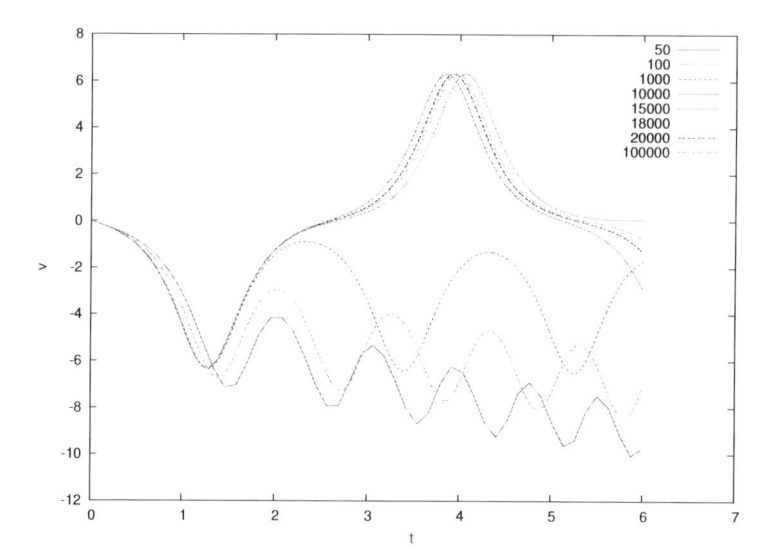

Figure 4.5: Convergence of Euler's method for the simple pendulum like in figure 4.2 for $\theta_0 = 3.0$, $\omega_0 = 0.0$. The behavior of the angular velocity is shown and we notice unstable behavior for $\text{Nt} \lesssim 1,000$.

orders of magnitude, adding the first one to the second is equivalent to adding zero and the contribution of the acceleration vanishes[4].

Writing programs that implement the methods discussed so far is quite simple. We will write a program that compares the results from all three methods Euler, Euler–Cromer and Euler–Verlet. The main program is mainly a user interface, and the computation is carried out by three functions euler, euler_cromer and euler_verlet. The user must provide the function accel(x) which gives the angular acceleration as a function of x. The variable x in our problem corresponds to the angle *theta*. For starters we take accel(x)= -10.0 * sin(x), the acceleration of the simple pendulum.

The data structure is very simple: Three double arrays T[P], X[P] and V[P] store the times t_n, the angles θ_n and the angular velocities ω_n for $n = 1, \ldots, \text{Nt}$. The user determines the time interval for the integration

---

[4] Numbers of type double have approximately sixteen significant digits. The accuracy of the operations described above is determined by the number ϵ, which is the smallest positive number such that $1 + \epsilon > 1$. For variables of type float, $\epsilon \approx 1.2 \times 10^{-7}$ and for variables of type double $\epsilon \approx 2.2 \times 10^{-16}$.

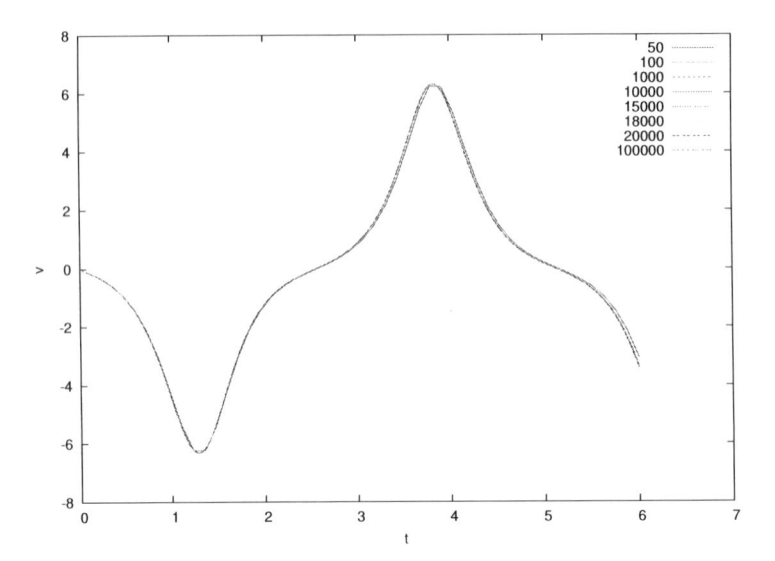

Figure 4.6: Convergence of Euler-Cromer's method, like in figure 4.5. We observe a faster convergence than for Euler's method.

from $t_i = 0$ to $t_f = $ Tfi and the number of discrete times Nt. The latter should be less than P, the size of the arrays. She also provides the initial conditions $\theta_0 = $ Xin and $\omega_0 = $ Vin. After this, we call the main integration functions which take as input the initial conditions, the time interval of the integration and the number of discrete times Xin,Vin,Tfi,Nt. The output of the routines is the arrays T,X,V which store the results for the time, position and velocity respectively. The results are printed to the files euler.dat, euler_cromer.dat and euler_verlet.dat.

After setting the initial conditions and computing the time step $\Delta t \equiv$ h = Tfi/(Nt - 1), the integration in each of the functions is performed in for loops which advance the solution by time Δt. The results are stored at each step in the arrays T,X,V. For example, the central part of the program for Euler's method is:

```
T[0] = 0.0;
X[0] = Xin;
V[0] = Vin;
h    = Tfi/(Nt-1);
for(i=1;i<Nt;i++){
```

4.2. PRELUDE: EULER METHODS

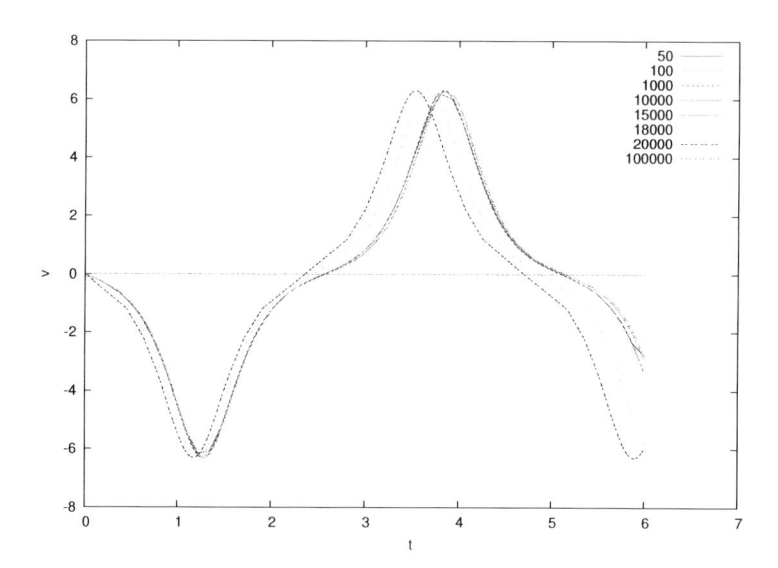

Figure 4.7: Convergence of the Euler–Verlet method, similarly to figure 4.5. We observe a faster convergence compared to Euler's method but that the roundoff errors make the results unstable for $Nt \gtrsim 18,000$. In this figure, float variables have been used instead of double in order to enhance the effect.

```
    T[i]   = T[i-1]+h;
    V[i]   = V[i-1]+accel(X[i-1])*h;
    X[i]   = X[i-1]+V[i]*h;
}
```

Some care has to be taken in the case of the Euler–Verlet method where one has to initialize the first two steps, as well as take special care for the last step for the velocity:

```
    T[0]     = 0.0;
    X[0]     = Xin;
    V[0]     = Vin;
    X0       = X[0] - V[0] * h + accel(X[0]) * h2 / 2.0;
    T[1]     = h;
    X[1]     = 2.0*X[0] - X0  + accel(X[0]) * h2;
    for(i=2;i<Nt;i++){
       .............
    }
    V[Nt-1]  = (X[Nt-1] - X[Nt-2])/h;
```

The full program can be found in the file euler.cpp and is listed below:

```cpp
//========================================================
//Program to integrate equations of motion for accelerations
//which are functions of x with the method of Euler,
//Euler-Cromer and Euler-Verlet.
//The user sets initial conditions and the functions return
//X[t] and V[t]=dX[t]/dt in arrays
//T[0..Nt-1],X[0..Nt-1],V[0..Nt-1]
//The user provides number of times Nt and the final
//time Tfi. Initial time is assumed to be t_i=0 and the
//integration step h = Tfi/(Nt-1)
//The user programs a real function accel(x) which gives the
//acceleration dV(t)/dt as function of X.
//NOTE: T[0] = 0 T[Nt-1] = Tfi
//========================================================
#include <iostream>
#include <fstream>
#include <cstdlib>
#include <string>
#include <cmath>
using namespace std;
//-------------------------------------------------------
const int P = 110000;
double T[P], X[P], V[P];
//-------------------------------------------------------
void euler       (const double& Xin, const double& Vin,
                  const double& Tfi, const int    & Nt );
void euler_cromer(const double& Xin, const double& Vin,
                  const double& Tfi, const int    & Nt );
void euler_verlet(const double& Xin, const double& Vin,
                  const double& Tfi, const int    & Nt );
double accel     (const double& x  );
//-------------------------------------------------------
int main(){
  double Xin, Vin, Tfi;
  int    Nt , i;
  string buf;
  //The user provides initial conditions X_0,V_0
  //final time t_f and Nt:
  cout << "Enter X_0,V_0,t_f,Nt (t_i=0):\n";
  cin  >> Xin >> Vin >> Tfi >> Nt; getline(cin,buf);
  //This check is necessary in order to avoid
  //memory access violations:
```

4.2. PRELUDE: EULER METHODS

```
  if(Nt>=P){cerr << "Error: Nt>=P\n";exit(1);}
  //Xin= X[0], Vin=V[0], T[0]=0 and the routine gives
  //evolution in T[1..Nt-1], X[1..Nt-1], V[1..Nt-1]
  //which we print in a file
  euler(Xin,Vin,Tfi,Nt);
  ofstream myfile("euler.dat");
  myfile.precision(17);
  for(i=0;i<Nt;i++)
    //Each line in file has time, position, velocity:
    myfile << T[i] << " " << X[i] << " " << V[i] << endl;
  myfile.close();//we close the stream to be reused below
  //----------------------------------------------------
  //We repeat everything for each method
  euler_cromer(Xin,Vin,Tfi,Nt);
  myfile.open("euler_cromer.dat");
  for(i=0;i<Nt;i++)
    myfile << T[i] << " " << X[i] << " " << V[i] << endl;
  myfile.close();
  //----------------------------------------------------
  euler_verlet(Xin,Vin,Tfi,Nt);
  myfile.open("euler_verlet.dat");
  for(i=0;i<Nt;i++)
    myfile << T[i] << " " << X[i] << " " << V[i] << endl;
  myfile.close();
}//main()
//====================================================
//Function which returns the value of acceleration at
//position x used in the integration functions
//euler, euler_cromer and euler_verlet
//====================================================
double accel(const double& x){
  return -10.0 * sin(x);
}
//====================================================
//Driver routine for integrating equations of motion
//using the Euler method
//Input:
//Xin=X[0], Vin=V[0] — initial condition at t=0,
//Tfi the final time and Nt the number of times
//Output:
//The arrays T[0..Nt-1], X[0..Nt-1], V[0..Nt-1] which
//gives x(t_k)=X[k-1], dx/dt(t_k)=V[k-1], t_k=T[k-1] k=1..Nt
//where for k=1 we have the initial condition.
//====================================================
void euler      (const double& Xin, const double& Vin,
```

```
                       const double& Tfi, const int   & Nt ){
  int    i;
  double h;
  //Initial conditions set here:
  T[0] = 0.0;
  X[0] = Xin;
  V[0] = Vin;
  //h is the time step Dt
  h    = Tfi/(Nt-1);
  for(i=1;i<Nt;i++){
    T[i] = T[i-1]+h;            //time advances by Dt=h
    X[i] = X[i-1]+V[i-1]*h;//advancement and storage of
    V[i] = V[i-1]+accel(X[i-1])*h;//position and velocity
  }
}//euler()
//=============================================================
//Driver routine for integrating equations of motion
//using the Euler-Cromer method
//Input:
//Xin=X[0], Vin=V[0] --- initial condition at t=0,
//Tfi the final time and Nt the number of times
//Output:
//The arrays T[0..Nt-1], X[0..Nt-1], V[0..Nt-1] which
//gives x(t_k)=X[k-1], dx/dt(t_k)=V[k-1], t_k=T[k-1] k=1..Nt
//where for k=1 we have the initial condition.
//=============================================================
void euler_cromer(const double& Xin, const double& Vin,
                  const double& Tfi, const int   & Nt ){
  int    i;
  double h;
  //Initial conditions set here:
  T[0] = 0.0;
  X[0] = Xin;
  V[0] = Vin;
  //h is the time step Dt
  h    = Tfi/(Nt-1);
  for(i=1;i<Nt;i++){
    T[i] = T[i-1]+h;
    V[i] = V[i-1]+accel(X[i-1])*h;
    X[i] = X[i-1]+V[i]*h; //note difference from Euler
  }
}//euler_cromer()
//=============================================================
//Driver routine for integrating equations of motion
```

4.2. PRELUDE: EULER METHODS

```cpp
//using the Euler-Verlet method
//Input:
//Xin=X[0], Vin=V[0] — initial condition at t=0,
//Tfi the final time and Nt the number of times
//Output:
//The arrays T[0..Nt-1], X[0..Nt-1], V[0..Nt-1] which
//gives x(t_k)=X[k-1], dx/dt(t_k)=V[k-1], t_k=T[k-1] k=1..Nt
//where for k=1 we have the initial condition.
//=========================================================
void euler_verlet(const double& Xin, const double& Vin,
                  const double& Tfi, const int    & Nt ){
  int    i;
  double h,h2,X0,o2h;
  //Initial conditions set here:
  T[0]     = 0.0;
  X[0]     = Xin;
  V[0]     = Vin;
  h        = Tfi/(Nt-1); //time step
  h2       = h*h;        //time step squared
  o2h      = 0.5/h;      // h/2
  //We have to initialize one more step:
  //X0 corresponds to 'X[-1]'
  X0       = X[0] - V[0] * h + accel(X[0]) * h2 / 2.0;
  T[1]     = h;
  X[1]     = 2.0*X[0] - X0   + accel(X[0]) * h2;
  //Now i starts from 2:
  for(i=2;i<Nt;i++){
    T[i]   = T[i-1] + h;
    X[i]   = 2.0*X[i-1] - X[i-2] + accel(X[i-1])*h2;
    V[i-1] = o2h * (X[i]- X[i-2]);
  }
  //we have one more step for the velocity:
  V[Nt-1]  = (X[Nt-1] - X[Nt-2])/h;
}//euler_verlet()
```

Compiling the running the program can be done with the commands:

```
> g++ euler.cpp -o euler
> ./euler
Enter X_0,V_0,t_f,Nt (t_i=0):
0.2 0.0 6.0 1000
> ls  euler*.dat
euler_cromer.dat   euler.dat   euler_verlet.dat
> head -n 5 euler.dat
```

```
0         0.20000   0
0.00600   0.20000  -0.01193
0.01201   0.19992  -0.02386
0.01801   0.19978  -0.03579
0.02402   0.19957  -0.04771
```

The last command shows the first 5 lines of the file euler.dat. We see the data for the time, the position and the velocity stored in 3 columns. We can graph the results using gnuplot:

```
gnuplot> plot "euler.dat" using 1:2 with lines
gnuplot> plot "euler.dat" using 1:3 with lines
```

These commands result in plotting the positions and the velocities as a function of time respectively. We can add the results of all methods to the last plot with the commands:

```
gnuplot> replot "euler_cromer.dat" using 1:3 with lines
gnuplot> replot "euler_verlet.dat" using 1:3 with lines
```

The results can be seen in figures 4.2–4.7. Euler's method is unstable unless we take a quite small time step. The Euler–Cromer method behaves impressively better. The results converge and remain constant for Nt~ 100,000. The Euler–Verlet method converges much faster, but roundoff errors kick in soon. This is more obvious in figure 4.7 where the initial angular position is large[5]. For small angles we can compare with the solution one obtains for the harmonic pendulum ($\sin(\theta) \approx \theta$):

$$\begin{aligned} \alpha(\theta) &= -\frac{g}{l}\theta \equiv -\Omega^2 \theta \\ \theta(t) &= \theta_0 \cos(\Omega t) + (\omega_0/\Omega)\sin(\Omega t) \\ \omega(t) &= \omega_0 \cos(\Omega t) - (\theta_0 \Omega)\sin(\Omega t)\,. \end{aligned} \quad (4.14)$$

In figures 4.2–4.4 we observe that the results agree with the above formulas for the values of Δt where the methods converge. This way we can check our program for bugs. The plot of the functions above can be

---

[5] In this figure, roundoff errors are enhanced by using float variables instead of double.

4.3. RUNGE–KUTTA METHODS

done with the following gnuplot commands[6]:

```
gnuplot> set dummy  t
gnuplot> omega2    = 10
gnuplot> X0        = 0.2
gnuplot> V0        = 0.0
gnuplot> omega     = sqrt(omega2)
gnuplot> x(t)      = X0 * cos(omega * t) +(V0/omega)*sin(omega*t)
gnuplot> v(t)      = V0 * cos(omega * t) -(omega*X0)*sin(omega*t)
gnuplot> plot x(t), v(t)
```

The results should not be compared only graphically since subtle differences can remain unnoticed. It is more desirable to plot the *differences* of the theoretical values from the numerically computed ones which can be done using the commands:

```
gnuplot> plot "euler.dat" using 1:($2-x($1)) with lines
gnuplot> plot "euler.dat" using 1:($3-v($1)) with lines
```

The command using 1:($2-x($1)) puts the values found in the first column on the x axis and the value found in the second column minus the value of the function x(t) for t equal to the value found in the first column on the y axis. This way, we can make the plots shown in[7] figures 4.11-4.14.

4.3 Runge–Kutta Methods

Euler's method is a one step finite difference method of first order. First order means that the total error introduced by the discretization of the integration interval $[t_i, t_f]$ by N discrete times is of order $\sim \mathcal{O}(h)$, where $h \equiv \Delta t = (t_f - t_i)/N$ is the time step of the integration. In this section we will discuss a generalization of this approach where the total error will be of higher order in h. This is the class of Runge-Kutta methods which are one step algorithms where the total discretization error is of order $\sim \mathcal{O}(h^p)$. The local error introduced at each step is of order $\sim \mathcal{O}(h^{p+1})$

---

[6]The command set dummy t sets the independent variable for functions to be t instead of x which is the default.

[7]A small modification is necessary in order to plot the *absolute* value of the differences.

leading after $N = (t_f - t_i)/\Delta t$ steps to a maximum error of order

$$\sim \mathcal{O}(h^{p+1}) \times N = \mathcal{O}(h^{p+1}) \times \frac{t_f - t_i}{\Delta t} \sim \mathcal{O}(h^{p+1}) \times \frac{1}{h} = \mathcal{O}(h^p). \quad (4.15)$$

In such a case we say that we have a Runge-Kutta method of pth order. The price one has to pay for the increased accuracy is the evaluation of the derivatives of the functions in more than one points in the interval $(t, t + \Delta t)$.

Let's consider for simplicity the problem with only one unknown function $x(t)$ which evolves in time according to the differential equation:

$$\frac{dx}{dt} = f(t, x). \quad (4.16)$$

Consider the first order method first. The most naive approach would

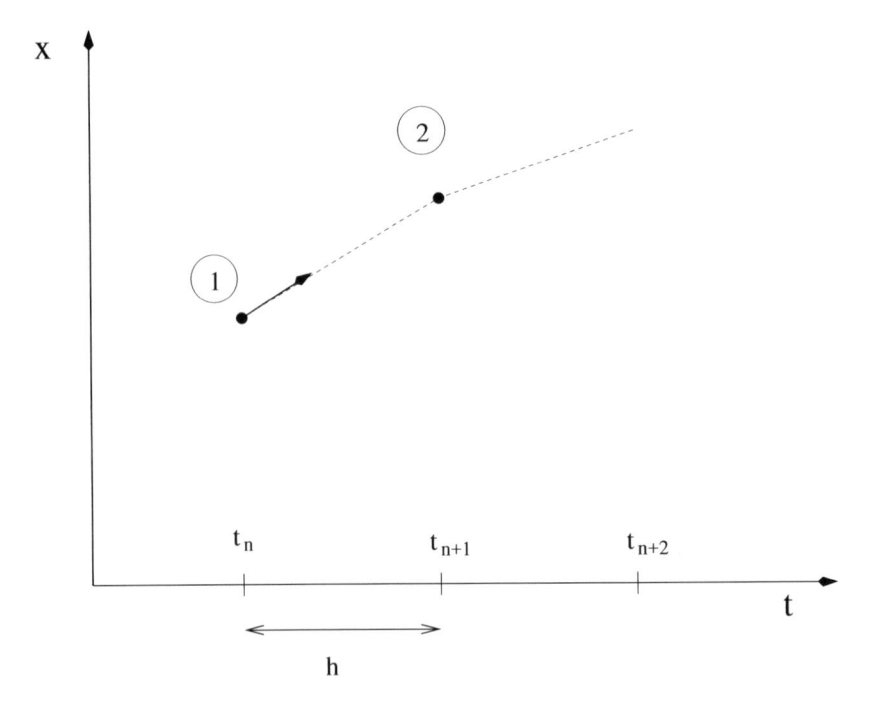

Figure 4.8: The geometry of the step of the Runge-Kutta method of 1st order given by equation (4.17).

be to take the derivative to be given by the finite difference

$$\frac{dx}{dt} \approx \frac{x_{n+1} - x_n}{\Delta t} = f(t_n, x_n) \Rightarrow x_{n+1} = x_n + hf(t_n, x_n). \quad (4.17)$$

4.3. RUNGE–KUTTA METHODS

By Taylor expanding, we see that the error at each step is $\mathcal{O}(h^2)$, therefore the error after integrating from $t_i \to t_f$ is $\mathcal{O}(h)$. Indeed,

$$x_{n+1} = x(t_n+h) = x_n + h\frac{dx}{dt}(x_n) + \mathcal{O}(h^2) = x_n + hf(t_n, x_n) + \mathcal{O}(h^2). \quad (4.18)$$

The geometry of the step is shown in figure 4.8. We start from point 1 and by linearly extrapolating in the direction of the derivative $k_1 \equiv f(t_n, x_n)$ we determine the point x_{n+1}.

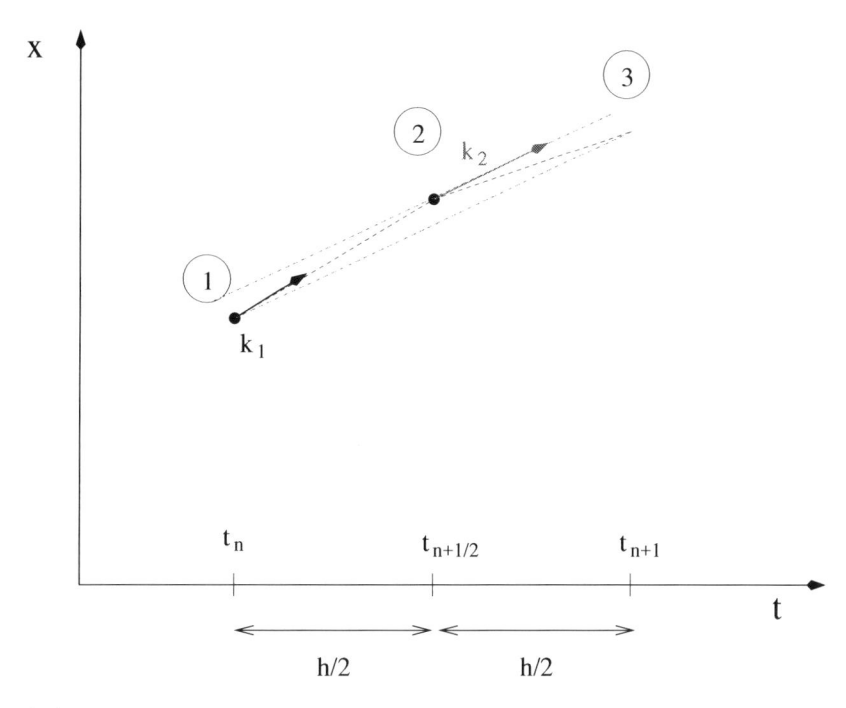

Figure 4.9: The geometry of an integration step of the 2nd order Runge-Kutta method given by equation (4.19).

We can improve the method above by introducing an intermediate point 2. This process is depicted in figure 4.9. We take the point 2 in the middle of the interval (t_n, t_{n+1}) by making a linear extrapolation from x_n in the direction of the derivative $k_1 \equiv f(t_n, x_n)$. Then we use the slope at point 2 as an *estimator* of the derivative within this interval, i.e. $k_2 \equiv f(t_{n+1/2}, x_{n+1/2}) = f(t_n + h/2, x_n + (h/2)k_1)$. We use k_2 to linearly extrapolate from x_n to x_{n+1}. Summarizing, we have that

$$k_1 = f(t_n, x_n)$$
$$k_2 \equiv f(t_n + \frac{h}{2}, x_n + \frac{h}{2} k_1)$$
$$x_{n+1} = x_n + h k_2. \quad (4.19)$$

For the procedure described above we have to evaluate f twice at each step, thereby doubling the computational effort. The error at each step (4.19) becomes $\sim \mathcal{O}(h^3)$, however, giving a total error of $\sim \mathcal{O}(h^2) \sim \mathcal{O}(1/N^2)$. So for given computational time, (4.19) is superior to (4.17).

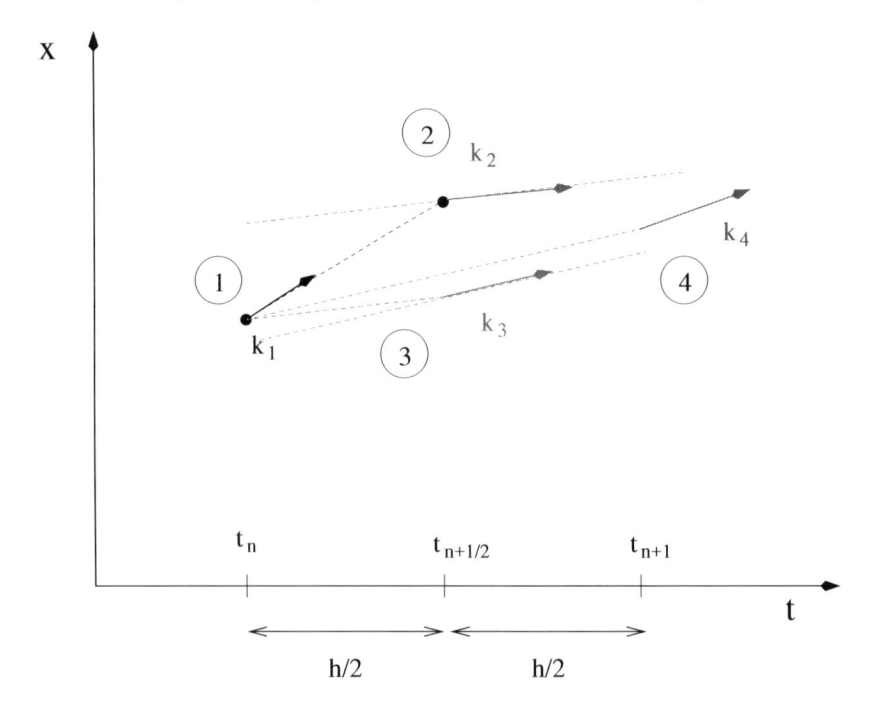

Figure 4.10: The geometry of an integration step of the Runge-Kutta method of 4th order given by equation (4.20).

We can further improve the accuracy gain by using the Runge–Kutta method of 4th order. In this case we have 4 evaluations of the derivative f per step, but the total error becomes now $\sim \mathcal{O}(h^4)$ and the method is superior to that of (4.19)[8]. The process followed is explained geometrically

---

[8] Not always though! Higher order does not necessarily mean higher accuracy, although this is true in the simple cases considered here.

4.3. RUNGE–KUTTA METHODS

in figure 4.10. We use 3 intermediate points for evolving the solution from x_n to x_{n+1}. Point 2 is determined by linearly extrapolating from x_n to the midpoint of the interval $(t_n, t_{n+1} = t_n + h)$ by using the direction given by the derivative $k_1 \equiv f(t_n, x_n)$, i.e. $x_2 = x_n + (h/2)k_1$. We calculate the derivative $k_2 \equiv f(t_n + h/2, x_n + (h/2)k_1)$ at the point 2 and we use it in order to determine point 3, also located at the midpoint of the interval (t_n, t_{n+1}). Then we calculate the derivative $k_3 \equiv f(t_n + h/2, x_n + (h/2)k_2)$ at the point 3 and we use it to linearly extrapolate to the *end* of the interval (t_n, t_{n+1}), thereby obtaining point 4, i.e. $x_4 = x_n + hk_3$. Then we calculate the derivative $k_4 \equiv f(t_n + h, x_n + hk_3)$ at the point 4, and we use all four derivative k_1, k_2, k_3 and k_4 as estimators of the derivative of the function in the interval (t_n, t_{n+1}). If each derivative contributes with a particular weight in this estimate, the discretization error can become $\sim \mathcal{O}(h^5)$. Such a choice is

$$
\begin{aligned}
k_1 &= f(t_n, x_n) \\
k_2 &= f(t_n + \frac{h}{2}, x_n + \frac{h}{2}k_1) \\
k_3 &= f(t_n + \frac{h}{2}, x_n + \frac{h}{2}k_2) \\
k_4 &= f(t_n + h, x_n + h\,k_3) \\
x_{n+1} &= x_n + \frac{h}{6}(k_1 + 2k_2 + 2k_3 + k_4).
\end{aligned}
\quad (4.20)
$$

We note that the second term of the last equation takes an average of the four derivatives with weights 1/6, 1/3, 1/3 and 1/6 respectively. A generic small change in these values will increase the discretization error to worse than h^5.

We remind to the reader the fact that by decreasing h the discretization errors decrease, but that roundoff errors will start showing up for small enough h. Therefore, a careful determination of h that minimizes the total error should be made by studying the dependence of the results as a function of h.

4.3.1 A Program for the 4th Order Runge–Kutta

Consider the problem of the motion of a particle in one dimension. For this, we have to integrate a system of two differential equations (4.5) for two unknown functions of time $x_1(t) \equiv x(t)$ and $x_2(t) \equiv v(t)$ so that

$$\frac{dx_1}{dt} = f_1(t, x_1, x_2) \qquad \frac{dx_2}{dt} = f_2(t, x_1, x_2) \qquad (4.21)$$

In this case, equations (4.20) generalize to:

$$\begin{aligned}
k_{11} &= f_1(t_n, x_{1,n}, x_{2,n}) \\
k_{21} &= f_2(t_n, x_{1,n}, x_{2,n}) \\
k_{12} &= f_1(t_n + \frac{h}{2}, x_{1,n} + \frac{h}{2} k_{11}, x_{2,n} + \frac{h}{2} k_{21}) \\
k_{22} &= f_2(t_n + \frac{h}{2}, x_{1,n} + \frac{h}{2} k_{11}, x_{2,n} + \frac{h}{2} k_{21}) \\
k_{13} &= f_1(t_n + \frac{h}{2}, x_{1,n} + \frac{h}{2} k_{12}, x_{2,n} + \frac{h}{2} k_{22}) \\
k_{23} &= f_2(t_n + \frac{h}{2}, x_{1,n} + \frac{h}{2} k_{12}, x_{2,n} + \frac{h}{2} k_{22}) \\
k_{14} &= f_1(t_n + h, x_{1,n} + h k_{13}, x_{2,n} + h k_{23}) \\
k_{24} &= f_2(t_n + h, x_{1,n} + h k_{13}, x_{1,n} + h k_{23}) \\
x_{1,n+1} &= x_{1,n} + \frac{h}{6}(k_{11} + 2k_{12} + 2k_{13} + k_{14}) \\
x_{2,n+1} &= x_{1,n} + \frac{h}{6}(k_{21} + 2k_{22} + 2k_{23} + k_{24}). \qquad (4.22)
\end{aligned}$$

Programming this algorithm is quite simple. The main program is an interface between the user and the driver routine of the integration. The user enters the initial and final times $t_i =$ Ti and $t_f =$ Tf and the number of discrete time points Nt. The initial conditions are $x_1(t_i) =$ X10, $x_2(t_i) =$ X20. The main data structure consists of three global double arrays T[P], X1[P], X2[P] which store the times $t_i \equiv t_1, t_2, \ldots, t_{Nt} \equiv t_f$ and the corresponding values of the functions $x_1(t_k)$ and $x_2(t_k)$, $k = 1, \ldots,$ Nt. The main program calls the driver routine RK(Ti,Tf,X10,X20,Nt) which "drives" the heart of the program, the function RKSTEP(t,x1,x2,dt) which performs one integration step using equations (4.22). RKSTEP evolves the functions x1, x2 at time t by one step $h =$ dt. The function RK stores the calculated values in the arrays T, X1 and X2 at each step. When RK returns the control to the main program, all the results are stored in T, X1 and X2, which are subsequently printed in the file rk.dat. The full program is listed below and can be found in the file rk.cpp:

4.3. RUNGE–KUTTA METHODS

```cpp
//========================================================
//Program to solve a 2 ODE system using Runge-Kutta Method
//User must supply derivatives
//dx1/dt=f1(t,x1,x2)  dx2/dt=f2(t,x1,x2)
//as real functions
//Output is written in file rk.dat
//========================================================
#include <iostream>
#include <fstream>
#include <cstdlib>
#include <string>
#include <cmath>
using namespace std;
//----------------------------------------------------
const int P = 110000;
double T[P], X1[P], X2[P];
//----------------------------------------------------
double
f1(const double& t  , const double& x1, const double& x2);
double
f2(const double& t  , const double& x1, const double& x2);
void
RK(const double& Ti , const double& Tf, const double& X10,
   const double& X20, const int   & Nt);
void
RKSTEP(double& t, double& x1, double& x2,
       const double& dt);
//----------------------------------------------------
int main(){
  double Ti,Tf,X10,X20;
  int    Nt;
  int    i;
  string buf;

  //Input:
  cout << "Runge-Kutta Method for 2-ODEs Integration\n";
  cout << "Enter Nt,Ti,TF,X10,X20:"    << endl;
  cin  >> Nt >> Ti >> Tf >> X10 >> X20;getline(cin,buf);
  cout << "Nt = "                 << Nt  << endl;
  cout << "Time: Initial Ti = " << Ti
       << " Final Tf = "        << Tf  << endl;
  cout << "          X1(Ti)= " << X10
       << " X2(Ti)= "           << X20 << endl;
  if(Nt >= P){cerr << "Error! Nt >= P\n";exit(1);}
  //Calculate:
```

```cpp
  RK(Ti,Tf,X10,X20,Nt);
  //Output:
  ofstream myfile("rk.dat");
  myfile.precision(17);
  for(i=0;i<Nt;i++)
    myfile << T [i] << " "
           << X1[i] << " "
           << X2[i] << '\n';
  myfile.close();
}//main()
//========================================================
//The functions f1,f2(t,x1,x2) provided by the user
//========================================================
double
f1(const double& t, const double& x1, const double& x2){
  return x2;
}
//--------------------------------------------------------
double
f2(const double& t, const double& x1, const double& x2){
  return -10.0*x1; //harmonic oscillator
}
//========================================================
//RK(Ti,Tf,X10,X20,Nt) is the driver
//for the Runge-Kutta integration routine RKSTEP
//Input: Initial and final times Ti,Tf
//        Initial values at t=Ti  X10,X20
//        Number of steps of integration: Nt-1
//Output: values in arrays T[Nt],X1[Nt],X2[Nt] where
//T[0]    = Ti X1[0]   = X10 X2[0] = X20
//         X1[k-1] = X1(at  t=T(k))
//         X2[k-1] = X2(at  t=T(k))
//T[Nt-1] = Tf
//========================================================
void
RK(const double& Ti , const double& Tf, const double& X10,
   const double& X20, const int    & Nt){
  double dt;
  double TS,X1S,X2S; //time and X1,X2 at given step
  int    i;
  //Initialize variables:
  dt       = (Tf-Ti)/(Nt-1);
  T [0]    = Ti;
  X1[0]    = X10;
  X2[0]    = X20;
```

4.3. RUNGE–KUTTA METHODS

```
  TS      = Ti;
  X1S     = X10;
  X2S     = X20;
  //Make RK steps: The arguments of RKSTEP are
  //replaced with the new ones!
  for(i=1;i<Nt;i++){
    RKSTEP(TS,X1S,X2S,dt);
    T [i]   = TS;
    X1[i]   = X1S;
    X2[i]   = X2S;
  }
}//RK()
//================================================
//Function RKSTEP(t,x1,x2,dt)
//Runge-Kutta Integration routine of ODE
//dx1/dt=f1(t,x1,x2)  dx2/dt=f2(t,x1,x2)
//User must supply derivative functions:
//real function f1(t,x1,x2)
//real function f2(t,x1,x2)
//Given initial point (t,x1,x2) the routine advances it
//by time dt.
//Input : Inital time t   and function values x1,x2
//Output: Final  time t+dt and function values x1,x2
//Careful!: values of t,x1,x2 are overwritten...
//================================================
void
RKSTEP(double& t, double& x1, double& x2,
       const double& dt){
  double k11,k12,k13,k14,k21,k22,k23,k24;
  double h,h2,h6;

  h  =dt;      //h =dt, integration step
  h2 =0.5*h;   //h2=h/2
  h6 =h/6.0;   //h6=h/6

  k11=f1(t,x1,x2);
  k21=f2(t,x1,x2);
  k12=f1(t+h2,x1+h2*k11,x2+h2*k21);
  k22=f2(t+h2,x1+h2*k11,x2+h2*k21);
  k13=f1(t+h2,x1+h2*k12,x2+h2*k22);
  k23=f2(t+h2,x1+h2*k12,x2+h2*k22);
  k14=f1(t+h ,x1+h *k13,x2+h *k23);
  k24=f2(t+h ,x1+h *k13,x2+h *k23);

  t  =t+h;
```

```
    x1 =x1+h6*(k11+2.0*(k12+k13)+k14);
    x2 =x2+h6*(k21+2.0*(k22+k23)+k24);

}//RKSTEP()
```

4.4 Comparison of the Methods

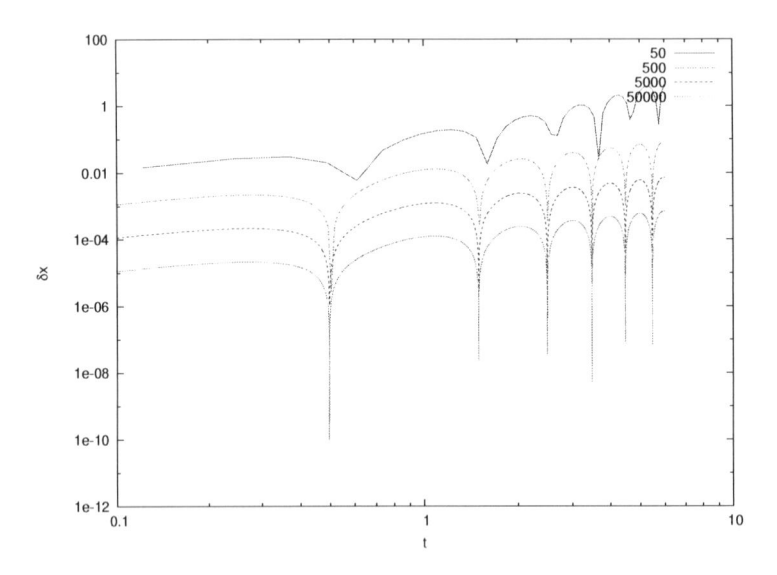

Figure 4.11: The discrepancy of the numerical results of the Euler method from the analytic solution for the simple harmonic oscillator. The parameters chosen are $\omega^2 = 10$, $t_i = 0$, $t_f = 6$, $x(0) = 0.2$, $v(0) = 0$ and the number of steps is $N = 50, 500, 5,000, 50,000$. Observe that the error becomes approximately ten times smaller each time according to the expectation of being of order $\sim \mathcal{O}(\Delta t)$.

In this section we will check our programs for correctness and accuracy w.r.t. discretization and roundoff errors. The simplest test is to check the results against a known analytic solution of a simple model. This will be done for the simple harmonic oscillator. We will change the functions that compute the acceleration of the particle to give $a = -\omega^2 x$. We will take $\omega^2 = 10$ ($T \approx 1.987$). Therefore the relevant part of the program in euler.cpp becomes

4.4. COMPARISON OF THE METHODS

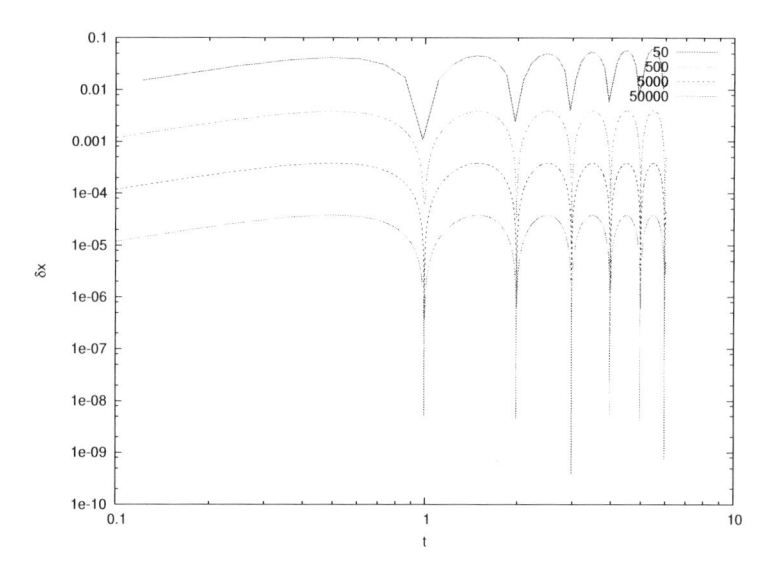

Figure 4.12: Like in figure 4.11 for the Euler-Cromer method. The error becomes approximately ten times smaller each time according to the expectation of being of order $\sim \mathcal{O}(\Delta t)$.

```
double accel(const double& x){
  return -10.0 * x;
}
```

and that of the program in rk.cpp becomes

```
double
f2(const double& t, const double& x1, const double& x2){
  return -10.0*x1;
}
```

The programs are run for a given time interval $t_i = 0$ to $t_f = 6$ with the initial conditions $x_0 = 0.2$, $v_0 = 0$. The time step Δt is varied by varying the number of steps Nt-1. The computed numerical solution is compared to the well known solution for the simple harmonic oscillator

$$\begin{aligned} a(x) &= -\omega^2 x \\ x_h(t) &= x_0 \cos(\omega t) + (v_0/\omega) \sin(\omega t) \\ v_h(t) &= v_0 \cos(\omega t) - (x_0 \omega) \sin(\omega t)\,, \end{aligned} \quad (4.23)$$

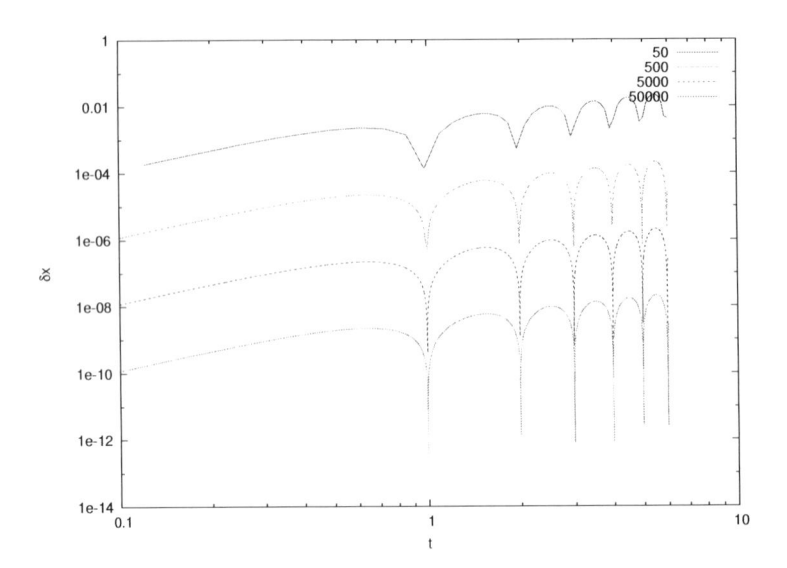

Figure 4.13: Like in figure 4.11 for the Euler-Verlet method. The error becomes approximately 100 times smaller each time according to the expectation of being of order $\sim \mathcal{O}(\Delta t^2)$.

We study the deviation $\delta x(t) = |x(t) - x_h(t)|$ and $\delta v(t) = |v(t) - v_h(t)|$ as a function of the time step Δt. The results are shown in figures 4.11–4.14. We note that for the Euler method and the Euler–Cromer method, the errors are of order $\mathcal{O}(\Delta t)$ as expected. However, the latter has smaller errors compared to the first one. For the Euler–Verlet method, the error turns out to be of order $\mathcal{O}(\Delta t^2)$ whereas for the 4th order Runge–Kutta is of order[9] $\mathcal{O}(\Delta t^4)$.

Another way for checking the numerical results is by looking at a conserved quantity, like the energy, momentum or angular momentum, and study its deviation from its original value. In our case we study the mechanical energy

$$E = \frac{1}{2}mv^2 + \frac{1}{2}m\omega^2 x^2 \qquad (4.24)$$

which is computed at each step. The deviation $\delta E = |E - E_0|$ is shown in figures 4.15–4.18.

---

[9]The reader should confirm these claims, initially by looking at the figures 4.11-4.14 and then by reproducing these results. A particular time t can be chosen and the errors can be plotted against Δt, Δt^2 and Δt^4 respectively.

4.5. THE FORCED DAMPED OSCILLATOR

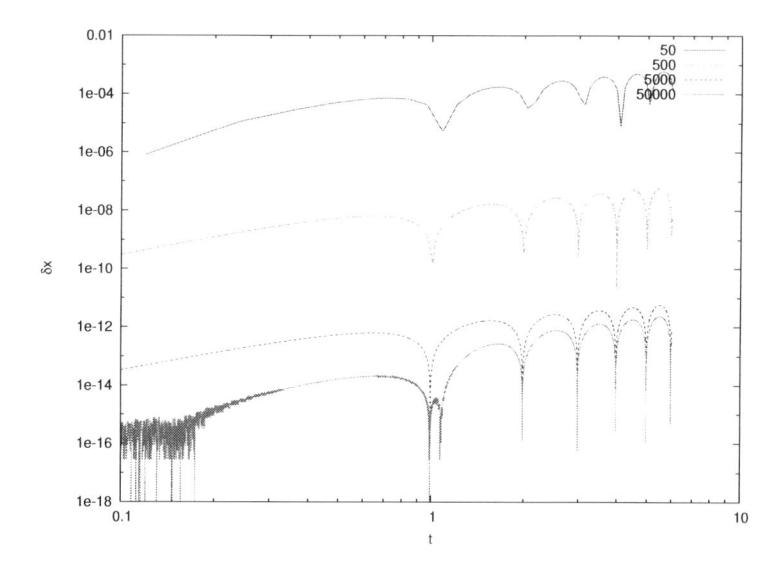

Figure 4.14: Like in figure 4.11 for the 4th order Runge–Kutta method. The error becomes approximately 10^{-4} times smaller each time according to the expectation of being of order $\sim \mathcal{O}(\Delta t^4)$. The roundoff errors become apparent for 50,000 steps.

4.5 The Forced Damped Oscillator

In this section we will study a simple harmonic oscillator subject to a damping force proportional to its velocity and an external periodic driving force, which for simplicity will be taken to have a sinusoidal dependence in time,

$$\frac{d^2x}{dt^2} + \gamma \frac{dx}{dt} + \omega_0^2 x = a_0 \sin \omega t, \qquad (4.25)$$

where $F(t) = m a_0 \sin \omega t$ and ω is the angular frequency of the driving force.

Consider initially the system without the influence of the driving force, i.e. with $a_0 = 0$. The real solutions of the differential equation[10] which are finite for $t \to +\infty$ are given by

$$x_0(t) = c_1 e^{-(\gamma + \sqrt{\gamma^2 - 4\omega_0^2})t/2} + c_2 e^{-(\gamma - \sqrt{\gamma^2 - 4\omega_0^2})t/2}, \quad \gamma^2 - 4\omega_0^2 > 0, \quad (4.26)$$

---

[10]These are easily obtained by substituting the ansatz $x(t) = A e^{-\Omega t}$ and solving for Ω.

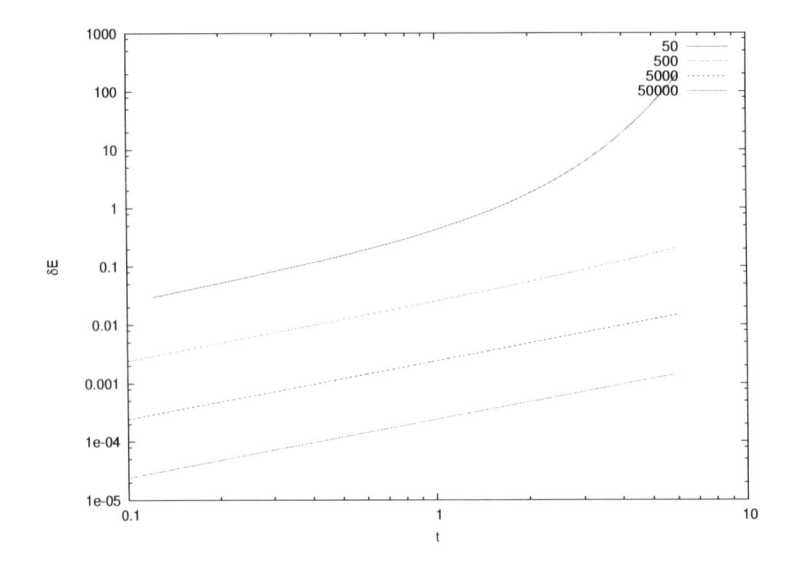

Figure 4.15: Like in figure 4.11 for the case of mechanical energy for the Euler method.

$$x_0(t) = c_1 e^{-\gamma t/2} + c_2 e^{-\gamma t/2} t, \quad \gamma^2 - 4\omega_0^2 = 0, \quad (4.27)$$

$$\begin{aligned} x_0(t) &= c_1 e^{-\gamma t/2} \cos\left(\sqrt{-\gamma^2 + 4\omega_0^2}\, t/2\right) \\ &+ c_2 e^{-\gamma t/2} \sin\left(\sqrt{-\gamma^2 + 4\omega_0^2}\, t/2\right), \quad \gamma^2 - 4\omega_0^2 < 0 \end{aligned} \quad (4.28)$$

In the last case, the solution oscillates with an amplitude decreasing exponentially with time.

In the $a_0 > 0$ case, the general solution is obtained from the sum of a special solution $x_s(t)$ and the solution of the homogeneous equation $x_0(t)$. A special solution can be obtained from the ansatz $x_s(t) = A \sin \omega t + B \cos \omega t$, which when substituted in (4.25) and solved for A and B we find that

$$x_s(t) = \frac{a_0 \left[(\omega_0^2 - \omega^2) \cos \omega t + \gamma \omega \sin \omega t\right]}{(\omega_0^2 - \omega^2)^2 + \omega^2 \gamma^2}, \quad (4.29)$$

and

$$x(t) = x_0(t) + x_s(t). \quad (4.30)$$

4.5. THE FORCED DAMPED OSCILLATOR

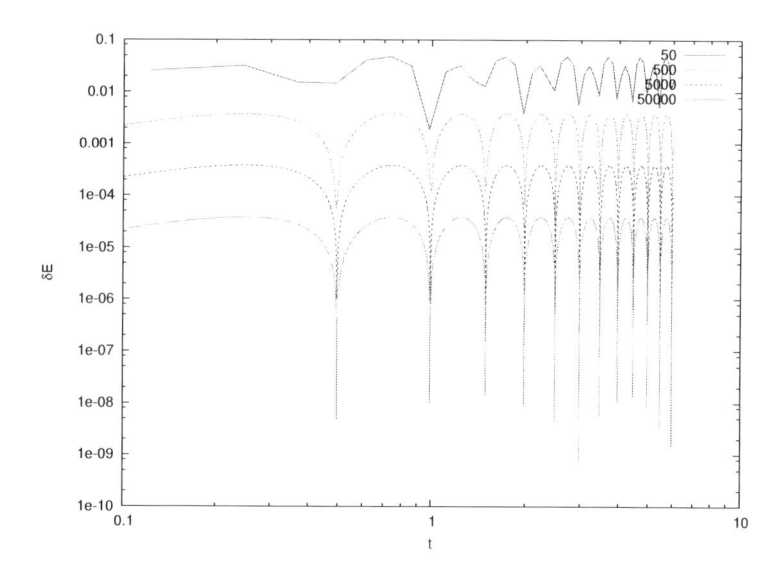

Figure 4.16: Like in figure 4.11 for the case of mechanical energy for the Euler–Cromer method.

The solution $x_0(t)$ decreases exponentially with time and eventually only $x_s(t)$ remains. The only case where this is not true, is when we have resonance without damping for $\omega = \omega_0$, $\gamma = 0$. In that case the solution is

$$x(t) = c_1 \cos \omega t + c_2 \sin \omega t + \frac{a_0}{4\omega^2}\left(\cos \omega t + 2(\omega t) \sin \omega t\right). \quad (4.31)$$

The first two terms are the same as that of the simple harmonic oscillator. The last one increases the amplitude linearly with time, which is a result of the influx of energy from the external force to the oscillator.

Our program will be a simple modification of the program in rk.cpp. The main routines RK(T0,TF,X10,X20,Nt) and RKSTEP(t,x1,x2,dt) remain as they are. We only change the user interface. The basic parameters ω_0, ω, γ, a_0 are entered interactively by the user from the standard input stdin. These parameters should be accessible also by the function f2(t,x1,x2) and they are declared within the global scope. Another point that needs our attention is the function f2(t,x1,x2) which now takes the velocity $v \to$ x2 in its arguments:

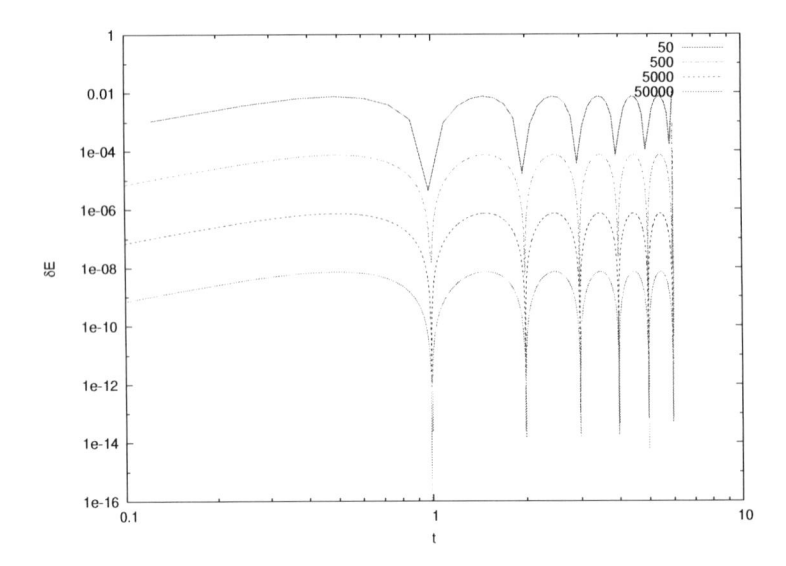

Figure 4.17: Like in figure 4.11 for the case of mechanical energy for the Euler–Verlet method.

```
double
f2(const double& t, const double& x1, const double& x2){
  double a;
  a = a_0*cos(omega*t);
  return -omega_02*x1-gam*x2+a;
}
```

The main program, found in the file dlo.cpp, is listed below. The functions RK, RKSTEP are the same as in rk.cpp and should also be included in the same file.

```
//===========================================================
//Program to solve Damped Linear Oscillator
//using 4th order Runge-Kutta Method
//Output is written in file dlo.dat
//===========================================================
#include <iostream>
#include <fstream>
#include <cstdlib>
#include <string>
#include <cmath>
using namespace std;
```

4.5. THE FORCED DAMPED OSCILLATOR

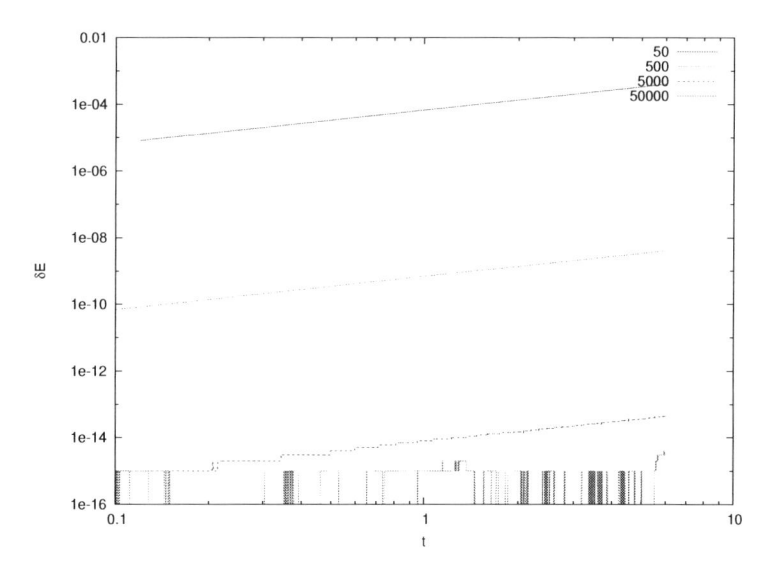

Figure 4.18: Like in figure 4.11 for the case of mechanical energy for the 4th order Runge–Kutta method. Roundoff errors appear for large enough number of steps.

```
//------------------------------------------------------------
const int P = 110000;
double T[P], X1[P], X2[P];
double omega_0,omega,gam,a_0,omega_02,omega2;
//------------------------------------------------------------
double
f1(const double& t   , const double& x1, const double& x2);
double
f2(const double& t   , const double& x1, const double& x2);
void
RK(const double& Ti , const double& Tf, const double& X10,
   const double& X20, const int    & Nt);
void
RKSTEP(double& t, double& x1, double& x2,
       const double& dt);
//------------------------------------------------------------
int main(){
  double Ti,Tf,X10,X20;
  double Energy;
  int    Nt;
  int    i;
  string buf;
```

```cpp
//Input:
  cout << "Runge-Kutta Method for DLO Integration\n";
  cout << "Enter omega_0, omega, gamma, a_0:\n";
  cin  >> omega_0>> omega>> gam>> a_0;getline(cin,buf);
  omega_02 = omega_0*omega_0;
  omega2   = omega *omega;
  cout << "omega_0= " << omega_0
       << " omega= "  << omega             << endl;
  cout << "gamma= "   << gamma
       << " a_0= "    << a_0               << endl;
  cout << "Enter Nt,Ti,TF,X10,X20:"        << endl;
  cin  >> Nt >> Ti >> Tf >> X10 >> X20;getline(cin,buf);
  cout << "Nt = "                << Nt << endl;
  cout << "Time: Initial Ti = " << Ti
       << " Final Tf = "         << Tf << endl;
  cout << "           X1(Ti)= " << X10
       << " X2(Ti)= "            << X20 << endl;
  if(Nt >= P){cerr << "Error! Nt >= P\n";exit(1);}
//Calculate:
  RK(Ti,Tf,X10,X20,Nt);
//Output:
  ofstream myfile("dlo.dat");
  myfile.precision(17);
  myfile << "# Damped Linear Oscillator - dlo\n";
  myfile << "# omega_0= " << omega_0 << " omega= " << omega
         << " gamma= " << gam       << " a_0= " << a_0 << ←
            endl;
  for(i=0;i<Nt;i++){
    Energy = 0.5*X2[i]*X2[i]+0.5*omega_02*X1[i]*X1[i];
    myfile << T [i]  << " "
           << X1[i]  << " "
           << X2[i]  << " "
           << Energy << '\n';
  }
  myfile.close();
}//main()
//========================================================
//The functions f1,f2(t,x1,x2) provided by the user
//========================================================
double
f1(const double& t, const double& x1, const double& x2){
  return x2;
}
//--------------------------------------------------------
double
```

4.5. THE FORCED DAMPED OSCILLATOR

```
f2(const double& t, const double& x1, const double& x2){
  double a;
  a = a_0*cos(omega*t);
  return -omega_02*x1-gam*x2+a;
}
```

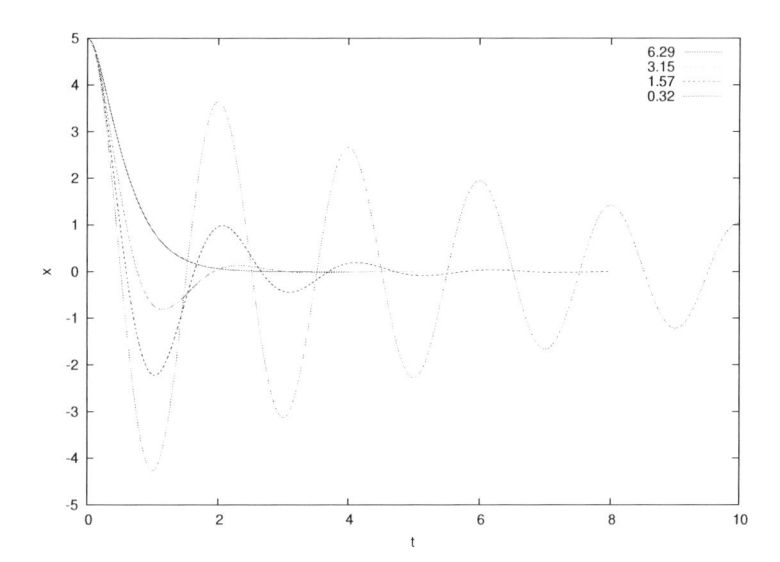

Figure 4.19: The position as a function of time for the damped oscillator for several values of γ and $\omega_0 = 3.145$.

The results are shown in figures 4.19–4.22. Figure 4.19 shows the transition from a damped motion for $\gamma > 2\omega_0$ to an oscillating motion with damping amplitude for $\gamma < 2\omega_0$. The exponential decrease of the amplitude is shown in figure 4.21, whereas the dependence of the period T from the damping coefficient γ is shown in figure 4.22. Motivated by equation (4.28), written in the form

$$4\omega_0^2 - \left(\frac{2\pi}{T}\right) = \gamma^2, \qquad (4.32)$$

we construct the plot in figure 4.22. The right hand side of the equation is put on the horizontal axis, whereas the left hand side on the vertical. Equation (4.32) predicts that both quantities are equal and all measurements should lie on a particular line, the diagonal $y = x$. The period T

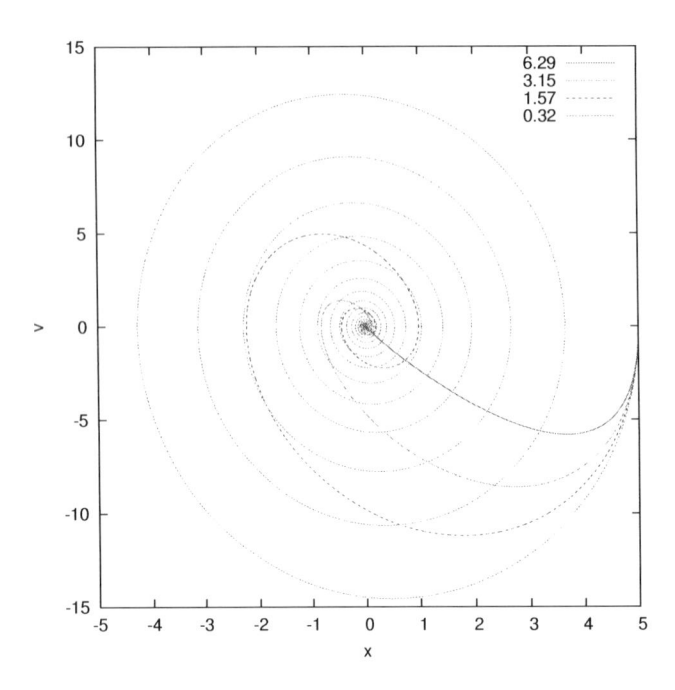

Figure 4.20: The phase space trajectory for the damped oscillator for several values of γ and $\omega_0 = 3.145$. Note the attractor at $(x, v) = (0, 0)$ where all trajectories are "attracted to" as $t \to +\infty$.

can be estimated from the time between two consecutive extrema of $x(t)$ or two consecutive zeros of the velocity $v(t)$ (see figure 4.19).

Finally it is important to study the trajectory of the system in phase space. This can be seen[11] in figure 4.20. A point in this space is a *state* of the system and a trajectory describes the evolution of the system's states in time. We see that all such trajectories end up as $t \to +\infty$ to the point $(0, 0)$, independently of the initial conditions. Such a point is an example of a system's *attractor*.

Next, we add the external force and study the response of the system to it. The system exhibits a *transient behavior* that depends on the initial conditions. For large enough times it approaches a *steady state* that does not depend on (almost all of) the initial conditions. This can be seen in figure 4.23. This is easily understood for our system by looking at equa-

---

[11] To be precise, phase space is the space of positions-momenta, but in our case the difference is trivial.

4.5. THE FORCED DAMPED OSCILLATOR

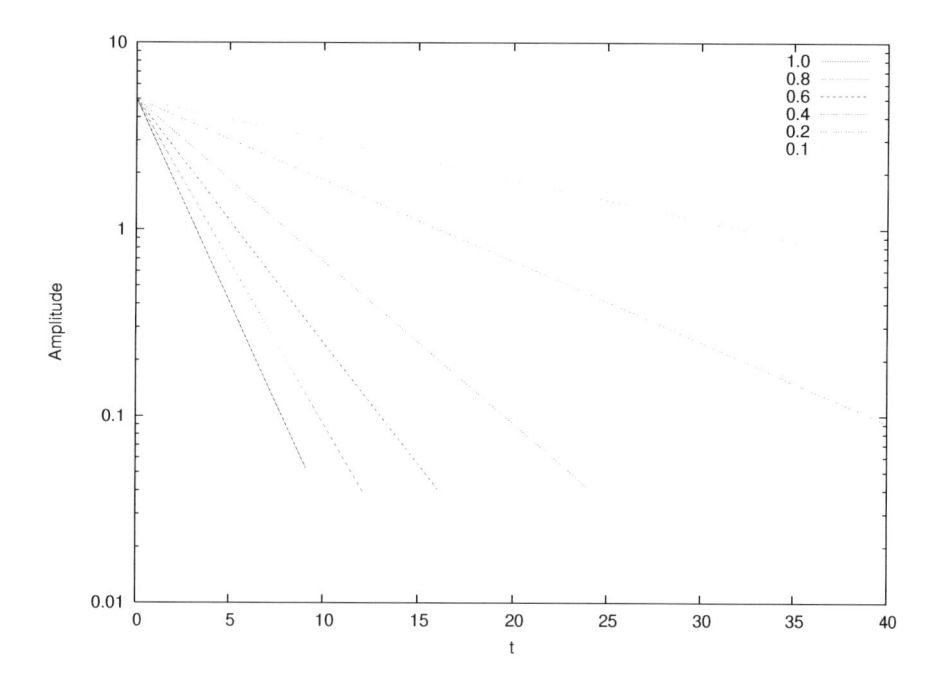

Figure 4.21: The amplitude of oscillation for the damped oscillator for several values of γ and $\omega_0 = 3.145$. Note the exponential damping of the amplitude with time.

tions (4.26)–(4.28). We see that the steady state $x_s(t)$ becomes dominant when the exponentials have damped away. $x_s(t)$ can be written in the form

$$x(t) = x_0(\omega) \cos(\omega t + \delta(\omega))$$
$$x_0(\omega) = \frac{a_0}{\sqrt{(\omega_0^2 - \omega^2)^2 + \gamma^2 \omega^2}}, \qquad \tan \delta(\omega) = \frac{\omega \gamma}{\omega^2 - \omega_0^2}. \quad (4.33)$$

These equations are verified in figure 4.24 where we study the dependence of the amplitude $x_0(\omega)$ on the angular frequency of the driving force. Finally we study the trajectory of the system in phase space. As we can see in figure 4.20, this time the attractor is an ellipse, which is a one dimensional curve instead of a zero dimensional point. For large enough times, all trajectories approach their attractor asymptotically.

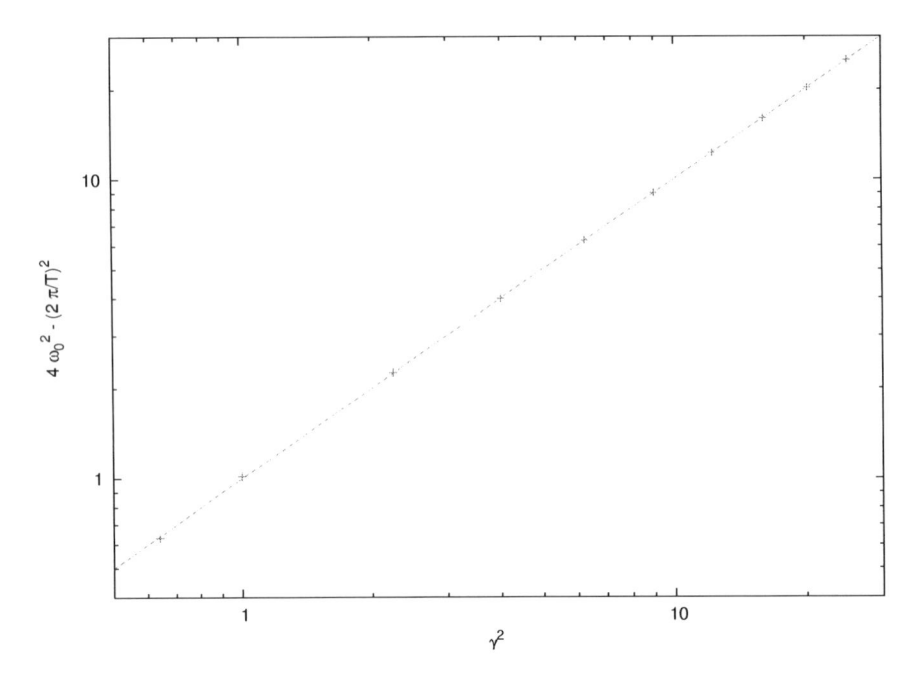

Figure 4.22: The period of oscillation of the damped oscillator for several values of γ and $\omega_0 = 3.145$. The axes are chosen so that equation (4.28) $(2\pi/T)^2 = 4\omega_0^2 - \gamma^2$ can be easily verified. The points in the plot are our measurements whereas the straight line is the theoretical prediction, the diagonal $y = x$

4.6 The Forced Damped Pendulum

In this section we will study a non-linear dynamical system which exhibits interesting chaotic behavior. This is a simple model which, despite its deterministic nature, the prediction of its future behavior becomes intractable after a short period of time. Consider a simple pendulum in a constant gravitational field whose motion is damped by a force proportional to its velocity and it is under the influence of a vertical, harmonic external driving force:

$$\frac{d^2\theta}{dt^2} + \gamma\frac{d\theta}{dt} + \omega_0^2 \sin\theta = -2A\cos\omega t \, \sin\theta. \qquad (4.34)$$

In the equation above, θ is the angle of the pendulum with the vertical axis, γ is the damping coefficient, $\omega_0^2 = g/L$ is the pendulum's natural angular frequency, ω is the angular frequency of the driving force and

4.6. THE FORCED DAMPED PENDULUM

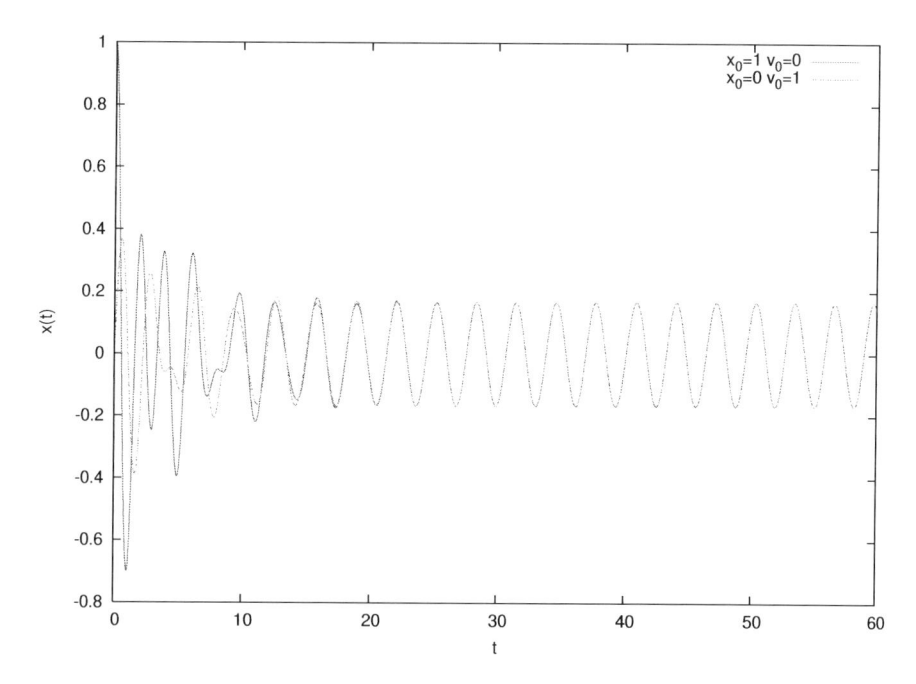

Figure 4.23: The period of oscillation for the forced damped oscillator for different initial conditions. We have chosen $\omega_0 = 3.145$, $\omega = 2.0$, $\gamma = 0.5$ and $a_0 = 1.0$. We note that after the transient behavior the system oscillates harmonically according to the relation $x(t) = x_0(\omega)\cos(\omega t + \delta)$.

$2A$ is the amplitude of the external angular acceleration caused by the driving force.

In the absence of the driving force, the damping coefficient drives the system to the point $(\theta, \dot{\theta}) = (0, 0)$, which is an attractor for the system. This continues to happen for small enough A, but for $A > A_c$ the behavior of the system becomes more complicated.

The program that integrates the equations of motion of the system can be obtained by making trivial changes to the program in the file dlo.cpp. This changes are listed in detail below, but we note that X1 $\leftrightarrow \theta$, X2 $\leftrightarrow \dot{\theta}$, a_0 $\leftrightarrow A$. The final program can be found in the file fdp.cpp. It is listed below, with the understanding that the commands in between the dots are the same as in the programs found in the files dlo.cpp, rk.cpp.

```
//===============================================
```

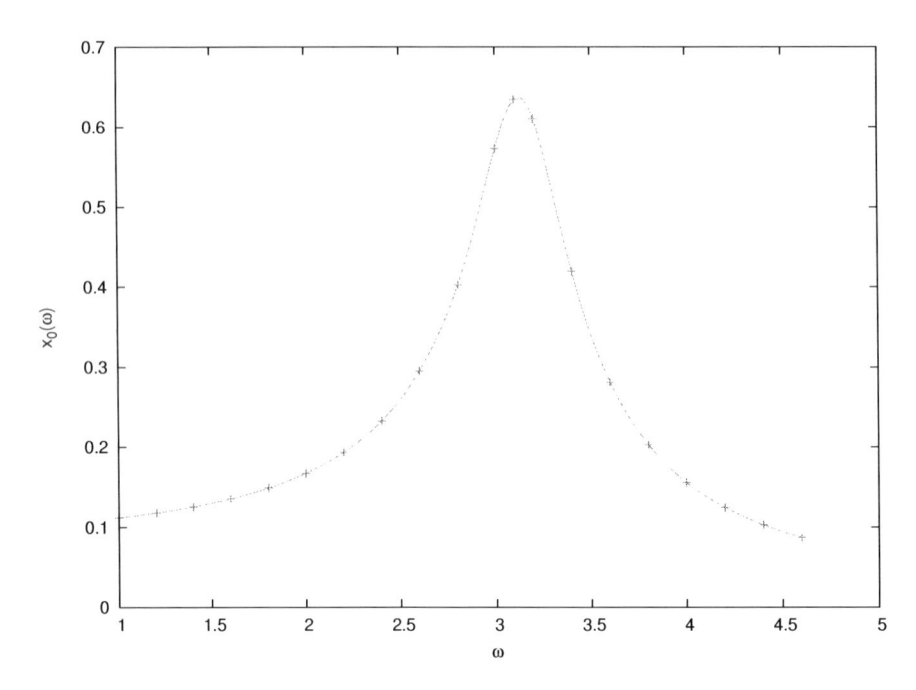

Figure 4.24: The oscillation amplitude $x_0(\omega)$ as a function of ω for the forced damped oscillator, where $\omega_0 = 3.145$, $\gamma = 0.5$ and $a_0 = 1.0$. We observe a resonance for $\omega \approx \omega_0$. The points of the plot are our measurements and the line is the theoretical prediction given by equation (4.33).

```
//Program to solve Forced Damped Pendulum
//using 4th order Runge-Kutta Method
//Output is written in file fdp.dat
//=================================================
..............................
const int P = 1010000;
..............................
    Energy = 0.5*X2[i]*X2[i]+omega_02*(1.0-cos(X1[i]));
..............................
double
f2(const double& t, const double& x1, const double& x2){
  return -(omega_02+2.0*a_0*cos(omega*t))*sin(x1)-gam*x2;
}
..............................
void
RKSTEP(double& t, double& x1, double& x2,
       const double& dt){
```

4.6. THE FORCED DAMPED PENDULUM

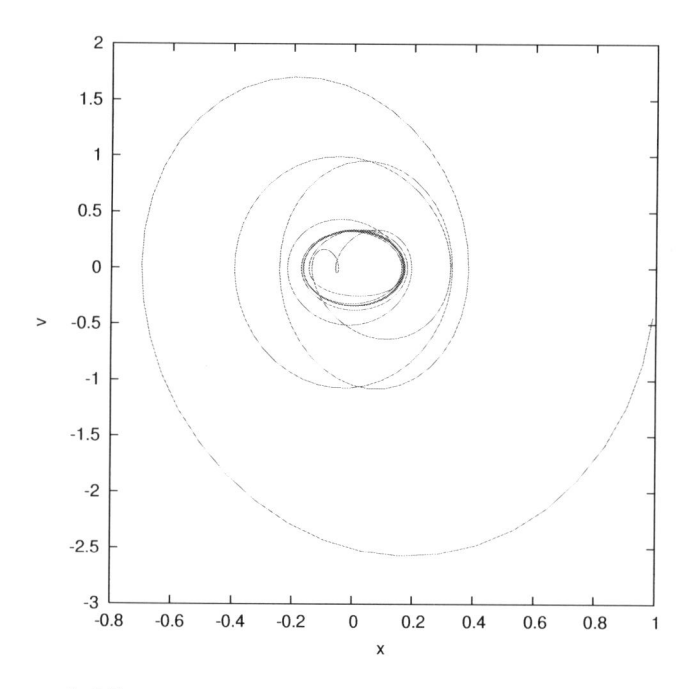

Figure 4.25: A phase space trajectory of the forced damped oscillator with $\omega_0 = 3.145$, $\omega = 2.0$, $\gamma = 0.5$ and $a_0 = 1.0$. The harmonic oscillation which is the steady state of the system is an ellipse, which is an attractor of all the phase space trajectories that correspond to different initial conditions.

```
................................
 const double pi = 3.14159265358979324;
 const double pi2= 6.28318530717958648;
 x1 =x1+h6*(k11+2.0*(k12+k13)+k14);
 x2 =x2+h6*(k21+2.0*(k22+k23)+k24);
  if( x1 >  pi ) x1 -= pi2;
  if( x1 < -pi ) x1 += pi2;
}//RKSTEP()
```

The final lines in the program are added so that the angle is kept within the interval $[-\pi, \pi]$.

In order to study the system's properties we will set $\omega_0 = 1$, $\omega = 2$, and $\gamma = 0.2$ unless we explicitly state otherwise. The natural period of the pendulum is $T_0 = 2\pi/\omega_0 = 2\pi \approx 6.28318530717958648$ whereas that of the driving force is $T = 2\pi/\omega = \pi \approx 3.14159265358979324$. For $A < A_c$, with $A_c \approx 0.18$, the point $(\theta, \dot\theta) = (0,0)$ is an attractor, which

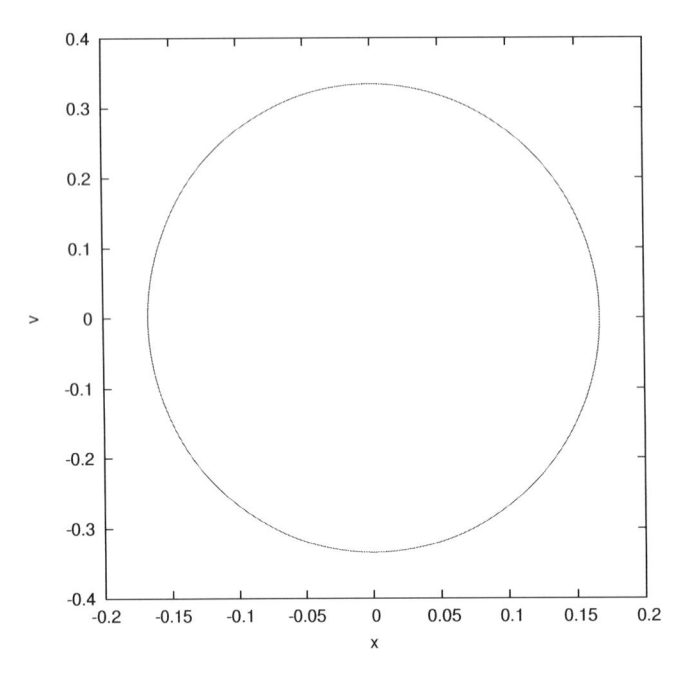

Figure 4.26: The trajectory shown in figure 4.25 for $t > 100$. The trajectory is almost on top of an ellipse corresponding to the steady state motion of the system. This ellipse is an attractor of the system.

means that the pendulum eventually stops at its stable equilibrium point. For $A_c < A < 0.71$ the attractor is a closed curve, which means that the pendulum at its steady state oscillates indefinitely without circling through its unstable equilibrium point at $\theta = \pm\pi$. The period of motion is found to be twice that of the driving force. For $0.72 < A < 0.79$ the attractor is an open curve, because at its steady state the pendulum crosses the $\theta = \pm\pi$ point. The period of the motion becomes equal to that of the driving force. For $0.79 < A \lesssim 1.033$ we have *period doubling* for critical values of A, but the trajectory is still periodic. For even larger values of A the system enters into a chaotic regime where the trajectories are non periodic. For $A \approx 3.1$ we find the system in a periodic steady state again, whereas for $A \approx 3.8 - 4.448$ we have period doubling. For $A \approx 4.4489$ we enter into a chaotic regime again etc. These results can be seen in figures 4.27–4.29. The reader should construct the bifurcation diagram of the system by solving problem 19 of this chapter.

4.6. THE FORCED DAMPED PENDULUM

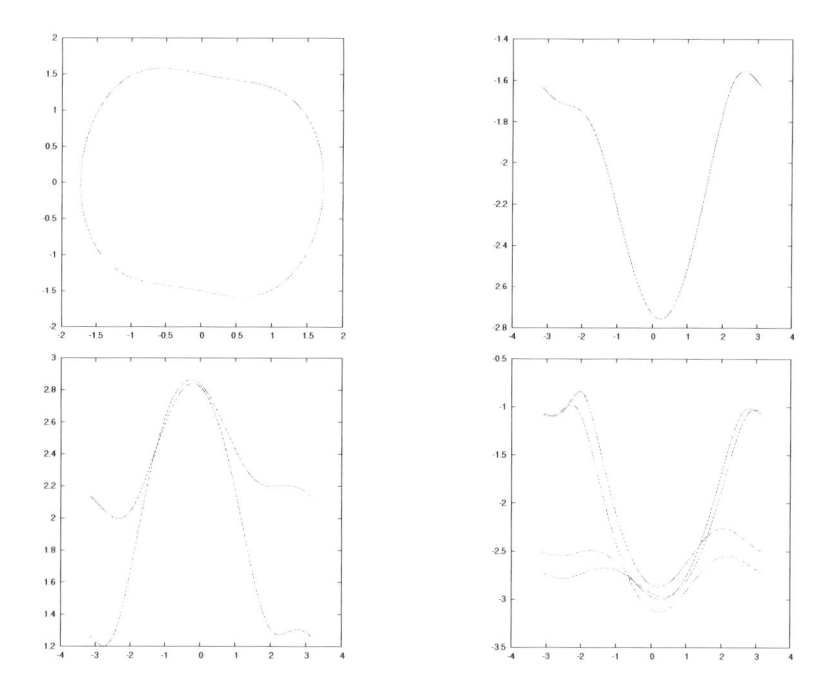

Figure 4.27: A phase space trajectory of the forced damped pendulum. The parameters chosen are $\omega_0 = 1.0$, $\omega = 2.0$, $\gamma = 0.2$ and $A = 0.60, 0.72, 0.85, 1.02$. We observe the phenomenon of period doubling.

We can also use the so called *Poincaré* diagrams in order to study the chaotic behavior of a system. These are obtained by placing a point in phase space when the time is an integer multiple of the period of the driving force. Then, if for example the period of the motion is equal to that of the period of the driving force, the Poincaré diagram consists of only one point. If the period of the motion is an n-multiple of the period of the driving force then the Poincaré diagram consists of only n points. Therefore, in the period doubling regime, the points of the Poincaré diagram double at each period doubling point. In the chaotic regime, the Poincaré diagram consists of an infinite number of points which belong to sets that have interesting fractal structure. One way to construct the Poincaré diagram numerically, is to process the data of the output file fdp.dat using awk[12]:

---

[12]The command can be written in one line without the final \ of the first and second lines.

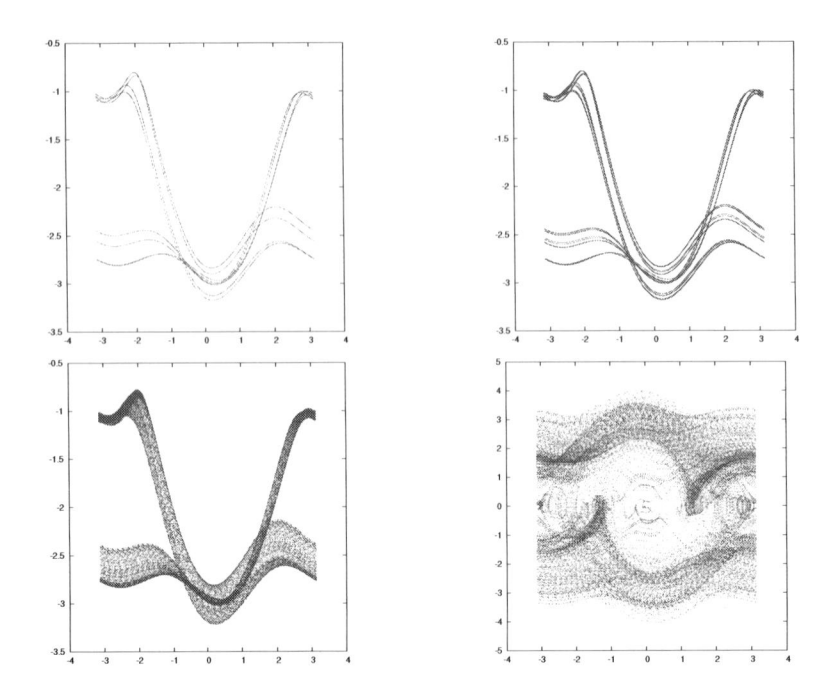

Figure 4.28: A phase space trajectory of the forced damped pendulum. The parameters chosen are $\omega_0 = 1.0$, $\omega = 2.0$, $\gamma = 0.2$ and $A = 1.031, 1.033, 1.04, 1.4$. We observe the chaotic behavior of the system.

```
awk -v o=$omega -v nt=$Nt -v tf=$TF \
  'BEGIN{T=6.283185307179/o;dt=tf/nt;} $1%T<dt{print $2,$3}'\
  fdp.dat
```

where $omega, $Nt, $TF are the values of the angular frequency ω, the number of points of time and the final time t_f. We calculate the period T and the time step dt in the program. Then we print those lines of the file where the time is an integer multiple of the period[13]. This is accomplished by the modulo operation $1 % T. The value of the expression $1 % T < dt is true when the remainder of the division of the first column ($1) of the file fdp.dat with the period T is smaller than dt. The results in the chaotic regime are displayed in figure 4.30.

We close this section by discussing another concept that helps us in

---

[13]The accuracy of this condition is limited by dt, which makes the points in the Poincaré diagram slightly fuzzy.

4.7. APPENDIX: ON THE EULER–VERLET METHOD

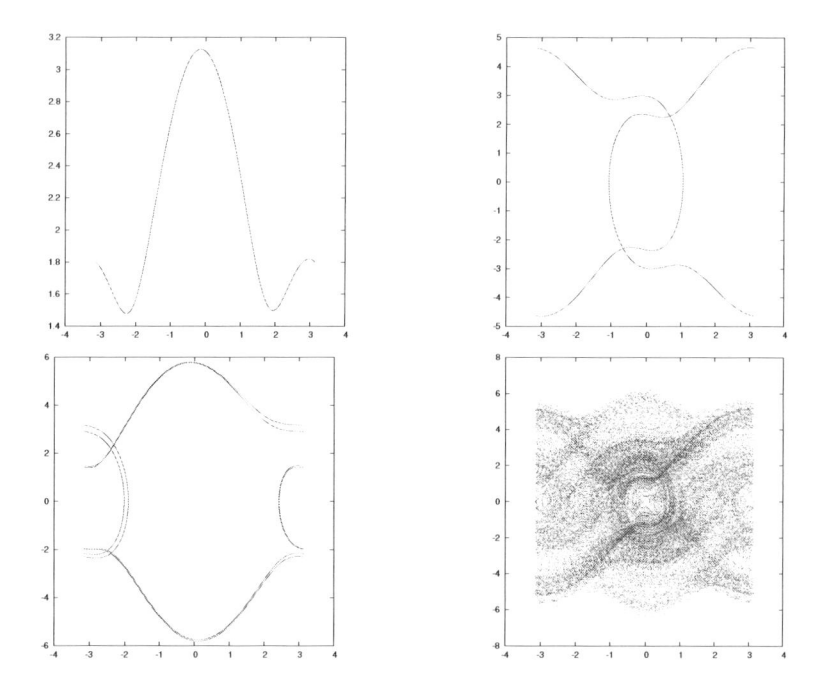

Figure 4.29: A phase space trajectory of the forced damped pendulum. The parameters chosen are $\omega_0 = 1.0$, $\omega = 2.0$, $\gamma = 0.2$ and $A = 1.568, 3.8, 4.44, 4.5$. We observe the system exiting and reentering regimes of chaotic behavior.

the analysis of the dynamical properties of the pendulum. This is the concept of the *basin of attraction* which is the set of initial conditions in phase space that lead the system to a specific attractor. Take for example the case for $A > 0.79$ in the regime where the pendulum at its steady state has a circular trajectory with a positive or negative direction. By taking a large sample of initial conditions and recording the direction of the resulting motion after the transient behavior, we obtain figure 4.31.

4.7 Appendix: On the Euler–Verlet Method

Equations (4.11) can be obtained from the Taylor expansion

$$\theta(t + \Delta t) = \theta(t) + (\Delta t)\theta'(t) + \frac{(\Delta t)^2}{2!}\theta''(t) + \frac{(\Delta t)^3}{3!}\theta'''(t) + \mathcal{O}((\Delta t)^4)$$

$$\theta(t - \Delta t) = \theta(t) - (\Delta t)\theta'(t) + \frac{(\Delta t)^2}{2!}\theta''(t) - \frac{(\Delta t)^3}{3!}\theta'''(t) + \mathcal{O}((\Delta t)^4).$$

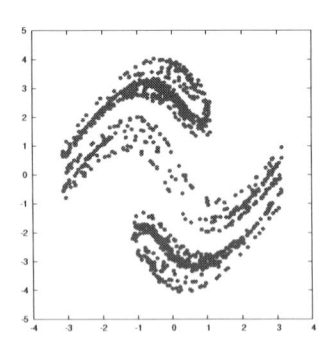

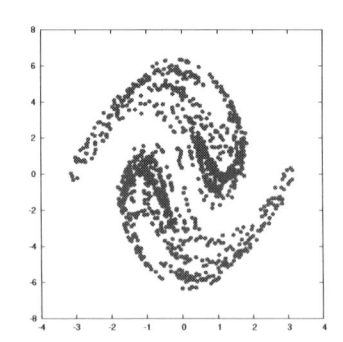

Figure 4.30: A Poincaré diagram for the forced damped pendulum in its chaotic regime. The parameters chosen are $\omega_0 = 1.0$, $\omega = 2.0$, $\gamma = 0.2$ and $A = 1.4, 4.5$.

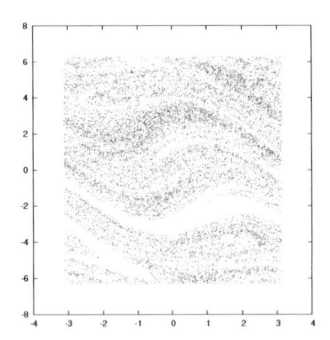

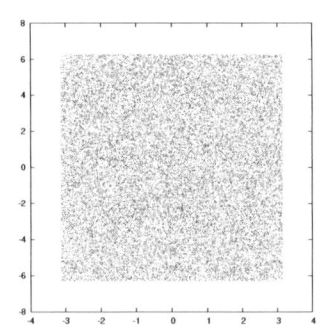

Figure 4.31: Basin of attraction for the forced damped pendulum. The parameters chosen are $\omega_0 = 1.0$, $\omega = 2.0$, $\gamma = 0.2$ and $A = 0.85, 1.4$.

By adding and subtracting the above equations we obtain

$$\begin{aligned} \theta(t + \Delta t) + \theta(t - \Delta t) &= 2\theta(t) + (\Delta t)^2 \theta''(t) + \mathcal{O}((\Delta t)^4) \\ \theta(t + \Delta t) - \theta(t - \Delta t) &= 2(\Delta t)\theta'(t) + \mathcal{O}((\Delta t)^3) \end{aligned} \quad (4.35)$$

which give equations (4.11)

$$\begin{aligned} \theta(t + \Delta t) &= 2\theta(t) - \theta(t - \Delta t) + (\Delta t)^2 \alpha(t) + \mathcal{O}((\Delta t)^4) \\ \omega(t) &= \frac{\theta(t + \Delta t) - \theta(t - \Delta t)}{2(\Delta t)} + \mathcal{O}((\Delta t)^2) \end{aligned} \quad (4.36)$$

From the first equation and equations (4.9) we obtain:

$$\theta(t + \Delta t) = \theta(t) + \omega(t)(\Delta t) + \mathcal{O}((\Delta t)^2) \quad (4.37)$$

4.7. APPENDIX: ON THE EULER–VERLET METHOD

When we perform a numerical integration, we are interested in the total error accumulated after $N-1$ integration steps. In this method, these errors must be studied carefully:

- The error in the velocity $\omega(t)$ does not accumulate because it is given by the *difference* of the positions $\theta(t+\Delta t) - \theta(t-\Delta t)$.

- The accumulation of the errors for the position is estimated as follows: Assume that $\delta\theta(t)$ is the *total* accumulated error from the integration from time t_0 to t. Then according to the expansions (4.36) the error for the first step is $\delta\theta(t_0 + \Delta t) = \mathcal{O}((\Delta t)^4)$. Then[14]

$$\begin{aligned}
\theta(t_0 + 2\Delta t) &= 2\theta(t_0 + \Delta t) - \theta(t_0) + \Delta t^2 \alpha(t_0 + \Delta t) + \mathcal{O}((\Delta t)^4) \Rightarrow \\
\delta\theta(t_0 + 2\Delta t) &= 2\delta\theta(t_0 + \Delta t) - \delta\theta(t_0) + \mathcal{O}((\Delta t)^4) \\
&= 2\mathcal{O}((\Delta t)^4) - 0 + \mathcal{O}((\Delta t)^4) \\
&= 3\mathcal{O}((\Delta t)^4).
\end{aligned}$$

For the next steps we obtain

$$\begin{aligned}
\theta(t_0 + 3\Delta t) &= 2\theta(t_0 + 2\Delta t) - \theta(t_0 + \Delta t) + \Delta t^2 \alpha(t_0 + 2\Delta t) + \mathcal{O}((\Delta t)^4) \Rightarrow \\
\delta\theta(t_0 + 3\Delta t) &= 2\delta\theta(t_0 + 2\Delta t) - \delta\theta(t_0 + \Delta t) + \mathcal{O}((\Delta t)^4) \\
&= 6\mathcal{O}((\Delta t)^4) - \mathcal{O}((\Delta t)^4) + \mathcal{O}((\Delta t)^4) \\
&= 6\mathcal{O}((\Delta t)^4),
\end{aligned}$$

$$\begin{aligned}
\theta(t_0 + 4\Delta t) &= 2\theta(t_0 + 3\Delta t) - \theta(t_0 + 2\Delta t) + \Delta t^2 \alpha(t_0 + 3\Delta t) + \mathcal{O}((\Delta t)^4) \Rightarrow \\
\delta\theta(t_0 + 4\Delta t) &= 2\delta\theta(t_0 + 3\Delta t) - \delta\theta(t_0 + 2\Delta t) + \mathcal{O}((\Delta t)^4) \\
&= 12\mathcal{O}((\Delta t)^4) - 3\mathcal{O}((\Delta t)^4) + \mathcal{O}((\Delta t)^4) \\
&= 10\mathcal{O}((\Delta t)^4).
\end{aligned}$$

Then, inductively, if $\delta\theta(t_0 + (n-1)\Delta t) = \frac{(n-1)n}{2}\mathcal{O}((\Delta t)^4)$, we obtain

$$\begin{aligned}
\theta(t_0 + n\Delta t) &= 2\theta(t_0 + (n-1)\Delta t) - \theta(t_0 + (n-2)\Delta t) + \Delta t^2 \alpha(t_0 + (n-1)\Delta t) \\
&\quad + \mathcal{O}((\Delta t)^4) \Rightarrow \\
\delta\theta(t_0 + n\Delta t) &= 2\delta\theta(t_0 + (n-1)\Delta t) - \delta\theta(t_0 + (n-2)\Delta t) + \mathcal{O}((\Delta t)^4) \\
&= 2\frac{(n-1)n}{2}\mathcal{O}((\Delta t)^4) - \frac{(n-2)(n-1)}{2}\mathcal{O}((\Delta t)^4) + \mathcal{O}((\Delta t)^4) \\
&= \frac{n(n+1)}{2}\mathcal{O}((\Delta t)^4).
\end{aligned}$$

---

[14] Remember that the acceleration $\alpha(t)$ is given, therefore $\delta\alpha(t) = 0$.

Finally

$$\delta\theta(t_0 + n\Delta t) = \frac{n(n+1)}{2}\mathcal{O}((\Delta t)^4) \sim \frac{1}{\Delta t^2}\mathcal{O}((\Delta t)^4) \sim \mathcal{O}((\Delta t)^2). \tag{4.38}$$

Therefore the total error is $\mathcal{O}((\Delta t)^2)$.

We also mention the Velocity Verlet method or the Leapfrog method. In this case we use the velocity explicitly:

$$\begin{aligned}
\theta_{n+1} &= \theta_n + \omega_n \Delta t + \frac{1}{2}\alpha_n \Delta t^2 \\
\omega_{n+\frac{1}{2}} &= \omega_n + \frac{1}{2}\alpha_n \Delta t \\
\omega_{n+1} &= \omega_{n+\frac{1}{2}} + \frac{1}{2}\alpha_{n+1}\Delta t.
\end{aligned} \tag{4.39}$$

The last step uses the acceleration α_{n+1} which should depend only on the position θ_{n+1} and not on the velocity.

The Verlet methods are popular in *molecular dynamics* simulations of many body systems. One of their advantages is that the constraints of the system of particles are easily encoded in the algorithm.

4.8 Appendix: 2nd order Runge–Kutta Method

In this appendix we will show how the choice of the intermediate point 2 in equation (4.17) reduces the error by a power of h. This choice is special, since by choosing another point (e.g. $t = t_n + 0.4h$) the result would have not been the same. Indeed, from the relation

$$\frac{dx}{dt} = f(t, x) \Rightarrow x_{n+1} = x_n + \int_{t_n}^{t_{n+1}} f(t, x)\, dx. \tag{4.40}$$

By Taylor expanding around the point $(t_{n+1/2}, x_{n+1/2})$ we obtain

$$f(t, x) = f(t_{n+1/2}, x_{n+1/2}) + (t - t_{n+1/2})\frac{df}{dt}(t_{n+1/2}) + \mathcal{O}(h^2). \tag{4.41}$$

4.8. APPENDIX: 2ND ORDER RUNGE–KUTTA METHOD

Therefore

$$\int_{t_n}^{t_{n+1}} f(t,x)\,dx$$

$$= f(t_{n+1/2}, x_{n+1/2})(t_{n+1} - t_n) + \frac{df}{dt}(t_{n+1/2}) \frac{(t - t_{n+1/2})^2}{2}\bigg|_{t_n}^{t_{n+1}}$$

$$+ \mathcal{O}(h^2)(t_{n+1} - t_n)$$

$$= f(t_{n+1/2}, x_{n+1/2})h + \frac{df}{dt}(t_{n+1/2}) \left\{ \frac{(t_{n+1} - t_{n+1/2})^2}{2} - \frac{(t_n - t_{n+1/2})^2}{2} \right\}$$

$$+ \mathcal{O}(h^2)h$$

$$= f(t_{n+1/2}, x_{n+1/2})h + \frac{df}{dt}(t_{n+1/2}) \left\{ \frac{h^2}{2} - \frac{(-h)^2}{2} \right\} + \mathcal{O}(h^3)$$

$$= f(t_{n+1/2}, x_{n+1/2})h + \mathcal{O}(h^3). \qquad (4.42)$$

Note that for the vanishing of the $\mathcal{O}(h)$ term it is necessary to place the intermediate point at time $t_{n+1/2}$.

This is not a unique choice. This can be most easily seen by a different analysis of the Taylor expansion. Expanding around the point (t_n, x_n) we obtain

$$\begin{aligned}
x_{n+1} &= x_n + (t_{n+1} - t_n)\frac{dx_n}{dt} + \frac{1}{2}(t_{n+1} - t_n)^2 \frac{d^2 x_n}{dt^2} + \mathcal{O}(h^3) \\
&= x_n + h f_n + \frac{h^2}{2}\frac{df_n}{dt} + \mathcal{O}(h^3) \\
&= x_n + h f_n + \frac{h^2}{2}\left(\frac{\partial f_n}{\partial t} + \frac{\partial f_n}{\partial x}\frac{dx_n}{dt}\right) + \mathcal{O}(h^3) \\
&= x_n + h f_n + \frac{h^2}{2}\left(\frac{\partial f_n}{\partial t} + \frac{\partial f_n}{\partial x}f_n\right) + \mathcal{O}(h^3), \qquad (4.43)
\end{aligned}$$

where we have set $f_n \equiv f(t_n, x_n)$, $\frac{dx_n}{dt} \equiv \frac{dx}{dt}(x_n)$ etc. We define

$$\begin{aligned}
k_1 &= f(t_n, x_n) = f_n \\
k_2 &= f(t_n + ah, x_n + bhk_1) \\
x_{n+1} &= x_n + h(c_1 k_1 + c_2 k_2). \qquad (4.44)
\end{aligned}$$

and we will determine the conditions so that the terms $\mathcal{O}(h^2)$ of the last equation in the error are identical with those of equation (4.43). By expanding k_2 we obtain

$$\begin{aligned} k_2 &= f(t_n + ah, x_n + bhk_1) \\ &= f(t_n, x_n + bhk_1) + ha\frac{\partial f}{\partial t}(t_n, x_n + bhk_1) + \mathcal{O}(h^2) \\ &= f(t_n, x_n) + hbk_1\frac{\partial f}{\partial x}(t_n, x_n) + ha\frac{\partial f}{\partial t}(t_n, x_n) + \mathcal{O}(h^2) \\ &= f_n + h\left\{a\frac{\partial f_n}{\partial t} + bk_1\frac{\partial f_n}{\partial x}\right\} + \mathcal{O}(h^2) \\ &= f_n + h\left\{a\frac{\partial f_n}{\partial t} + bf_n\frac{\partial f_n}{\partial x}\right\} + \mathcal{O}(h^2) \end{aligned} \quad (4.45)$$

Substituting in (4.44) we obtain

$$\begin{aligned} x_{n+1} &= x_n + h(c_1 k_1 + c_2 k_2) \\ &= x_n + h\left\{c_1 f_n + c_2 f_n + c_2 h\left(a\frac{\partial f_n}{\partial t} + bf_n\frac{\partial f_n}{\partial x}\right) + \mathcal{O}(h^2)\right\} \\ &= x_n + h(c_1 + c_2)f_n + \frac{h^2}{2}\left((2c_2 a)\frac{\partial f_n}{\partial t} + (2c_2 b)f_n\frac{\partial f_n}{\partial x}\right) \\ &\quad + \mathcal{O}(h^3). \end{aligned} \quad (4.46)$$

All we need is to choose

$$\begin{aligned} c_1 + c_2 &= 1 \\ 2c_2 a &= 1 \\ 2c_2 b &= 1. \end{aligned} \quad (4.47)$$

The choice $c_1 = 0$, $c_2 = 1$, $a = b = 1/2$ leads to equation (4.19). Some other choices in the bibliography are $c_2 = 1/2$ and $c_2 = 3/4$.

4.9 Problems

4.1 Prove that the total error in the Euler–Cromer method is of order Δt.

4.2 Reproduce the results in figures 4.11–4.18

4.3 Improve your programs so that there is no accumulation of roundoff error in the calculation of time when h is very small for the methods Euler, Euler-Cromer, Euler-Verlet and Runge-Kutta. Repeat the analysis of the previous problem.

4.4 Compare the results obtained from the Euler, Euler-Cromer, Euler-Verlet, Runge-Kutta methods for the following systems where the analytic solution is known:

(a) Particle falling in a constant gravitational field. Consider the case $v(0) = 0$, $m = 1$, $g = 10$.

(b) Particle falling in a constant gravitational field moving in a fluid from which exerts a force $F = -kv$ on the particle. Consider the case $v(0) = 0$, $m = 1$, $g = 10$ $k = 0.1, 1.0, 2.0$. Calculate the limiting velocity of the particle numerically and compare the value obtained to the theoretical expectation.

(c) Repeat for the case of a force of resistance of magnitude $|F| = kv^2$.

4.5 Consider the damped harmonic oscillator

$$\frac{d^2x}{dt^2} + \gamma\frac{dx}{dt} + \omega_0^2 x = 0. \qquad (4.48)$$

Take $\omega_0 = 3.145$, $\gamma = 0.5$ and calculate its mechanical energy as a function of time. Is it monotonic? Why? (show that $d(E/m)/dt = -\gamma v^2$). Repeat for $\gamma = 4, 5, 6, 7, 8$. When is the system oscillating and when it's not? Calculate numerically the critical value of γ for which the system passes from a non oscillating to an oscillating regime. Compare your results with the theoretical expectations.

4.6 Reproduce the results of figures 4.19–4.22.

4.7 Reproduce the results of figures 4.23–4.26. Calculate the phase $\delta(\omega)$ numerically and compare with equation (4.33).

4.8 Consider a simple model for a swing. Take the damped harmonic oscillator and a driving force which periodically exerts a momentary push with angular frequency ω. Define "momentary" to be an impulse given by the acceleration a_0 by an appropriately small time interval Δt. The acceleration is 0 for all other times. Calculate the amplitude $x_0(\omega)$ for $\omega_0 = 3.145$ and $\gamma = 0.5$.

4.9 Consider a "half sine" driving force on a damped harmonic oscillator
$$a(t) = \begin{cases} a_0 \cos \omega t & \cos \omega t > 0 \\ 0 & \cos \omega t \leq 0 \end{cases}$$
Study the transient behavior of the system for several initial conditions and calculate its steady state motion for $\omega_0 = 3.145$ and $\gamma = 0.5$. Calculate the amplitude $x_0(\omega)$.

4.10 Consider the driving force on a damped oscillator given by
$$a(t) = \frac{1}{\pi} + \frac{1}{2}\cos\omega + \frac{2}{3\pi}\cos 2\omega t - \frac{2}{15\pi}\cos 4\omega t$$
Study the transient behavior of the system for several initial conditions and calculate its steady state motion for $\omega_0 = 3.145$ and $\gamma = 0.5$. Calculate the amplitude $x_0(\omega)$. Compare your results with those of the previous problem and comment about.

4.11 Write a program that simulates N identical, independent harmonic oscillators. Take $N = 20$ and choose random initial conditions for each one of them. Study their trajectories in phase space and check whether they cross each other. Comment on your results.

4.12 Place the $N = 20$ harmonic oscillators of the previous problem in a small square in phase space whose center is at the origin of the axes. Consider the evolution of the system in time. Does the shape of the rectangle change in time? Does the area change in time? Explain...

4.13 Repeat the previous problem when each oscillator is damped with $\gamma = 0.5$. Take $\omega_0 = 3.145$.

4.9. PROBLEMS

4.14 Consider the forced damped oscillator with $\omega = 2$, $\omega_0 = 1.0$, $\gamma = 0.2$. Study the transient behavior of the system in the plots of $\theta(t)$, $\dot\theta(t)$ for $A = 0.1, 0.5, 0.79, 0.85, 1.03, 1.4$.

4.15 Consider the forced damped pendulum with $\omega = 2$, $\omega_0 = 1.0$, $\gamma = 0.2$ and study the phase space trajectories for $A = 0.1$, 0.19, 0.21, 0.25, 0.5, 0.71, 0.79, 0.85, 1.02, 1.031, 1.033, 1.05, 1.08, 1.1, 1.4, 1.8, 3.1, 3.5, 3.8, 4.2, 4.42, 4.44, 4.445, 4.447, 4.4488. Consider both the transient behavior and the steady state motion.

4.16 Reproduce the results in figures 4.30.

4.17 Reproduce the results in figures 4.31.

4.18 Consider the forced damped oscillator with

$$\omega_0 = 1, \quad \omega = 2, \quad \gamma = 0.2$$

After the transient behavior, the motion of the system for $A = 0.60$, $A = 0.75$ and $A = 0.85$ is periodic. Measure the period of the motion with an accuracy of three significant digits and compare it with the natural period of the pendulum and with the period of the driving force. Take as initial conditions the following pairs: $(\theta_0, \dot\theta_0) = (3.1, 0.0)$, $(2.5, 0.0)$, $(2.0, 0.0)$, $(1.0, 0.0)$, $(0.2, 0.0)$, $(0.0, 1.0)$, $(0.0, 3.0)$, $(0.0, 6.0)$. Check if the period is independent of the initial conditions.

4.19 Consider the forced damped pendulum with

$$\omega_0 = 1, \quad \omega = 2, \quad \gamma = 0.2$$

Study the motion of the pendulum when the amplitude A takes values in the interval $[0.2, 5.0]$. Consider specific discrete values of A by splitting the interval above in subintervals of width equal to $\delta A = 0.002$. For each value of A, record in a file the value of A, the angular position and the angular velocity of the pendulum when $t_k = k\pi$ with $k = k_{trans}, k_{trans} + 1, k_{trans} + 2, \ldots, k_{max}$:

$$A \quad \theta(t_k) \quad \dot\theta(t_k)$$

The choice of k_{trans} is made so that the transient behavior will be discarded and study only the steady state of the pendulum. You may take $k_{max} = 500$, $k_{trans} = 400$, $t_i = 0$, $t_f = 500\pi$, and split the intervals $[t_k, t_k + \pi]$ to 50 subintervals. Choose $\theta_0 = 3.1$, $\dot{\theta}_0 = 0$.

(a) Construct the bifurcation diagram by plotting the points $(A, \theta(t_k))$.

(b) Repeat by plotting the points $(A, \dot{\theta}(t_k))$.

(c) Check whether your results depend on the choice of θ_0, $\dot{\theta}_0$. Repeat your analysis for $\theta_0 = 0$, $\dot{\theta}_0 = 1$.

(d) Study the onset of chaos: Take $A \in [1.0000, 1.0400]$ with $\delta A = 0.0001$ and $A \in [4.4300, 4.4500]$ with $\delta A = 0.0001$ and compute with the given accuracy the value A_c where the system enters into the chaotic behavior regime.

(e) The plot the points $(\theta(t_k), \dot{\theta}(t_k))$ for $A = 1.034$, 1.040, 1.080, 1.400, 4.450, 4.600. Put 2000 points for each value of A and commend on the strength of the chaotic behavior of the pendulum.

Chapter 5

Planar Motion

In this chapter we will study the motion of a particle moving on the plane under the influence of a dynamical field. Special emphasis will be given to the study of the motion in a central field, like in the problem of planetary motion and scattering. We also study the motion of two or more interacting particles moving on the plane, which requires the solution of a larger number of dynamical equations. These problems can be solved numerically by using Runge–Kutta integration methods, therefore this chapter extends and applies the numerical methods studied in the previous chapter.

5.1 Runge–Kutta for Planar Motion

In two dimensions, the initial value problem that we are interested in, is solving the system of equations (4.6)

$$\frac{dx}{dt} = v_x \qquad \frac{dv_x}{dt} = a_x(t, x, v_x, y, v_y)$$
$$\frac{dy}{dt} = v_y \qquad \frac{dv_y}{dt} = a_y(t, x, v_x, y, v_y). \qquad (5.1)$$

The 4th order Runge-Kutta method can be programmed by making small modifications of the program in the file rk.cpp. In order to facilitate the study of many different dynamical fields, for each field we put the code of the respective acceleration in a different file. The code which is common for all the forces, namely the user interface and the implementation of the Runge–Kutta method, will be put in the file rk2.cpp.

The program that computes the acceleration will be put in a file named rk_XXX.cpp, where XXX is a string of characters that identifies the force. For example, the file rk2_hoc.cpp contains the program computing the acceleration of the simple harmonic oscillator, the file rk2_g.cpp the acceleration of a constant gravitational field $\vec{g} = -g\hat{y}$ etc.

Different force fields will require the use of one or more coupling constants which need to be accessible to the code in the main program and some subroutines. For this reason, we will provide two variables k1, k2 in the global scope which will be accessed by the acceleration functions f3 and f4, the function energy and the main program where the user will enter their. The initial conditions are stored in the variables X10 $\leftrightarrow x_0$, X20 $\leftrightarrow y_0$, V10 $\leftrightarrow v_{x0}$, V20 $\leftrightarrow v_{y0}$, and the values of the functions of time will be stored in the arrays X1[P] $\leftrightarrow x(t)$, X2[P] $\leftrightarrow y(t)$, V1[P] $\leftrightarrow v_x(t)$, V2[P] $\leftrightarrow v_y(t)$. The integration is performed by a call to the function RK(Ti,Tf,X10,X20,V10,V20,Nt) The results are written to the file rk2.dat. Each line in this file contains the time, position, velocity and the total mechanical energy, where the energy is calculated by the function energy(t,x1,x2,v1,v2). The code for the function energy, which is different for each force field, is written in the same file with the acceleration. The code for the function RKSTEP(t,x1,x2,x3,x4,dt) should be extended in order to integrate four instead of two functions. The full code is listed below:

```
//========================================================
//Program to solve a 4 ODE system using Runge-Kutta Method
//User must supply derivatives
//dx1/dt=f1(t,x1,x2,x3,x4) dx2/dt=f2(t,x1,x2,x3,x4)
//dx3/dt=f3(t,x1,x2,x3,x4) dx4/dt=f4(t,x1,x2,x3,x4)
//as double functions
//Output is written in file rk2.dat
//========================================================
#include <iostream>
#include <fstream>
#include <cstdlib>
#include <string>
#include <cmath>
using namespace std;
//--------------------------------------------------------
const int P = 1010000;
double T[P], X1[P], X2[P], V1[P], V2[P];
```

5.1. RUNGE–KUTTA FOR PLANAR MOTION

```
double k1,k2;
//----------------------------------------
double
f1(const double& t  , const double& x1, const double& x2,
   const double& v1 , const double& v2);
double
f2(const double& t  , const double& x1, const double& x2,
   const double& v1 , const double& v2);
double
f3(const double& t  , const double& x1, const double& x2,
   const double& v1 , const double& v2);
double
f4(const double& t  , const double& x1, const double& x2,
   const double& v1 , const double& v2);
double
energy
  (const double& t  , const double& x1, const double& x2,
   const double& v1 , const double& v2);
void
RK(const double& Ti , const double& Tf ,
   const double& X10, const double& X20,
   const double& V10, const double& V20,
   const int    & Nt);
void
RKSTEP(double& t ,
       double& x1, double& x2,
       double& x3, double& x4,
       const    double& dt);
//----------------------------------------
int main(){
  string buf;
  double Ti,Tf,X10,X20,V10,V20;
  int    Nt,i ;
  double E0,EF,DE;
  //Input:
  cout << "Runge-Kutta Method for 4-ODEs Integration\n";
  cout << "Enter coupling constants:\n";
  cin  >> k1 >> k2;getline(cin,buf);
  cout << "k1= " << k1 << " k2= " << k2 << endl;
  cout << "Enter Nt,Ti,Tf,X10,X20,V10,V20:\n";
  cin  >> Nt >> Ti >> Tf>> X10 >> X20 >> V10 >> V20;
  getline(cin,buf);
  cout << "Nt = " << Nt << endl;
  cout << "Time: Initial Ti = " << Ti
       << " Final Tf= "         << Tf << endl;
```

```cpp
  cout <<  "               X1(Ti)= " << X10
       << " X2(Ti)="              << X20 << endl;
  cout <<  "               V1(Ti)= " << V10
       << " V2(Ti)="              << V20 << endl;
  //Calculate:
  RK(Ti,Tf,X10,X20,V10,V20,Nt);
  ofstream myfile("rk2.dat");
  myfile.precision(17);
  for(i=0;i<Nt;i++)
    myfile << T [i] << " "
           << X1[i] << " " << X2[i] << " "
           << V1[i] << " " << V2[i] << " "
           << energy(T[i],X1[i],X2[i],V1[i],V2[i])
           << endl;
  myfile.close();
  //Rutherford scattering angles:
  cout.precision(17);
  cout <<"v-angle: "<< atan2(V2[Nt-1],V1[Nt-1])  << endl;
  cout <<"b-angle: "<< 2.0*atan(k1/(V10*V10*X20))<< endl;
  E0=energy(Ti     ,X10     ,X20     ,V10     ,V20     );
  EF=energy(T[Nt-1],X1[Nt-1],X2[Nt-1],V1[Nt-1],V2[Nt-1]);
  DE = abs(0.5*(EF-E0)/(EF+E0));
  cout << "E0,EF, DE/E= " << E0
       << " "             << EF
       << " "             << DE << endl;
}//main()
//========================================================
//The velocity functions f1,f2(t,x1,x2,v1,v2)
//========================================================
double
f1(const double& t , const double& x1, const double& x2,
   const double& v1, const double& v2){
  return v1;
}
//--------------------------------------------------------
double
f2(const double& t , const double& x1, const double& x2,
   const double& v1, const double& v2){
  return v2;
}
//========================================================
//RK(Ti,Tf,X10,X20,V10,V20,Nt) is the driver
//for the Runge-Kutta integration routine RKSTEP
//Input: Initial and final times Ti,Tf
//       Initial values at t=Ti   X10,X20,V10,V20
```

5.1. RUNGE–KUTTA FOR PLANAR MOTION

```
//          Number of steps of integration: Nt-1
//          Size of arrays T,X1,X2,V1,V2
//Output: real arrays T[Nt],X1[Nt],X2[Nt],
//                    V1[Nt],V2[Nt] where
//T[0] = Ti  X1[0] = X10  X2[0] = X20  V1[0] = V10  V2[0] = V20
//          X1[k] = X1(at t=T[k])  X2[k] = X2(at t=T[k])
//          V1[k] = V1(at t=T[k])  V2[k] = V2(at t=T[k])
//T[Nt-1]= Tf
//========================================================
void
RK(const double& Ti , const double& Tf ,
   const double& X10, const double& X20,
   const double& V10, const double& V20,
   const int    & Nt){

  double dt;
  double TS,X1S,X2S; //values of time and X1,X2 at given step
  double     V1S,V2S;
  int i;
  //Initialize:
  dt    = (Tf-Ti)/(Nt-1);
  T [0] = Ti;
  X1[0] = X10; X2[0] = X20;
  V1[0] = V10; V2[0] = V20;
  TS    = Ti;
  X1S   = X10; X2S   = X20;
  V1S   = V10; V2S   = V20;
  //Make RK steps: The arguments of RKSTEP are
  //replaced with the new ones
  for(i=1;i<Nt;i++){
    RKSTEP(TS,X1S,X2S,V1S,V2S,dt);
    T [i] = TS;
    X1[i] = X1S; X2[i] = X2S;
    V1[i] = V1S; V2[i] = V2S;
  }
}//RK()
//========================================================
//Subroutine RKSTEP(t,x1,x2,dt)
//Runge-Kutta Integration routine of ODE
//dx1/dt=f1(t,x1,x2,x3,x4) dx2/dt=f2(t,x1,x2,x3,x4)
//dx3/dt=f3(t,x1,x2,x3,x4) dx4/dt=f4(t,x1,x2,x3,x4)
//User must supply derivative functions:
//real function f1(t,x1,x2,x3,x4)
//real function f2(t,x1,x2,x3,x4)
//real function f3(t,x1,x2,x3,x4)
```

```cpp
//real function f4(t,x1,x2,x3,x4)
//Given initial point (t,x1,x2) the routine advances it
//by time dt.
//Input : Inital time t    and function values x1,x2,x3,x4
//Output: Final  time t+dt and function values x1,x2,x3,x4
//Careful: values of t,x1,x2,x3,x4 are overwritten...
//========================================================
void
RKSTEP(double& t ,
       double& x1, double& x2,
       double& x3, double& x4,
       const    double& dt){
  double k11,k12,k13,k14,k21,k22,k23,k24;
  double k31,k32,k33,k34,k41,k42,k43,k44;
  double h,h2,h6;

  h =dt;       // h  = dt, integration step
  h2=0.5*h;    // h2 = h/2
  h6=h/6.0;    // h6 = h/6

  k11=f1(t,x1,x2,x3,x4);
  k21=f2(t,x1,x2,x3,x4);
  k31=f3(t,x1,x2,x3,x4);
  k41=f4(t,x1,x2,x3,x4);

  k12=f1(t+h2,x1+h2*k11,x2+h2*k21,x3+h2*k31,x4+h2*k41);
  k22=f2(t+h2,x1+h2*k11,x2+h2*k21,x3+h2*k31,x4+h2*k41);
  k32=f3(t+h2,x1+h2*k11,x2+h2*k21,x3+h2*k31,x4+h2*k41);
  k42=f4(t+h2,x1+h2*k11,x2+h2*k21,x3+h2*k31,x4+h2*k41);

  k13=f1(t+h2,x1+h2*k12,x2+h2*k22,x3+h2*k32,x4+h2*k42);
  k23=f2(t+h2,x1+h2*k12,x2+h2*k22,x3+h2*k32,x4+h2*k42);
  k33=f3(t+h2,x1+h2*k12,x2+h2*k22,x3+h2*k32,x4+h2*k42);
  k43=f4(t+h2,x1+h2*k12,x2+h2*k22,x3+h2*k32,x4+h2*k42);

  k14=f1(t+h ,x1+h *k13,x2+h *k23,x3+h *k33,x4+h *k43);
  k24=f2(t+h ,x1+h *k13,x2+h *k23,x3+h *k33,x4+h *k43);
  k34=f3(t+h ,x1+h *k13,x2+h *k23,x3+h *k33,x4+h *k43);
  k44=f4(t+h ,x1+h *k13,x2+h *k23,x3+h *k33,x4+h *k43);

  t =t+h;
  x1=x1+h6*(k11+2.0*(k12+k13)+k14);
  x2=x2+h6*(k21+2.0*(k22+k23)+k24);
  x3=x3+h6*(k31+2.0*(k32+k33)+k34);
  x4=x4+h6*(k41+2.0*(k42+k43)+k44);
```

```
} //RKSTEP()
```

5.2 Projectile Motion

Consider a particle in the constant gravitational field near the surface of the earth which moves with constant acceleration $\vec{g} = -g\hat{y}$ so that

$$\begin{aligned} x(t) &= x_0 + v_{0x}t &, & y(t) &= y_0 + v_{0y}t - \tfrac{1}{2}gt^2 \\ v_x(t) &= v_{0x} &, & v_y(t) &= v_{0y} - gt \\ a_x(t) &= 0 &, & a_y(t) &= -g \end{aligned} \quad (5.2)$$

The particle moves on a parabolic trajectory that depends on the initial conditions

$$\begin{aligned} (y - y_0) &= \left(\frac{v_{0y}}{v_{0x}}\right)(x - x_0) - \frac{1}{2}\frac{g}{v_{0x}^2}(x - x_0)^2 \\ &= \tan\theta \, (x - x_0) - \frac{\tan^2\theta}{4h_{\max}}(x - x_0)^2, \end{aligned} \quad (5.3)$$

where $\tan\theta = v_{0y}/v_{0x}$ is the direction of the initial velocity and $h_{\max}$ is the maximum height of the trajectory.

The acceleration $a_x(t) = 0$ $a_y(t) = -g$ ($a_x \leftrightarrow$ f3, $a_y \leftrightarrow$ f4) and the mechanical energy is coded in the file rk2_g.cpp:

```
//==========================================================
//The acceleration functions f3,f4(t,x1,x2,v1,v2) provided
//by the user
//==========================================================
#include <iostream>
#include <fstream>
#include <cstdlib>
#include <string>
#include <cmath>
using namespace std;
extern double k1,k2;
//----------------------------------------------------------
//Free fall in constant gravitational field with
//g = -k2
double
```

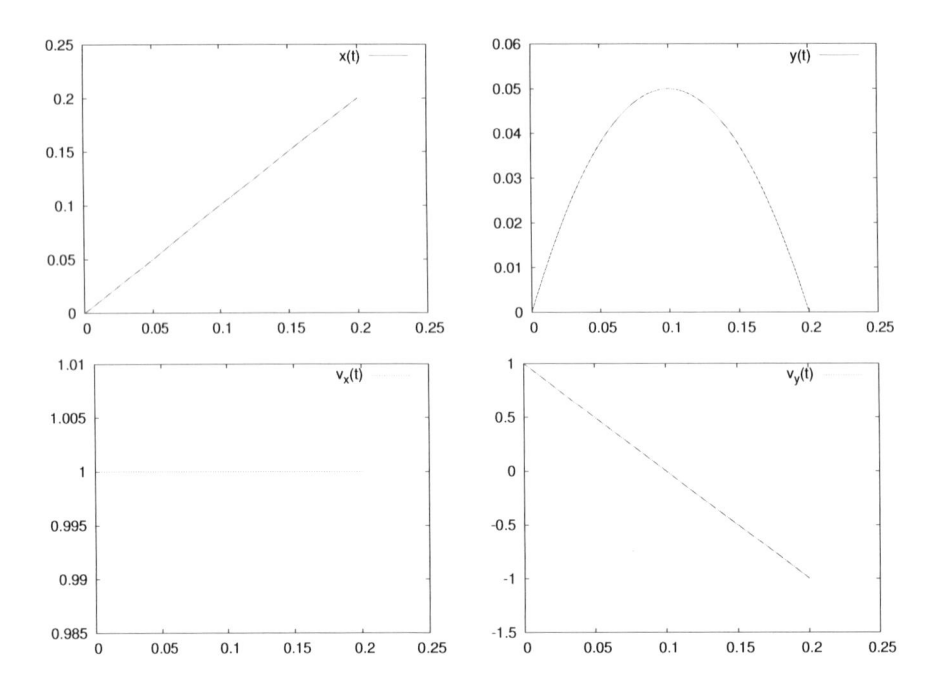

Figure 5.1: Plots of $x(t)$, $y(t)$, $v_x(t)$, $v_y(t)$ for a projectile fired in a constant gravitational field $\vec{g} = -10.0\,\hat{y}$ with initial velocity $\vec{v}_0 = \hat{x} + \hat{y}$.

```
f3(const double& t , const double& x1, const double& x2,
   const double& v1, const double& v2){
  return 0.0;    // dx3/dt=dv1/dt=a1
}
//--------------------------------------------------------
double
f4(const double& t , const double& x1, const double& x2,
   const double& v1, const double& v2){
  return -k1;    // dx4/dt=dv2/dt=a2
}
//--------------------------------------------------------
double
energy
  (const double& t , const double& x1, const double& x2,
   const double& v1, const double& v2){
  return 0.5*(v1*v1+v2*v2) + k1*x2;
}
```

In order to calculate a projectile's trajectory you may use the following

5.2. PROJECTILE MOTION

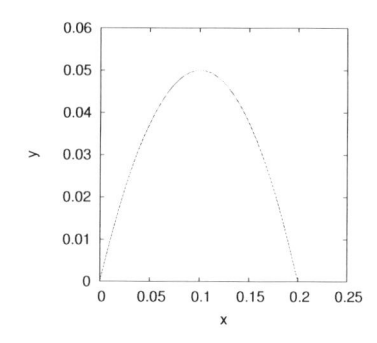

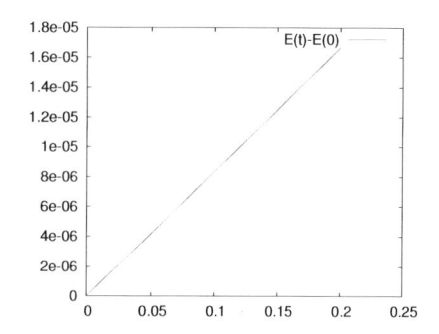

Figure 5.2: (Left) The parabolic trajectory of a projectile fired in a constant gravitational field $\vec{g} = -10.0\,\hat{y}$ with initial velocity $\vec{v}_0 = \hat{x} + \hat{y}$. (Right) The deviation of the projectile's energy from its initial value is due to numerical errors.

commands:

```
> g++ -O2 rk2.cpp rk2_g.cpp -o rk2
> ./rk2
Runge-Kutta Method for 4-ODEs Integration
Enter coupling constants:
10.0 0.0
k1= 10 k2= 0
Enter Nt,Ti,Tf,X10,X20,V10,V20:
20000 0.0 0.2 0.0 0.0 1.0 1.0
Nt = 20000
Time: Initial Ti = 0 Final Tf= 0.2
        X1(Ti)= 0 X2(Ti)=0
        V1(Ti)= 1 V2(Ti)=1
```

The analysis of the results contained in the file rk2.dat can be done using gnuplot:

```
gnuplot> set terminal x11 1
gnuplot> plot "rk2.dat" using 1:2 with lines title "x(t)"
gnuplot> set terminal x11 2
gnuplot> plot "rk2.dat" using 1:3 with lines title "y(t)"
gnuplot> set terminal x11 3
gnuplot> plot "rk2.dat" using 1:4 with lines title "vx(t)"
gnuplot> set terminal x11 4
gnuplot> plot "rk2.dat" using 1:5 with lines title "vy(t)"
gnuplot> set terminal x11 5
```

```
gnuplot> plot "rk2.dat" using 1:($6-1.0) w lines t "E(t)E-(0)"
gnuplot> set terminal x11 6
gnuplot> set size square
gnuplot> set title "Trajectory"
gnuplot> plot "rk2.dat" using 2:3 with lines notit
```

The results can be seen in figures 5.1 and 5.2. We note a small increase in the mechanical energy which is due to the accumulation of numerical errors.

We can animate the trajectory by writing a script of gnuplot commands in a file rk2_animate.gpl

```
icount = icount+skip
plot   "<cat -n rk2.dat" \
  using 3:($1<= icount ? $4: 1/0) with lines notitle
# pause 1
if(icount < nlines ) reread
```

Before calling the script, the user must set the values of the variables icount, skip and nlines. Each time gnuplot reads the script, it plots icount number of lines from rk2.dat. Then the script is read again and a new plot is made with skip lines more than the previous one, unless icount < nlines. The plotted "file" "<cat -n rk2.dat" is the standard output (stdout) of the command cat -n rk2.dat which prints to the stdout the contents of the file rk2.dat line by line, together with the line number. Therefore the plot command reads data which are the line number, the time, the coordinate x, the coordinate y etc. The keyword using in

```
using 3:($1<= icount ? $4: 1/0)
```

instructs the plot command to use the 3rd column on the horizontal axis and if the first column is less than icount ($1<= icount) put on the vertical axis the value of the 4th column if the first column is less than icount. Otherwise ($1 > icount) it prints an undefined number (1/0) which makes gnuplot print nothing at all. You may also uncomment the command pause if you want to make the animation slower. In order to run the script from gnuplot, issue the commands

5.2. PROJECTILE MOTION

```
gnuplot> icount = 10
gnuplot> skip   = 200
gnuplot> nlines = 20000
gnuplot> load "rk2_animate.gpl"
```

The scripts shown above can be found in the accompanying software. More scripts can be found there that automate many of the boring procedures. The usage of two of these is explained below. The first one is in the file rk2_animate.csh:

```
> ./rk2_animate.csh -h
Usage: rk2_animate.csh -t [sleep time] -d [skip points] <file>
Default file is rk2.dat
Other options:
  -x: set lower value in xrange
  -X: set lower value in xrange
  -y: set lower value in yrange
  -Y: set lower value in yrange
  -r: automatic determination of x-y range
> ./rk2_animate.csh -r -d 500 rk2.dat
```

The last line is a command that animates a trajectory read from the file rk2.dat. Each animation frame contains 500 more points than the previous one. The option -r calculates the plot range automatically. The option -h prints a short help message.

A more useful script is in the file rk2.csh.

```
> ./rk2.csh -h
Usage: rk2.csh -f <force> k1 k2 x10 x20 v10 v20 STEPS t0 tf
Other Options:
  -n Do not animate trajectory
Available forces (value of <force>):
1: ax=-k1           ay= -k2 y           Harmonic oscillator
2: ax= 0            ay= -k1             Free fall
3: ax= -k2     vx   ay= -k2      vy - k1  Free fall + \
                                          air resistance ~ v
4: ax= -k2 |v| vx   ay= -k2 |v|vy - k1   Free fall + \
                                          air resistance ~ v^2
5: ax= k1*x1/r^3    ay= k1*x2/r^3        Coulomb Force
....
```

The option -h prints operating instructions. A menu of forces is available, and a choice can be made using the option -f. The rest of the command line consists of the parameters read by the program in rk2.cpp, i.e. the coupling constants k1, k2, the initial conditions x10, x20, v10, v20 and the integration parameters STEPS, t0 and tf. For example, the commands

```
> rk2.csh -f 2 -- 10.0 0.0 0.0 0.0 1.0 1.0 20000 0.0 0.2
> rk2.csh -f 1 -- 16.0 1.0 0.0 1.0 1.0 0.0 20000 0.0 6.29
> rk2.csh -f 5 -- 10.0 0.0 -10 0.2 10. 0.0 20000 0.0 3.00
```

compute the trajectory of a particle in the constant gravitational field discussed above, the trajectory of an anisotropic harmonic oscillator (k1 = a_x = $-\omega_1^2 x$, k2 = a_y = $-\omega_2^2 y$) and the scattering of a particle in a Coulomb field – try them! I hope that you will have enough curiosity to look "under the hood" of the scripts and try to modify them or create new ones. Some advise to the lazy guys: If you need to program your own force field follow the recipe: Write the code of your acceleration field in a file named e.g. rk2_myforce.cpp as we did with rk2_g.cpp. Edit the file rk2.csh and modify the line

```
set forcecode = (hoc g vg v2g cb)
```

to

```
set forcecode = (hoc g vg v2g cb myforce)
```

(the variable $forcecode may have more entries than the ones shown above). Count the order of the string myforce, which is 6 in our case. In order to access this force field from the command line, use the option -f 6:

```
> rk2.csh -f 6 -- .......
```

Now, we will study the effect of the air resistance on the motion of the projectile. For small velocities this is a force proportional to the velocity $\vec{F}_r = -mk\vec{v}$, therefore

$$\begin{aligned} a_x &= -kv_x \\ a_y &= -kv_y - g\,. \end{aligned} \quad (5.4)$$

5.2. PROJECTILE MOTION

By taking

$$\begin{aligned}
x(t) &= x_0 + \frac{v_{0x}}{k}\left(1 - e^{-kt}\right) \\
y(t) &= y_0 + \frac{1}{k}\left(v_{0y} + \frac{g}{k}\right)\left(1 - e^{-kt}\right) - \frac{g}{k}t \\
v_x(t) &= v_{0x}e^{-kt} \\
v_y(t) &= \left(v_{0y} + \frac{g}{k}\right)e^{-kt} - \frac{g}{k},
\end{aligned} \tag{5.5}$$

we obtain the motion of a particle with terminal velocity $v_y(+\infty) = -g/k$ ($x(+\infty) = $ const., $y(+\infty) \sim t$).

The acceleration caused by the air resistance is programmed in the file ($\text{k1} \leftrightarrow g$, $\text{k2} \leftrightarrow k$) rk2_vg.cpp:

```cpp
//==========================================================
//The acceleration functions f3,f4(t,x1,x2,v1,v2) provided
//by the user
//==========================================================
//Free fall in constant gravitational filled with
// ax = -k2 vx     ay = -k2 vy - k1
#include <iostream>
#include <fstream>
#include <cstdlib>
#include <string>
#include <cmath>
using namespace std;
extern double k1,k2;
//--------------------------------------------------------
double
f3(const double& t , const double& x1, const double& x2,
   const double& v1, const double& v2){
  return -k2*v1;    // dx3/dt=dv1/dt=a1
}
//--------------------------------------------------------
double
f4(const double& t , const double& x1, const double& x2,
   const double& v1, const double& v2){
  return -k2*v2-k1; // dx4/dt=dv2/dt=a2
}
//--------------------------------------------------------
double
energy
```

```
  (const double& t , const double& x1, const double& x2,
    const double& v1, const double& v2){
  return 0.5*(v1*v1+v2*v2) + k1*x2;
}
```

The results are shown in figure 5.3 where we see the effect of an increasing air resistance on the particle trajectory. The effect of a resistance force of the form $\vec{F}_r = -mkv^2\hat{v}$ is shown in figure 5.4.

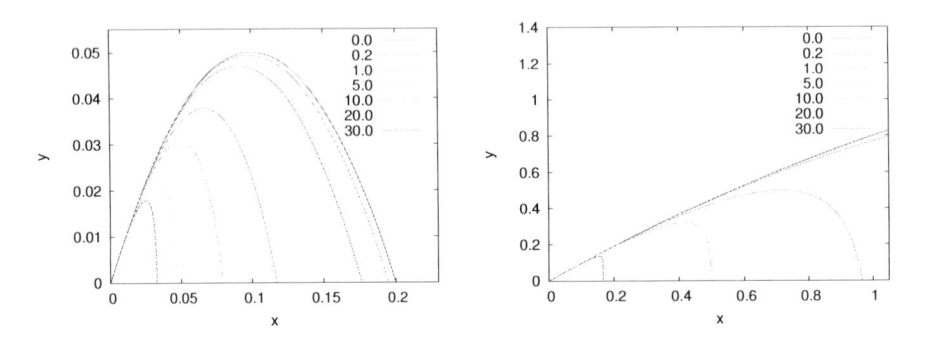

Figure 5.3: The trajectory of a projectile moving in a constant gravitational field $\vec{g} = -10\,\hat{y}$ with air resistance causing acceleration $\vec{a}_r = -k\vec{v}$ for $k = 0, 0.2, 1, 5, 10, 20, 30$. The left plot has $\vec{v}(0) = \hat{x} + \hat{y}$ and the right plot has $\vec{v}(0) = 5\hat{x} + 5\hat{y}$.

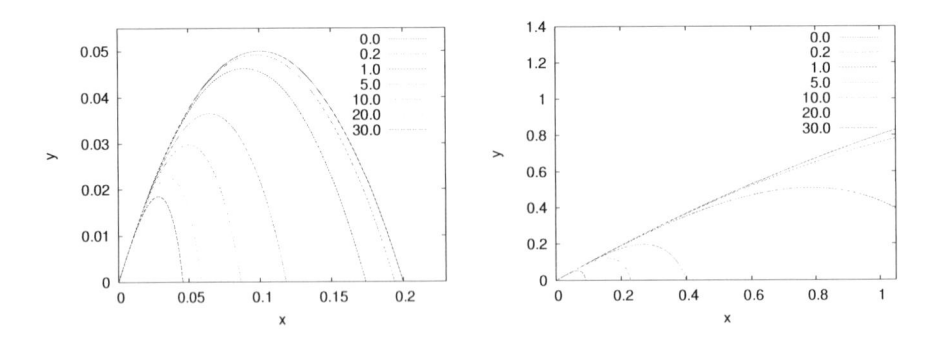

Figure 5.4: The trajectory of a projectile moving in a constant gravitational field $\vec{g} = -10\,\hat{y}$ with air resistance causing acceleration $\vec{a}_r = -kv^2\hat{v}$ for $k = 0, 0.2, 1, 5, 10, 20, 30$. The left plot has $\vec{v}(0) = \hat{x} + \hat{y}$ and the right plot has $\vec{v}(0) = 5\hat{x} + 5\hat{y}$.

5.3 Planetary Motion

Consider the simple planetary model of a "sun" of mass M and a planet "earth" at distance r from the sun and mass m such that $m \ll M$. According to Newton's law of gravity, the earth's acceleration is

$$\vec{a} = \vec{g} = -\frac{GM}{r^2}\hat{r} = -\frac{GM}{r^3}\vec{r}, \tag{5.6}$$

where $G = 6.67 \times 10^{-11} \frac{\text{m}^3}{\text{kgr} \cdot \text{sec}^2}$, $M = 1.99 \times 10^{30}\text{kgr}$, $m = 5.99 \times 10^{24}\text{kgr}$. When the hypothesis $m \ll M$ is not valid, the two body problem is reduced to that of the one body problem with the mass replaced by the reduced mass μ

$$\frac{1}{\mu} = \frac{1}{m} + \frac{1}{M}.$$

The force of gravity is a central force. This implies conservation of the angular momentum $\vec{L} = \vec{r} \times \vec{p}$ with respect to the center of the force, which in turn implies that the motion is confined on one plane. We choose the z axis so that

$$\vec{L} = L_z\hat{k} = m(xv_y - yv_x)\hat{k}. \tag{5.7}$$

The force of gravity is conservative and the mechanical energy

$$E = \frac{1}{2}mv^2 - \frac{GmM}{r} \tag{5.8}$$

is conserved. If we choose the origin of the coordinate axes to be the center of the force, the equations of motion (5.6) become

$$\begin{aligned} a_x &= -\frac{GM}{r^3}x \\ a_y &= -\frac{GM}{r^3}y, \end{aligned} \tag{5.9}$$

where $r^2 = x^2 + y^2$. This is a system of two coupled differential equations for the functions $x(t)$, $y(t)$. The trajectories are conic sections which are either an ellipse (bound states - "planet"), a parabola (e.g. escape to infinity when the particle starts moving with speed equal to the escape velocity) or a hyperbola (e.g. scattering).

Kepler's third law of planetary motion states that the orbital period T of a planet satisfies the equation

$$T^2 = \frac{4\pi^2}{GM} a^3, \qquad (5.10)$$

where a is the semi-major axis of the elliptical trajectory. The eccentricity is a measure of the deviation of the trajectory from being circular

$$e = \sqrt{1 - \frac{b^2}{a^2}}, \qquad (5.11)$$

where b is the semi-minor axis. The eccentricity is 0 for the circle and tends to 1 as the ellipse becomes more and more elongated. The foci F_1 and F_2 are located at a distance ea from the center of the ellipse. They have the property that for every point on the ellipse

$$PF_1 + PF_2 = 2a. \qquad (5.12)$$

The acceleration given to the particle by Newton's force of gravity is programmed in the file rk2_cb.cpp:

```
//================================================
//The acceleration functions f3,f4(t,x1,x2,v1,v2) provided
//by the user
//================================================
#include <iostream>
#include <fstream>
#include <cstdlib>
#include <string>
#include <cmath>
using namespace std;
extern double k1,k2;
//------------------------------------------------
//Motion in Coulombic potential:
//ax= k1*x1/r^3 ay= k1*x2/r^3
double
f3(const double& t , const double& x1, const double& x2,
   const double& v1, const double& v2){
  double r2,r3;
  r2=x1*x1+x2*x2;
  r3=r2*sqrt(r2);
  if(r3>0.0)
```

5.3. PLANETARY MOTION

```
      return k1*x1/r3;  // dx3/dt=dv1/dt=a1
    else
      return 0.0;
}
//----------------------------------------------------------------
double
f4(const double& t , const double& x1, const double& x2,
    const double& v1, const double& v2){
  double r2,r3;
  r2=x1*x1+x2*x2;
  r3=r2*sqrt(r2);
  if(r3>0.0)
    return k1*x2/r3;  // dx4/dt=dv4/dt=a4
  else
    return 0.0;
}
//----------------------------------------------------------------
double
energy
  (const double& t , const double& x1, const double& x2,
    const double& v1, const double& v2){
  double r;
  r=sqrt(x1*x1+x2*x2);
  if( r > 0.0)
    return 0.5*(v1*v1+v2*v2) + k1/r;
  else
    return 0.0;
}
```

We set k1= $-GM$ and take special care to avoid hitting the center of the force, the singular point at $(0,0)$. The same code can be used for the electrostatic Coulomb field with k1= $qQ/4\pi\epsilon_0 m$.

At first we study trajectories which are bounded. We set $GM = 10$, $x(0) = 1.0$, $y(0) = 0$, $v_{0x} = 0$ and vary v_{0y}. We measure the period T and the length of the semi axes of the resulting ellipse. The results can be found in table 5.1. Some of the trajectories are shown in figure 5.5. There we can see the dependence of the size of the ellipse on the period. Figure 5.6 confirms Kepler's third law of planetary motion given by equation (5.10).

In order to confirm Kepler's third law of planetary motion numeri-

v_{0x}	$T/2$	$2a$
3.2	1.030	2.049
3.4	1.281	2.370
3.6	1.682	2.841
3.8	2.396	3.597
4.0	3.927	5.000
4.1	5.514	6.270
4.2	8.665	8.475
4.3	16.931	13.245
4.3	28.088	18.561
4.38	42.652	24.522
4.40	61.359	31.250
4.42	99.526	43.141

Table 5.1: The results for the period T and the length of the semi-major axis a of the trajectory of planetary motion for $GM = 10$, $x(0) = 1.0$, $y(0) = 0$, $v_{0y} = 0$.

cally, we take the logarithm of both sides of equation (5.10)

$$\ln T = \frac{3}{2} \ln a + \frac{1}{2} \ln \left(\frac{4\pi^2}{GM} \right). \tag{5.13}$$

Therefore, the points $(\ln a, \ln T)$ lie on a straight line. Using a linear least squares fit we calculate the slope and the intercept which should be equal to $\frac{3}{2}$ and $1/2 \ln (4\pi^2/GM)$ respectively. This is left as an exercise.

In the case where the initial velocity of the particle becomes larger than the escape velocity v_e, the particle escapes from the influence of the gravitational field to infinity. The escape velocity corresponds to zero mechanical energy, which gives

$$v_e^2 = \frac{2GM}{r}. \tag{5.14}$$

When $GM = 10$, $x(0) = 1.0$, $y(0) = 0$, we obtain $v_e \approx 4.4721\ldots$. The numerical calculation of v_e is left as an exercise.

5.4. SCATTERING

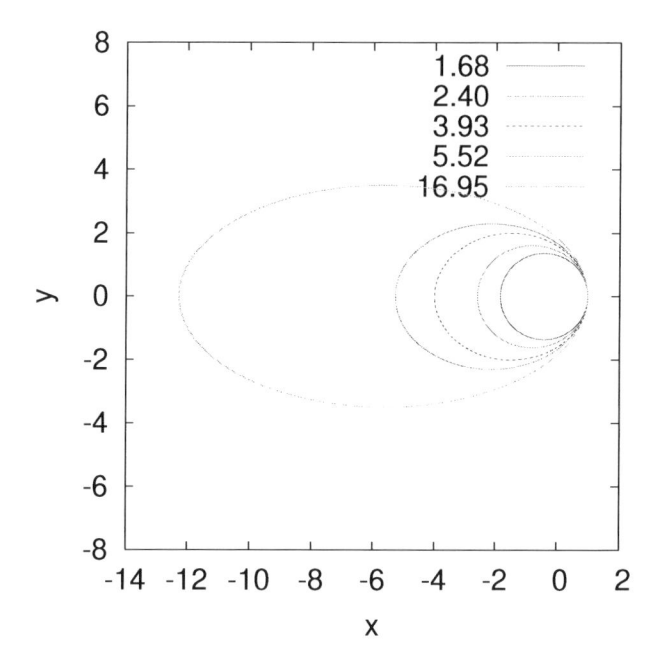

Figure 5.5: Planetary trajectories for $GM = 10$, $x(0) = 1.0$, $y(0) = 0$, $v_{0y} = 0$ and $v_{0x} = 3.6, 3.8, 4.0, 4.1, 4.3$. The numbers are the corresponding half periods.

5.4 Scattering

In this section we consider scattering of particles from a central potential[1]. We assume particles that follow unbounded trajectories that start from infinity and move almost free from the influence of the force field towards its center. When they approach the region of interaction they get deflected and get off to infinity in a new direction. We say that the particles have been scattered and that the angle between their original and final direction is the scattering angle θ. Scattering problems are interesting because we can infer to the properties of the scattering potential from the distribution of the scattering angle. This approach is heavily used in today's particle accelerators for the study of fundamental interactions between elementary particles.

First we will discuss scattering of small hard spheres of radius r_1 by

---
[1] We refer the reader to [40], chapter 4.

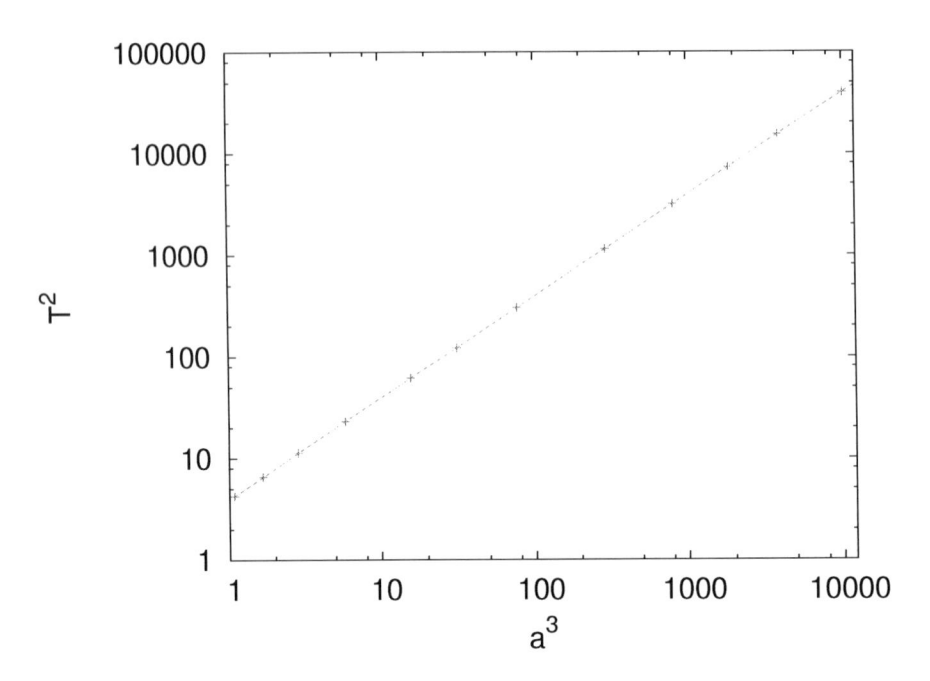

Figure 5.6: Kepler's third law of planetary motion for $GM = 10$. The points are the measurements taken from table 5.1. The solid line is the known analytic solution (5.10).

other hard spheres or radius R_2. The interaction potential[2] is given by

$$V(r) = \begin{cases} 0 & r > R_2 + r_1 \\ \infty & r < R_2 + r_1 \end{cases}, \qquad (5.15)$$

where r is the distance between the center of r_1 from the center of R_2. Assume that the particles in the beam do not interact with each other and that there is only one collision per scattering. Let J be the intensity of the beam[3] and A its cross sectional area. Assume that the target has n particles per unit area. The cross sectional area of the interaction is $\sigma = \pi(r_1 + R_2)^2$ where r_1 and R_2 are the radii of the scattered particles and targets respectively (see figure (5.8)): All the spheres of the beam which lie outside this area are not scattered by the particular target. The

---

[2] The so called hard core potential.

[3] The number of particles crossing a surface perpendicular to the beam per unit time and unit area.

5.4. SCATTERING

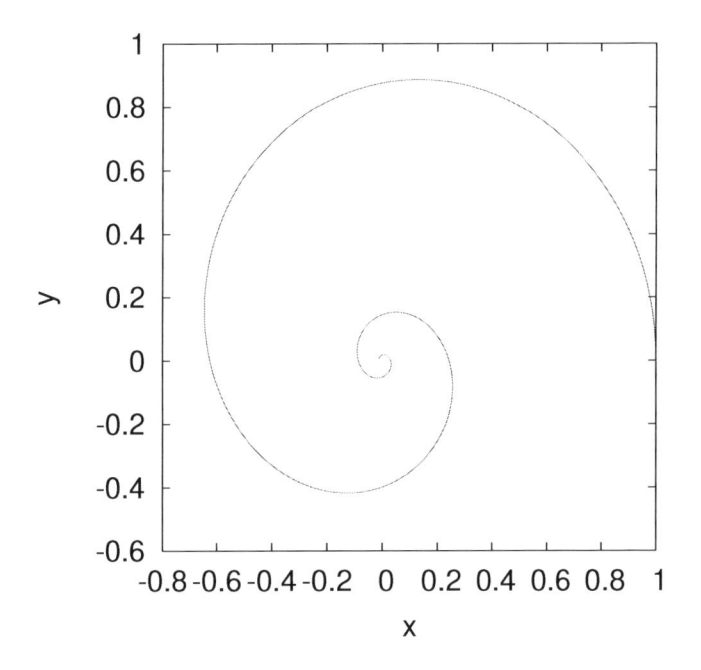

Figure 5.7: The spiral orbit of a particle moving under the influence of a central force $\vec{F} = -k/r^3 \hat{r}$.

total interaction cross section is

$$\Sigma = nA\sigma, \tag{5.16}$$

where nA is the total number of target spheres which lie within the beam. On the average, the scattering rate is

$$N = J\Sigma = JnA\sigma. \tag{5.17}$$

The above equation is the definition of the total scattering cross section σ of the interaction. The differential cross section $\sigma(\theta)$ is defined by the relation

$$dN = JnA\sigma(\theta) \, d\Omega, \tag{5.18}$$

where dN is the number of particles per unit time scattered within the solid angle $d\Omega$. The total cross section is

$$\sigma_{tot} = \int_\Omega \sigma(\theta) \, d\Omega = \int \sigma(\theta) \sin\theta \, d\theta d\phi = 2\pi \int \sigma(\theta) \sin\theta \, d\theta. \tag{5.19}$$

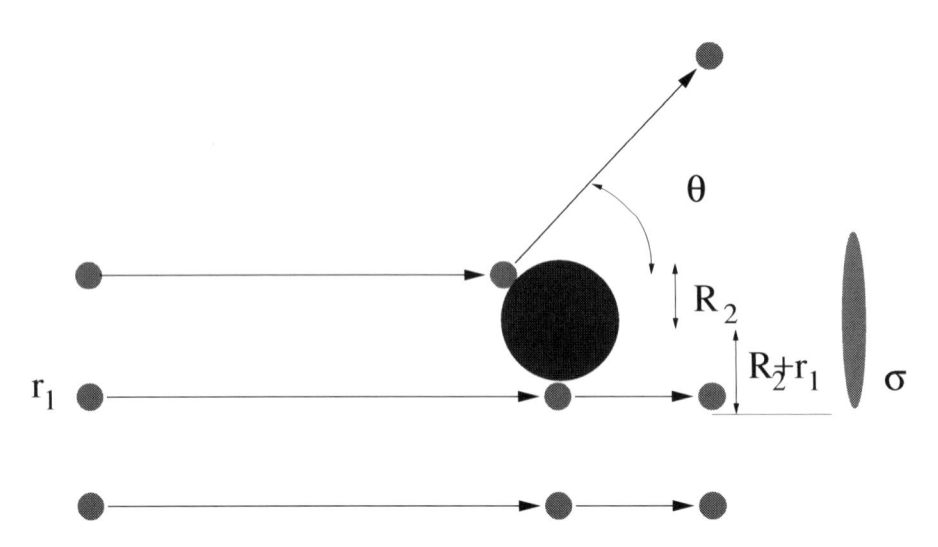

Figure 5.8: Scattering of hard spheres. The scattering angle is θ. The cross sectional area σ is shown to the right.

In the last relation we used the cylindrical symmetry of the interaction with respect to the axis of the collision. Therefore

$$\sigma(\theta) = \frac{1}{nAJ}\frac{dN}{2\pi \sin\theta\, d\theta}. \qquad (5.20)$$

This relation can be used in experiments for the measurement of the differential cross section by measuring the rate of detection of particles within the space contained in between two cones defined by the angles θ and $\theta + d\theta$. This is the relation that we will use in the numerical calculation of $\sigma(\theta)$.

Generally, in order to calculate the differential cross section we shoot a particle at a target as shown in figure 5.9. The scattering angle θ depends on the impact parameter b. The part of the beam crossing the ring of radius $b(\theta)$, thickness db and area $2\pi b\, db$ is scattered in angles between θ and $\theta + d\theta$. Since there is only one particle at the target we have that $nA = 1$. The number of particles per unit time crossing the ring is $J2\pi b\, db$, therefore

$$2\pi b(\theta)\, db = -2\pi\sigma(\theta)\sin\theta\, d\theta \qquad (5.21)$$

(the $-$ sign is because as b increases, θ decreases). From the potential we

5.4. SCATTERING

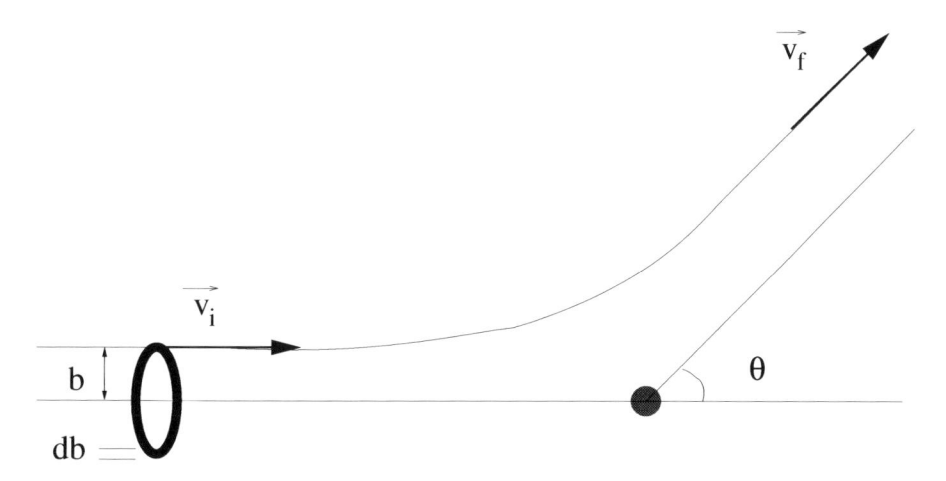

Figure 5.9: Beam particles passing through the ring $2\pi b\, db$ are scattered within the solid angle $d\Omega = 2\pi \sin\theta\, d\theta$.

can calculate $b(\theta)$ and from $b(\theta)$ we can calculate $\sigma(\theta)$. Conversely, if we measure $\sigma(\theta)$, we can calculate $b(\theta)$.

5.4.1 Rutherford Scattering

The scattering of a charged particle with charge q ("electron") in a Coulomb potential of a much heavier charge Q ("nucleus") is called Rutherford scattering. In this case, the interaction potential is given by

$$V(r) = \frac{1}{4\pi\epsilon_0} \frac{Q}{r}, \qquad (5.22)$$

which accelerates the particle with acceleration

$$\vec{a} = \frac{qQ}{4\pi\epsilon_0 m} \frac{\hat{r}}{r^2} \equiv \alpha \frac{\vec{r}}{r^3}. \qquad (5.23)$$

The energy of the particle is $E = \frac{1}{2}mv^2$ and the magnitude of its angular momentum is $l = mvb$, where $v \equiv |\vec{v}|$. The dependence of the impact parameter on the scattering angle is [40]

$$b(\theta) = \frac{\alpha}{v^2} \cot\frac{\theta}{2}. \qquad (5.24)$$

Using equation (5.21) we obtain

$$\sigma(\theta) = \frac{\alpha^2}{4} \frac{1}{v^4} \sin^{-4}\frac{\theta}{2}. \tag{5.25}$$

Consider the scattering trajectories. The results for same charges are

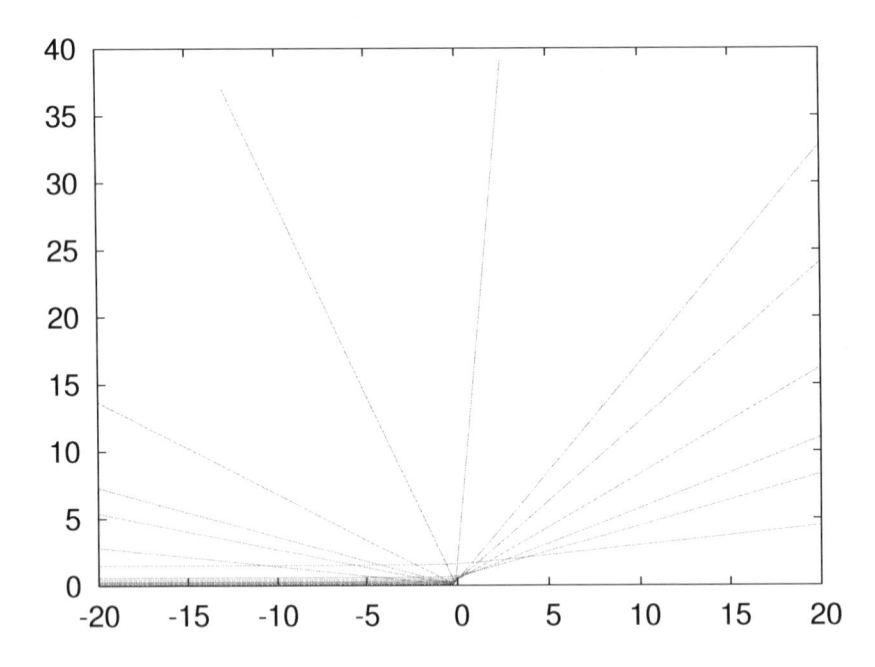

Figure 5.10: Rutherford scattering trajectories. We set $k1 \equiv \frac{qQ}{4\pi\epsilon_0 m} = 1$ (see code in the file rk2_cb.cpp) and $b = 0.08, 0.015, 0.020, 0.035, 0.080, 0.120, 0.200, 0.240, 0.320, 0.450, 0.600, 1.500$. The initial position of the particle is at $x(0) = -50$ and its initial velocity is $v = 3$ in the x direction. The number of integration steps is 1000, the initial time is 0 and the final time is 30.

shown in figure 5.10. A similar figure is obtained in the case of opposite charges. In the latter case we have to take special care for small impact parameters $b < 0.2$ where the scattering angle is ≈ 1. A large number of integration steps is needed in order to obtain the desired accuracy. A useful monitor of the accuracy of the calculation is the measurement of the energy of the particle which should be conserved. The results are shown in table 5.2. We will now describe a method for calculating the cross section by using equation (5.20). Alternatively we could have used

5.4. SCATTERING

b	θ_n	θ_a	$\Delta E/E$	Nt
0.008	2.9975	2.9978	$2.8\,10^{-9}$	5000
0.020	2.7846	2.7854	$2.7\,10^{-9}$	5000
0.030	2.6131	2.6142	$2.5\,10^{-9}$	5000
0.043	2.4016	2.4031	$2.3\,10^{-9}$	5000
0.056	2.2061	2.2079	$2.0\,10^{-9}$	5000
0.070	2.0152	2.0172	$1.7\,10^{-9}$	5000
0.089	1.7887	1.7909	$1.4\,10^{-9}$	5000
0.110	1.5786	1.5808	$1.0\,10^{-9}$	5000
0.130	1.4122	1.4144	$0.8\,10^{-9}$	5000
0.160	1.2119	1.2140	$0.5\,10^{-9}$	5000
0.200	1.0123	1.0142	$0.3\,10^{-9}$	5000
0.260	0.8061	0.8077	$0.1\,10^{-9}$	5000
0.360	0.5975	0.5987	$2.9\,10^{-11}$	5000
0.560	0.3909	0.3917	$0.3\,10^{-11}$	5000
1.160	0.1905	0.1910	$5.3\,10^{-14}$	5000

Table 5.2: Scattering angles of Rutherford scattering. We set $\mathtt{k1} \equiv \frac{qQ}{4\pi\epsilon_0 m} = 1$ (see file rk2_cb.cpp) and study the resulting trajectories for the values of b shown in column 1. θ_n is the numerically calculated scattering angle and θ_a is the one calculated from equation (5.24). The ratio $\Delta E/E$ shows the change in the particle's energy due to numerical errors. The last column is the number of integration steps. The particle's initial position is at $x(0) = -50$ and initial velocity $\vec{v} = 3\hat{x}$.

equation (5.21) and perform a numerical calculation of the derivatives. This is left as an exercise for the reader. Our calculation is more like an experiment. We place a "detector" that "detects" particles scattered within angles θ and $\theta + \delta\theta$. For this reason we split the interval $[0, \pi]$ in N_b bins so that $\delta\theta = \pi/N_b$. We perform "scattering experiments" by varying $b \in [b_m, b_M]$ with step δb. Due to the symmetry of the problem we fix ϕ to be a constant, therefore a given θ corresponds to a cone with an opening angle θ and an apex at the center of scattering. For given b we measure the scattering angle θ and record the number of particles per unit time $\delta N \propto b \delta b$. The latter is proportional to the area of the ring of radius b. All we need now is the beam intensity J which is the total number of particles per unit time $J \propto \sum_i b \delta b$ (note than in the ratio $\delta N/J$ the proportionality constant and δb cancel) and the solid angle $2\pi \sin(\theta)\,\delta\theta$.

b	θ_n	θ_a	$\Delta E/E$	STEPS
0.020	2.793	2.785	0.02	1 000 000
0.030	2.620	2.614	$8.2\,10^{-3}$	300 000
0.043	2.405	2.403	$7.2\,10^{-4}$	150 000
0.070	2.019	2.017	$3.2\,10^{-7}$	150 000
0.089	1.793	1.791	$8.2\,10^{-7}$	60 000
0.110	1.583	1.581	$1.2\,10^{-6}$	30 000
0.130	1.417	1.414	$9.4\,10^{-7}$	20 000
0.160	1.216	1.214	$6.0\,10^{-5}$	5 000
0.200	1.016	1.014	$4.1\,10^{-6}$	5 000
0.260	0.8093	0.8077	$2.2\,10^{-7}$	5 000
0.360	0.6000	0.5987	$7.6\,10^{-9}$	5 000
0.560	0.3926	0.3917	$1.2\,10^{-10}$	5 000
1.160	0.1913	0.1910	$2.9\,10^{-13}$	5 000

Table 5.3: Rutherford scattering of opposite charges with $\frac{qQ}{4\pi\epsilon_0 m} = -1$. The table is similar to table 5.2. We observe the numerical difficulty for small impact parameters.

Finally we can easily use equation (5.19) in order to calculate the total cross section σ_{tot}. The program that performs this calculation is in the file scatter.cpp and it is a simple modification of the program in rk2.cpp:

```cpp
//========================================================
//Program that computes scattering cross-section of a central
//force on the plane. The user should first check that the
//parameters used, lead to a free state in the end.
//     ** X20 is the impact parameter b **
//A 4 ODE system is solved using Runge-Kutta Method
//User must supply derivatives
//dx1/dt=f1(t,x1,x2,x3,x4) dx2/dt=f2(t,x1,x2,x3,x4)
//dx3/dt=f3(t,x1,x2,x3,x4) dx4/dt=f4(t,x1,x2,x3,x4)
//as real(8) functions
//Output is written in file scatter.dat
//========================================================
#include <iostream>
#include <fstream>
#include <cstdlib>
#include <string>
#include <cmath>
```

5.4. SCATTERING

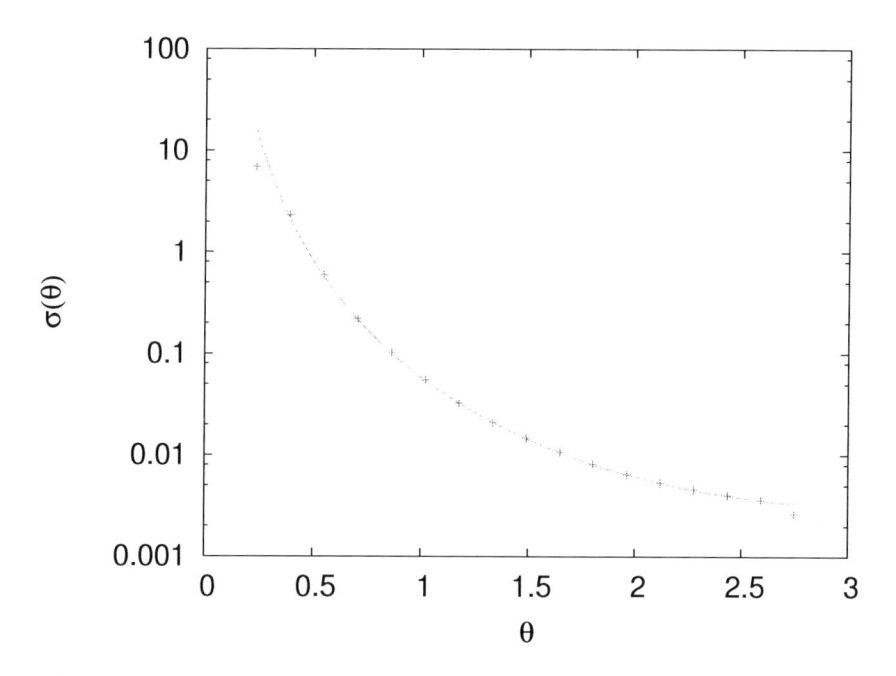

Figure 5.11: Differential cross section of the Rutherford scattering. The solid line is the function (5.25) for $\alpha = 1$, $v = 3$. We set $\frac{qQ}{4\pi\epsilon_0 m} = 1$. The particle's initial position is $x(0) = -50$ and its initial velocity is $\vec{v} = 3\hat{x}$. We used 5000 integration steps, initial time equal to 0 and final time equal to 30. The impact parameter varies between 0.02 and 1 with step equal to 0.0002.

```
using namespace std;
//————————————————————————————
const int P = 1010000;
double T[P], X1[P], X2[P], V1[P], V2[P];
double k1,k2;
//————————————————————————————
double
f1(const double& t  , const double& x1, const double& x2,
   const double& v1 , const double& v2);
double
f2(const double& t  , const double& x1, const double& x2,
   const double& v1 , const double& v2);
double
f3(const double& t  , const double& x1, const double& x2,
   const double& v1 , const double& v2);
double
```

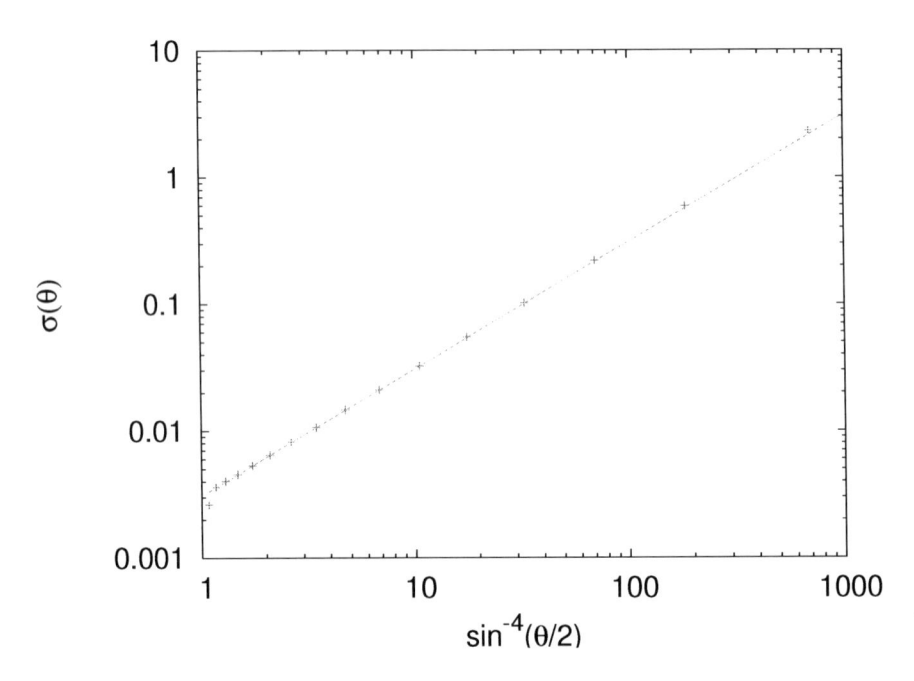

Figure 5.12: Differential cross section of the Rutherford scattering like in figure 5.11. The solid line is the function $1/(4\times 3^4)x$ from which we can deduce the functional form of $\sigma(\theta)$.

```
f4(const double& t   , const double& x1, const double& x2,
   const double& v1 , const double& v2);
double
energy
  (const double& t   , const double& x1, const double& x2,
   const double& v1 , const double& v2);
void
RK(const double& Ti , const double& Tf ,
   const double& X10, const double& X20,
   const double& V10, const double& V20,
   const int    & Nt);
void
RKSTEP(double& t ,
       double& x1, double& x2,
       double& x3, double& x4,
       const    double& dt);
//------------------------------------------------------
int main(){
```

5.4. SCATTERING

```
  string buf;
  double Ti,Tf,X10,X20,V10,V20;
  double X20F,dX20; //max impact parameter and step
  int    Nt,i ;
  const int Nbins=20;
  int index;
  double angle,bins[Nbins],Npart;
  const double PI     =3.14159265358979324;
  const double rad2deg=180.0/PI;
  const double dangle =PI/Nbins;
  double R,density,dOmega,sigma,sigmatot;

//Input:
  cout << "Runge-Kutta Method for 4-ODEs Integration\n";
  cout << "Enter coupling constants:\n";
  cin  >> k1 >> k2;getline(cin,buf);
  cout << "k1= " << k1 << " k2= " << k2 << endl;
  cout << "Enter Nt,Ti,Tf,X10,X20,V10,V20:\n";
  cin  >> Nt >> Ti >> Tf>> X10 >> X20 >> V10 >> V20;
  getline(cin,buf);
  cout << "Enter final impact parameter X20F and step dX20:\n";
  cin  >> X20F >> dX20;
  cout << "Nt = " << Nt << endl;
  cout << "Time: Initial Ti = " << Ti
       << " Final Tf= "         << Tf   << endl;
  cout << "             X1(Ti)= " << X10
       << " X2(Ti)="              << X20  << endl;
  cout << "             V1(Ti)= " << V10
       << " V2(Ti)="              << V20  << endl;
  cout << "Impact par X20F =" << X20F
       << " dX20 ="            << dX20 << endl;
  ofstream myfile("scatter.dat");
  myfile.precision(17);
  for(i=0;i<Nbins;i++) bins[i] = 0.0;
//Calculate:
  Npart = 0.0;
  X20   = X20 + dX20/2.0; //starts in middle of first interval
  while( X20 < X20F ){
    RK(Ti,Tf,X10,X20,V10,V20,Nt);
    //Take absolute value due to symmetry:
    angle = abs(atan2(V2[Nt-1],V1[Nt-1]));
    //Output: The final angle. Check if almost constant
    myfile << "@ " << X20 << " " << angle
           << " "  << abs(atan2(V2[Nt-51],V1[Nt-51]))
           << " "  << k1/(V10*V10)/tan(angle/2.0) << endl;
```

```
   //Update histogram:
   index        = int(angle/dangle);
   //Number of incoming particles per unit time
   //is proportional to radius of ring
   //of radius X20, the impact parameter:
   bins[index] += X20 ;   // db is cancelled from density
   Npart       += X20 ;   // <-- i.e. from here
   X20         += dX20;
 }//while( X20 < X20F )
 //Print scattering cross section:
 R          = X20;                 //beam radius
 density    = Npart/(PI*R*R);      //beam flux density J
 sigmatot   = 0.0;                 //total cross section
 for(i=0;i<Nbins;i++){
    angle   = (i+0.5)*dangle;
    dOmega  = 2.0*PI*sin(angle)*dangle; //d(Solid Angle)
    sigma   = bins[i]/(density*dOmega);
    if(sigma>0.0)
       myfile << "ds= " << angle
              << " "    << angle*rad2deg
              << " "    << sigma << endl;
    sigmatot += sigma*dOmega;
 }//for(i=0;i<Nbins;i++)
 myfile << "sigmatot= " << sigmatot << endl;
 myfile.close();
}//main()
```

The results are recorded in the file scatter.dat. An example session that reproduces figures 5.11 and 5.12 is

```
> g++ scatter.cpp rk2_cb.cpp -o scatter
> ./scatter
Runge-Kutta Method for 4-ODEs Integration
Enter coupling constants:
1.0 0.0
k1= 1 k2= 0
Enter Nt,Ti,Tf,X10,X20,V10,V20:
5000 0 30 -50 0.02 3 0
Enter final impact parameter X20F and step dX20:
1 0.0002
Nt = 5000
Time: Initial Ti = 0 Final Tf= 30
         X1(Ti)= -50 X2(Ti)=0.02
         V1(Ti)= 3   V2(Ti)=0
```

5.4. SCATTERING

```
Impact par X20F  =1 dX20  =0.0002
```

The results can be plotted with the gnuplot commands:

```
gnuplot> set log
gnuplot> plot [:1000] "<grep ds= scatter.dat" \
   u ((sin($2/2))**(-4)):($4) notit,\
        (1./(4.*3.**4))*x notit
gnuplot> unset log
gnuplot> set log y
gnuplot> plot [:] "<grep ds= scatter.dat" u 2:4 notit, \
   (1./(4.*3.**4))*(sin(x/2))**(-4) notit
```

The results are in a very good agreement with the theoretical ones given by (5.25). The next step will be to study other central potentials whose solution is not known analytically.

5.4.2 More Scattering Potentials

Consider scattering from a force field

$$\vec{F} = f(r)\hat{r}, \qquad f(r) = \begin{cases} \frac{1}{r^2} - \frac{r}{a^3} & r \leq a \\ 0 & r > a \end{cases}. \tag{5.26}$$

This is a very simple classical model of the scattering of a positron e^+ by the hydrogen atom. The positron has positive charge $+e$ and the hydrogen atom consists of a positively charged proton with charge $+e$ in an electron cloud of opposite charge $-e$. We set the scales so that $m_{e^+} = 1$ and $e^2/4\pi\epsilon_0 = 1$. We will perform a numerical calculation of $b(\theta)$, $\sigma(\theta)$ and σ_{tot}.

The potential energy is given by

$$f(r) = -\frac{dV(r)}{dr} \Rightarrow V(r) = \frac{1}{r} + \frac{r^2}{2a^2} - \frac{3}{2a}. \tag{5.27}$$

where $V(r) = 0$ for $r \geq a$. The program containing the calculation of the acceleration caused by this force can be found in the file rk_hy.cpp:

```
//===========================================================
//The acceleration functions f3,f4(t,x1,x2,v1,v2) provided
```

```cpp
//by the user
//================================================
#include <iostream>
#include <fstream>
#include <cstdlib>
#include <string>
#include <cmath>
using namespace std;
extern double k1,k2;
//------------------------------------------------
//Motion in hydrogen atom + positron:
//f(r) = 1/r^2-r/k1^3
//ax= f(r)*x1/r  ay= f(r)*x2/r
double
f3(const double& t , const double& x1, const double& x2,
   const double& v1, const double& v2){
  double r2,r,fr;
  r2=x1*x1+x2*x2;
  r =sqrt(r2);
  if(r <= k1 && r2 > 0.0)
    fr = 1.0/r2-r/(k1*k1*k1);
  else
    fr = 0.0;

  if(fr > 0.0 && r > 0.0)
    return fr*x1/r; // dx3/dt=dv1/dt=a1
  else
    return 0.0;

}
//------------------------------------------------
double
f4(const double& t , const double& x1, const double& x2,
   const double& v1, const double& v2){
  double r2,r,fr;
  r2=x1*x1+x2*x2;
  r =sqrt(r2);
  if(r <= k1 && r2 > 0.0)
    fr = 1.0/r2-r/(k1*k1*k1);
  else
    fr = 0.0;

  if(fr > 0.0 && r > 0.0)
    return fr*x2/r; // dx4/dt=dv2/dt=a2
  else
```

```
      return 0.0;
}
//------------------------------------------------------------
double
energy
  (const double& t , const double& x1, const double& x2,
   const double& v1, const double& v2){
  double r,Vr;
  r=sqrt(x1*x1+x2*x2);
  if(r <= k1 && r > 0.0)
    Vr = 1/r + 0.5*r*r/(k1*k1*k1) - 1.5 / k1;
  else
    Vr = 0.0;
  return 0.5*(v1*v1+v2*v2) + Vr;
}
```

The results are shown in figures 5.13–5.14. We find that $\sigma_{tot} = \pi a^2$ (see problem 5.10).

Another interesting dynamical field is given by the Yukawa potential. This is a phenomenological model of nuclear interactions:

$$V(r) = k\frac{e^{-r/a}}{r}. \tag{5.28}$$

This field can also be used as a model of the effective interaction of electrons in metals (Thomas–Fermi) or as the Debye potential in a classic plasma. The resulting force is

$$\vec{F}(r) = f(r)\hat{r}, \qquad f(r) = k\frac{e^{-r/a}}{r^2}\left(1 + \frac{r}{a}\right) \tag{5.29}$$

The program of the resulting acceleration can be found in the file rk2_yu.cpp. The results are shown in figures 5.15–5.16.

5.5 More Particles

In this section we will generalize the discussion of the previous paragraphs in the case of a dynamical system with more degrees of freedom. The number of dynamical equations that need to be solved depends on the number of degrees of freedom and we have to write a program that

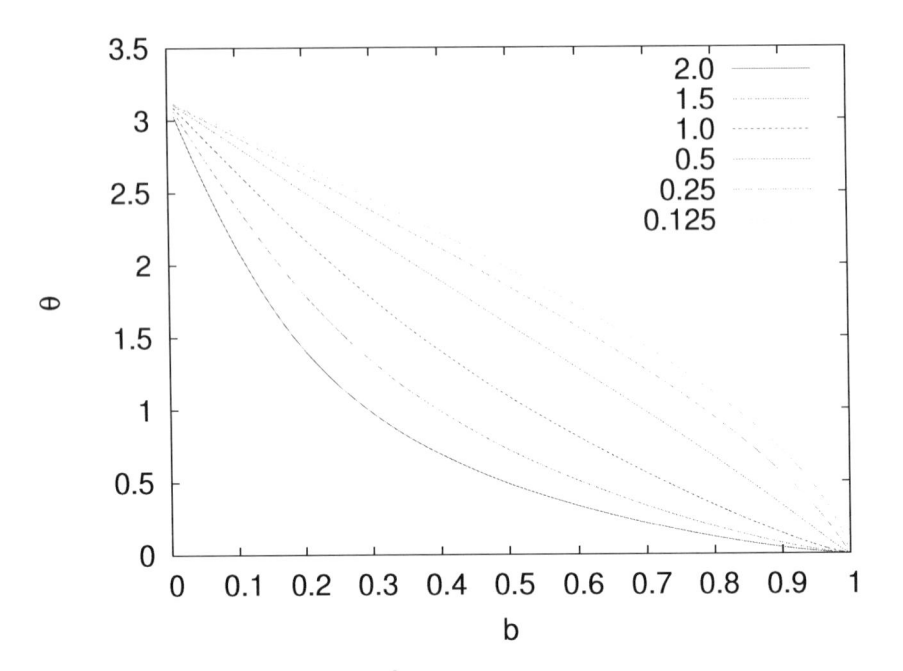

Figure 5.13: The impact parameter $b(\theta)$ for the potential given by equation (5.27) for different values of the initial velocity v. We set $a = 1$, $x(0) = -5$ and made 4000 integration steps from $t_i = 0$ to $t_f = 40$.

implements the 4th order Runge–Kutta method for an arbitrary number of equations NEQ. We will explain how to allocate memory *dynamically*, in which case the necessary memory storage space, which depends on NEQ, is allocated at the time of running the program and not at compilation time.

Until now, memory has been allocated *statically*. This means that arrays have sizes which are known at compile time. For example, in the program rk2.cpp the integer parameter P had a given value which determined the size of all arrays using the declarations:

```
const int P = 1010000;
double T[P], X1[P], X2[P], V1[P], V2[P];
```

Changing P after compilation is impossible and if this becomes necessary we have to edit the file, change the value of P and recompile. Dynamical memory allocation allows us to read in Nt and NEQ at execution time and

5.5. MORE PARTICLES

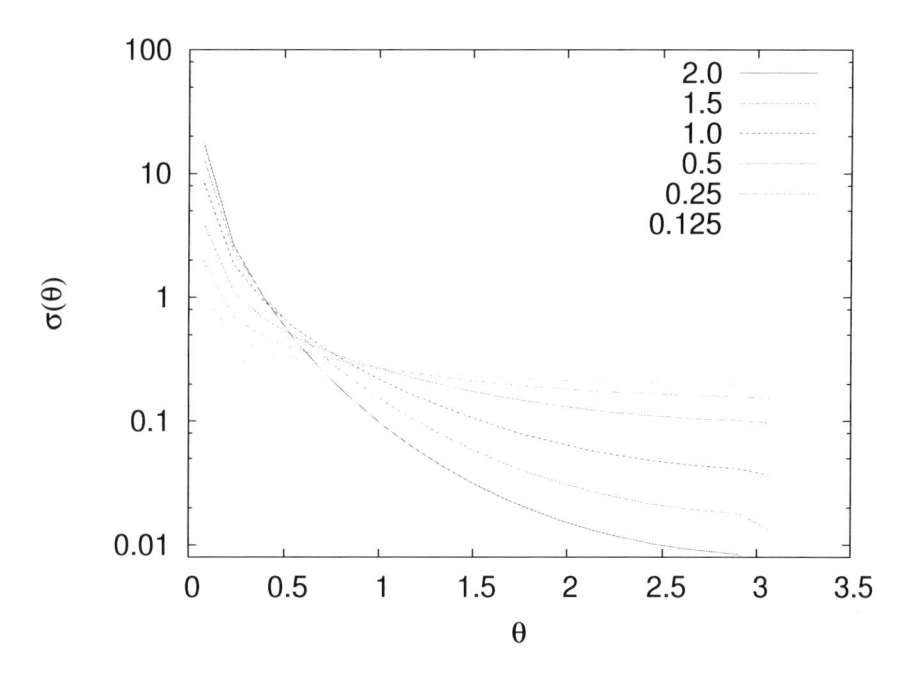

Figure 5.14: The function $\sigma(\theta)$ for the potential given by equation (5.27) for different values of the initial velocity v. We set $a = 1$, $x(0) = -5$ and the integration is performed by making 4000 steps from $t_i = 0$ to $t_f = 40$.

then ask from the operating system to allocate the necessary memory. The needed memory can be asked for at execution time by using the `new` operator. Here is an example:

```
double   *T;   //Declare 1–Dim arrays as pointers
double  **X;   //Declare 2–Dim arrays as pointers to pointers
int NEQ,Nt;    //Variables of array sizes
//----------------------------------------------------------------
finit(NEQ);    //function that sets NEQ at run time
cin  >> Nt;    //read Nt at run time
//----------------------------------------------------------------
// allocates 1–Dim array of Nt doubles
T  = new double [Nt];
// allocates 1–Dim array of Nt pointers to doubles
X  = new double*[Nt];
//for each i, allocate an array of NEQ doubles at X[i]:
for(int i=0;i<Nt;i++) X[i] = new double[NEQ];
```

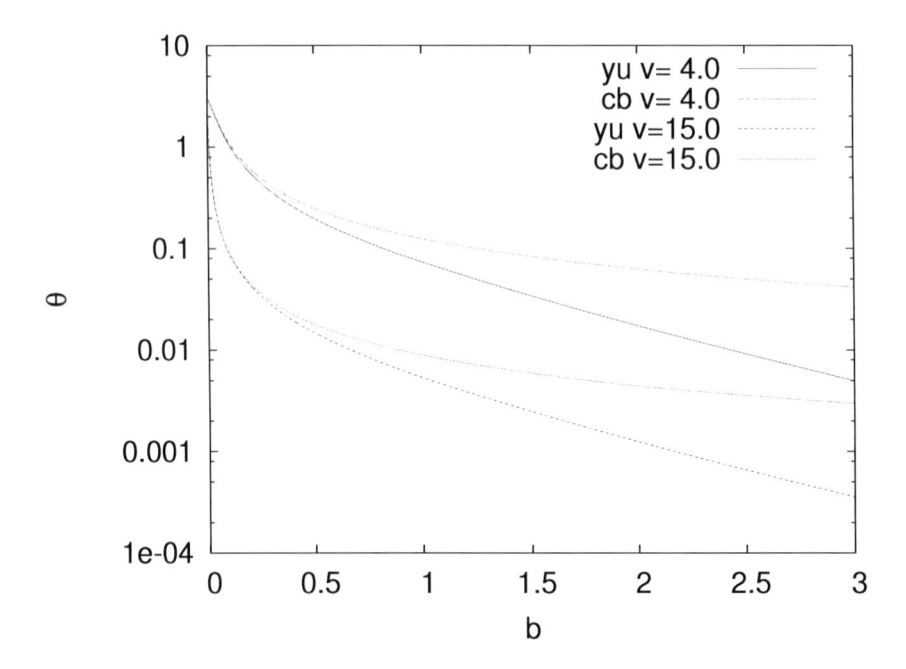

Figure 5.15: The function $b(\theta)$ for the Yukawa scattering for several values of the initial velocity v. We set $a=1$, $k=1$, $x(0)=-50$ and the integration is performed with 5000 steps from $t_i=0$ to $t_f=30$. The lines marked as cb are equation (5.24) of the Rutherford scattering.

```
.....
(compute with T[Nt], X[Nt][NEQ])
.....
//return allocated memory back to the system:
delete [] T; // deallocate an array
//for each i, deallocate the array of doubles X[i][NEQ]
for(int i=0;i<NEQ;i++) delete [] X[i];
//and then deallocate the array of pointers to doubles X[Nt]:
delete[] X ;
....................
void finit(int& NEQ){//function that sets value of NEQ
  NEQ = 4;
}
```

In this program, we should remember the fact that in C++, the name T of an array T[Nt] is a pointer to T[0], which is denoted by &T[0]. This is the address of the first element of the array in the memory. Therefore,

5.5. MORE PARTICLES

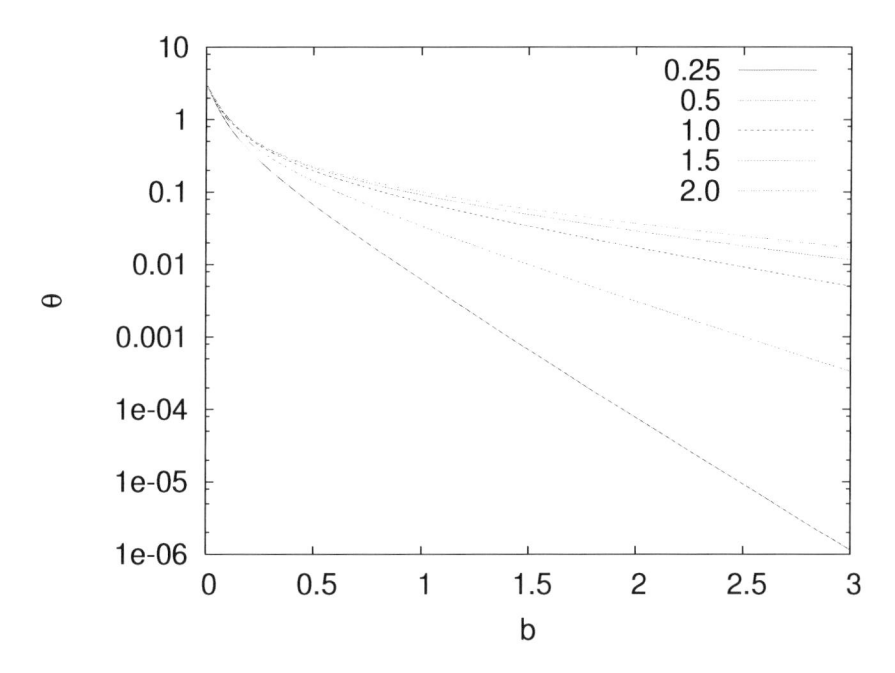

Figure 5.16: The function $b(\theta)$ for the Yukawa scattering for several values of the range a of the force. We set $v = 4.0$, $k = 1$, $x(0) = -50$ and the integration is performed with 5000 steps from $t_i = 0$ to $t_f = 30$.

if the array is `double T[Nt]`, then `T` is a pointer to a double: `double *T`. Then `T+i` is a pointer to the address of the `(i+1)`-th element of the array `T[i]` and

```
T+i = &T[i]
```

as well as

```
T[i] = *(T+i)
```

We can use pointers with the same notation as we do with arrays: If we declare a pointer to a double `*T1`, and assign `T1=T`, then `T1[0]`, `T1[1]`, ... are the same as `T[0]`, `T[1]`, ... For example, `T1[0] = 2.0` assigns also `T[0]` and vice-versa. Conversely, we can declare a pointer to a double `*T`, as we do in our program, and make it point to a region in the memory where we have reserved space for `Nt` doubles. This is what the operator new does. It asks for the memory for Nt doubles and returns a pointer to it. Then we assign this pointer to T:

```
double *T;
T = new double[Nt];
```

Then we can use T[0], T[1], ..., T[Nt-1] as we do with ordinary arrays.

Two dimensional arrays are slightly trickier: For a two dimensional array, double X[Nt][NEQ], X is a pointer to the value X[0][0], which is &X[0][0]. Then X[i] is a *pointer* to the one dimensional array X[i][NEQ], therefore X is a pointer to a pointer of a double!

```
X[i][jeq] is:   double
X[i]      is:   double *
X         is:   double **
```

Conversely, we can declare a double **X, and use the operator new to return a pointer to an array of Nt pointers to doubles, and then for each element of the array, use new to return a pointer to NEQ doubles:

```
double **X
X        = new double*[Nt];
X[0]     = new double[NEQ];
X[1]     = new double[NEQ];
...
X[Nt-1]  = new double[NEQ];
```

Then we can use the notation T[i][jeq], i= 0, 1, ..., Nt-1 and jeq= 0, 1, ..., NEQ-1, as we do with statically defined arrays.

The memory that we ask to be allocated dynamically is a finite resource that can easily be exhausted (heard of *memory leaks*?). Therefore, we should be careful to return unused memory to the system, so that it can be recycled. This should happen especially within functions that we call many times, which allocate large memory dynamically. The operator delete can be used to deallocate memory that has been allocated with the operator new. For one dimensional arrays, this is particularly simple:

```
double *T;
```

5.5. MORE PARTICLES

```
T = new double[Nt];
....
(use T[i])
...
delete [] T;
...
(cannot use T after delete)
```

For "two dimensional arrays" that have been allocated as we described above, first we have to delete the arrays pointed by X[i] for i=0, ... , Nt-1, and then the arrays of pointers pointed by X:

```
X   = new double*[Nt ];
for(int i=0;i<Nt;i++) X[i] = new double[NEQ];
.... use X[i][jeq] ....
for(int i=0;i<NEQ;i++) delete [] X[i]; //delete all arrays X[i]
delete[] X ;                           //delete array of pointers
```

The main program will be written in the file rkA.cpp, whereas the force-dependent part of the code will be written in files with names of the form rkA_XXX.cpp. In the latter, the user must program a *function* f(t,X,dXdt) which takes as input the time t and the values of the functions X[NEQ] and outputs the values of their derivatives dXdt[NEQ] at time t. The function finit(NEQ) sets the number of functions in f and it is called once during the initialization phase of the program.

The program in the file rkA.cpp is listed below:

```
//===============================================================
//Program to solve an ODE system using the
//4th order Runge–Kutta Method
//NEQ: Number of equations
//User supplies two functions:
//f(t,x,xdot): with double t,x[NEQ],xdot[NEQ] which
//given the time t and current values of functions x[NEQ]
//it returns the values of derivatives: xdot = dx/dt
//The values of two coupling constants k1,k2 may be used
//in f which are read in the main program
//finit(NEQ) : sets the value of NEQ
//
//User Interface:
//double k1,k2:   coupling constants in global scope
//Nt,Ti,Tf: Nt-1 integration steps , initial/final time
```

```cpp
//double X0[NEQ]: initial conditions
//Output:
//rkA.dat with Nt lines consisting of: T[Nt],X[Nt][NEQ]
//============================================================
#include <iostream>
#include <fstream>
#include <cstdlib>
#include <string>
#include <cmath>
using namespace std;
//------------------------------------------------------------
double  *T;  // T[Nt]       stores the values of times
double **X;  // X[Nt][NEQ] stores the values of functions
double k1,k2;
//------------------------------------------------------------
void RK(const double& Ti, const double& Tf, double* X0,
        const int  & Nt, const int  & NEQ);
void RKSTEP(  double& t ,        double* x ,
        const double& dt, const int  & NEQ);
//The following functions are defined in rkA_XX.cpp:
void finit(   int  & NEQ); //Sets number of equations
//Derivative and energy functions:
void  f      (const double& t, double* X,double* dXdt);
double energy(const double& t, double* X);
//------------------------------------------------------------
int main(){
  string   buf;
  int      NEQ,Nt;
  double*  X0;
  double Ti,Tf;
  //Get number of equations and allocate memory for X0:
  finit(NEQ);
  X0 = new double [NEQ];
  //Input:
  cout << "Runge-Kutta Method for ODE Integration.\n";
  cout << "NEQ= " << NEQ << endl;
  cout << "Enter coupling constants:\n";
  cin  >> k1 >> k2;getline(cin,buf);
  cout << "k1= " << k1 << " k2= " << k2 << endl;
  cout << "Enter Nt, Ti, Tf, X0:\n";
  cin  >>       Nt>>Ti>>Tf;
  for(int i=0;i<NEQ;i++) cin >> X0[i];getline(cin,buf);
  cout << "Nt = " << Nt << endl;
  cout << "Time: Initial Ti =" << Ti << " "
       << "Final Tf="          << Tf << endl;
```

5.5. MORE PARTICLES

```cpp
  cout << "                X0 =";
  for(int i=0;i<NEQ;i++) cout   << X0[i] << " ";
  cout << endl;
  //Allocate memory for data arrays:
  T  = new double [Nt];
  X  = new double*[Nt];
  for(int i=0;i<Nt;i++) X[i] = new double[NEQ];
  //The Calculation:
  RK(Ti,Tf,X0,Nt,NEQ);
  //Output:
  ofstream myfile("rkA.dat");
  myfile.precision(16);
  for(int i=0;i<Nt;i++){
    myfile    << T[i]       << " ";
    for(int jeq=0;jeq<NEQ;jeq++)
      myfile  << X[i][jeq] << " ";
    myfile    << energy(T[i],X[i]) << '\n';
  }
  myfile.close();
  //------------------------------------------------
  //Cleaning up dynamic memory: delete[] each array
  //created with the new operator (Not necessary in this
  //program,it is done at the end of the program anyway)
  delete[] X0;
  delete[] T ;
  for(int i=0;i<NEQ;i++) delete [] X[i];
  delete[] X ;
}//main()
//================================================
// Driver of the RKSTEP routine
//================================================
void RK(const double& Ti, const double& Tf, double* X0,
        const int  & Nt, const int   & NEQ){
  double  dt;
  double  TS;
  double* XS;

  XS      = new double[NEQ];
  //Initialize variables:
  dt    = (Tf-Ti)/(Nt-1);
  T [0] = Ti;
  for(int ieq=0;ieq<NEQ;ieq++) X[0][ieq]=X0[ieq];
  TS    = Ti;
  for(int ieq=0;ieq<NEQ;ieq++) XS  [ieq]=X0[ieq];
  //Make RK steps: The arguments of RKSTEP are
```

```cpp
  //replaced with the new ones
  for(int i=1;i<Nt;i++){
    RKSTEP(TS,XS,dt,NEQ);
    T[i] = TS;
    for(int ieq=0;ieq<NEQ;ieq++) X[i][ieq] = XS[ieq];
  }
  // Clean up memory:
  delete [] XS;
}//RK()
//==========================================================
//Function RKSTEP(t,X,dt)
//Runge-Kutta Integration routine of ODE
//==========================================================
void RKSTEP(double& t, double* x, const double& dt ,
                                  const int   & NEQ){
  double  tt;
  double *k1, *k2, *k3, *k4, *xx;
  double  h,h2,h6;

  k1 = new double[NEQ];
  k2 = new double[NEQ];
  k3 = new double[NEQ];
  k4 = new double[NEQ];
  xx = new double[NEQ];

  h =dt;       // h =dt, integration step
  h2=0.5*h;    // h2=h/2
  h6=h/6.0;    // h6=h/6
  //1st step:
  f(t ,x ,k1);
  //2nd step:
  for(int ieq=0;ieq<NEQ;ieq++)
    xx[ieq] = x[ieq] + h2*k1[ieq];
  tt =t+h2;
  f(tt,xx,k2);
  //3rd step:
  for(int ieq=0;ieq<NEQ;ieq++)
    xx[ieq] = x[ieq] + h2*k2[ieq];
  tt =t+h2;
  f(tt,xx,k3);
  //4th step:
  for(int ieq=0;ieq<NEQ;ieq++)
    xx[ieq] = x[ieq] + h *k3[ieq];
  tt=t+h ;
  f(tt,xx,k4);
```

5.5. MORE PARTICLES

```
// Update:
t += h;
for(int ieq=0;ieq<NEQ;ieq++)
  x[ieq] += h6*(k1[ieq]+2.0*(k2[ieq]+k3[ieq])+k4[ieq]);
// Clean up memory:
delete[] k1;
delete[] k2;
delete[] k3;
delete[] k4;
delete[] xx;
}//RKSTEP()
```

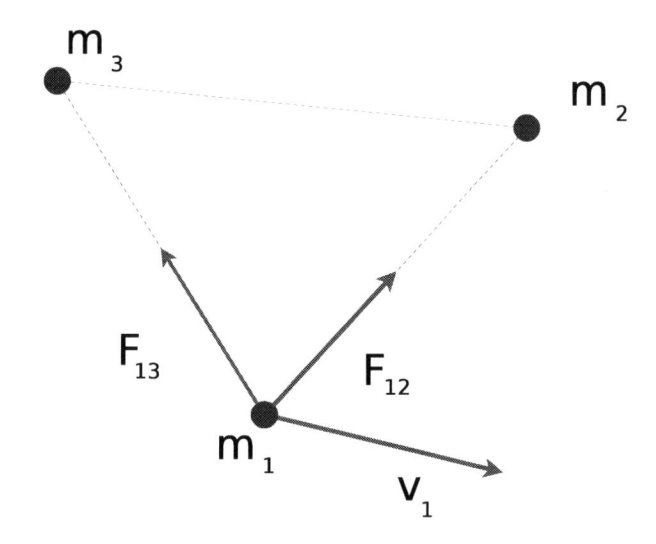

Figure 5.17: Three particles of equal mass interact via their mutual gravitational attraction. The problem is solved numerically using the program in the files rkA.cpp, rkA_3pcb.cpp. The same program can be used in order to study the motion of three equal charges under the influence of their attractive or repulsive electrostatic force.

Consider three particles of equal mass exerting a force of gravitational attraction on each other[4] like the ones shown in figure 5.17. The forces exerting on each other are given by

$$\vec{F}_{ij} = \frac{mk_1}{r_{ij}^3}\vec{r}_{ij}, \quad i,j = 1,2,3, \tag{5.30}$$

---

[4]The same program can be used for three equal charges exerting an electrostatic force on each other, which can be either attractive or repulsive.

where $k_1 = -Gm$ and the equations of motion become ($i = 1, 2, 3$)

$$\frac{dx_i}{dt} = v_{ix} \qquad \frac{dv_{ix}}{dt} = k_1 \sum_{j=1, j \neq i}^{3} \frac{x_i - x_j}{r_{ij}^3}$$

$$\frac{dy_i}{dt} = v_{iy} \qquad \frac{dv_{iy}}{dt} = k_1 \sum_{j=1, j \neq i}^{3} \frac{y_i - y_j}{r_{ij}^3}, \qquad (5.31)$$

where $r_{ij}^2 = (x_i - x_j)^2 + (y_i - y_j)^2$. The total energy of the system is

$$E/m = \frac{1}{2}(v_1^2 + v_2^2) + \sum_{i,j=1, j<i}^{3} \frac{k_1}{r_{ij}}. \qquad (5.32)$$

The relations shown above are programmed in the file rkA_3pcb.cpp listed below:

```cpp
#include <iostream>
#include <fstream>
#include <cstdlib>
#include <string>
#include <cmath>
using namespace std;
//-----------------------------------------------------------
extern double k1,k2;
//-----------------------------------------------------------
//Sets number of equations:
void finit(int& NEQ){
  NEQ = 12;
}
//===========================================================
//Three particles of the same
//mass on the plane interacting
//via Coulombic force
//===========================================================
void f(const double& t, double* X, double* dXdt){
  double x11,x12,x21,x22,x31,x32;
  double v11,v12,v21,v22,v31,v32;
  double r12,r13,r23;
  //---------------------------------------
  x11 = X[0];x21 = X[4];x31 = X[8];
  x12 = X[1];x22 = X[5];x32 = X[9];
```

5.5. MORE PARTICLES

```
    v11 = X[2];v21 = X[6];v31 = X[10];
    v12 = X[3];v22 = X[7];v32 = X[11];
  //------------------------
    r12 = pow((x11-x21)*(x11-x21)+(x12-x22)*(x12-x22),-3.0/2.0);
    r13 = pow((x11-x31)*(x11-x31)+(x12-x32)*(x12-x32),-3.0/2.0);
    r23 = pow((x21-x31)*(x21-x31)+(x22-x32)*(x22-x32),-3.0/2.0);
  //------------------------
    dXdt[0]  = v11;
    dXdt[1]  = v12;
    dXdt[2]  = k1*(x11-x21)*r12+k1*(x11-x31)*r13; // a11=dv11/dt
    dXdt[3]  = k1*(x12-x22)*r12+k1*(x12-x32)*r13; // a12=dv12/dt
  //------------------------
    dXdt[4]  = v21;
    dXdt[5]  = v22;
    dXdt[6]  = k1*(x21-x11)*r12+k1*(x21-x31)*r23; // a21=dv21/dt
    dXdt[7]  = k1*(x22-x12)*r12+k1*(x22-x32)*r23; // a22=dv22/dt
  //------------------------
    dXdt[8]  = v31;
    dXdt[9]  = v32;
    dXdt[10] = k1*(x31-x11)*r13+k1*(x31-x21)*r23; // a31=dv31/dt
    dXdt[11] = k1*(x32-x12)*r13+k1*(x32-x22)*r23; // a32=dv32/dt
}
//=====================================
double energy(const double& t, double* X){
    double x11,x12,x21,x22,x31,x32;
    double v11,v12,v21,v22,v31,v32;
    double r12,r13,r23;
    double e;
  //------------------------
    x11 = X[0];x21 = X[4];x31 = X[8];
    x12 = X[1];x22 = X[5];x32 = X[9];
    v11 = X[2];v21 = X[6];v31 = X[10];
    v12 = X[3];v22 = X[7];v32 = X[11];
  //------------------------
    r12 = pow((x11-x21)*(x11-x21)+(x12-x22)*(x12-x22),-0.5);
    r13 = pow((x11-x31)*(x11-x31)+(x12-x32)*(x12-x32),-0.5);
    r23 = pow((x21-x31)*(x21-x31)+(x22-x32)*(x22-x32),-0.5);
  //------------------------
    e  = 0.5*(v11*v11+v12*v12+v21*v21+v22*v22+v31*v31+v32*v32);
    e += k1*(r12+r13+r23);
    return e;
}
```

In order to run the program and see the results, look at the commands in the shell script in the file rkA_3pcb.csh. In order to run the script use the command

```
> rkA_3pcb.csh -0.5 4000 1.5 -1 0.1 1 0 1 -0.1 -1 0 0.05 1 0 -1
```

which will run the program setting $k_1 = -0.5$, $\vec{r}_1(0) = -\hat{x}+0.1\hat{y}$, $\vec{v}_1(0) = \hat{x}$, $\vec{r}_2(0) = \hat{x} - 0.1\hat{y}$, $\vec{v}_2(0) = -\hat{x}$, $\vec{r}_3(0) = 0.05\hat{x} + \hat{y}$, $\vec{v}_3(0) = -\hat{y}$, Nt= 4000 and $t_f = 1.5$.

5.6 Problems

5.1 Reproduce the results shown in figures 5.3 and 5.4. Compare your results to the known analytic solution.

5.2 Write a program for the force on a charged particle in a constant magnetic field $\vec{B} = B\hat{k}$ and compute its trajectory for $\vec{v}(0) = v_{0x}\hat{x} + v_{0y}\hat{y}$. Set $x(0) = 1, y(0) = 0, v_{0y} = 0$ and calculate the resulting radius of the trajectory. Plot the relation between the radius and v_{0x}. Compare your results to the known analytic solution. (assume non relativistic motion)

5.3 Consider the anisotropic harmonic oscillator $a_x = -\omega_1^2 x$, $a_y = -\omega_2^2 y$. Construct the Lissajous curves by setting $x(0) = 0, y(0) = 1, v_x(0) = 1, v_y(0) = 0$, $t_f = 2\pi$, $\omega_2^2 = 1$, $\omega_1^2 = 1, 2, 4, 9, 16, \ldots$. What happens when $\omega_1^2 \neq n\omega_2^2$?

5.4 Reproduce the results displayed in table 5.1 and figures 5.5 and 5.6. Plot $\ln a$ vs $\ln T$ and calculate the slope of the resulting straight line by using the linear least squares method. Is it what you expect? Calculate the intercept and compare your result with the expected one.

5.5 Calculate the angular momentum with respect to the center of the force at each integration step of the planetary motion and check whether it is conserved. Show analytically that conservation of angular momentum implies that the position vector sweeps areas at constant rate.

5.6 Calculate the escape velocity of a planet v_e for $GM = 10.0$, $y(0) = 0.0$, $x_0 = x(0) = 1$ using the following steps: First show that $v_0^2 = -GM(1/a) + v_e^2$. Then set $v_x(0) = 0$, $v_y(0) = v_0$. Vary $v_y(0) = v_0$ and measure the resulting semi-major axis a. Determine the intercept of the resulting straight line in order to calculate v_e.

5.7 Repeat the previous problem for $x_0 = 0.5, 1.0, 1.5, 2.0, 2.5, 3.0, 3.5, 4.0$. From the $v_e = f(1/x_0)$ plot confirm the relation (5.14).

5.8 Check that for the bound trajectory of a planet with $GM = 10.0$, $x(0) = 1$, $y(0) = 0.0$, $v_x(0) = 0$, $v_y(0) = 4$ you obtain that $F_1 P +$

$F_2P = 2a$ for each point P of the trajectory. The point F_1 is the center of the force. After determining the semi-major axis a numerically, the point F_2 will be taken symmetric to F_1 with respect to the center of the ellipse.

5.9 Consider the planetary motion studied in the previous problem. Apply a momentary push in the tangential direction after the planet has completed 1/4 of its elliptical orbit. How stable is the particle trajectory (i.e. what is the dependence of the trajectory on the magnitude and the duration of the push?)? Repeat the problem when the push is in the vertical direction.

5.10 Consider the scattering potential of the positron-hydrogen system given by equation (5.26). Plot the functions $f(r)$ and $V(r)$ for different values of a. Calculate the total cross section σ_{tot} numerically and show that it is equal to πa^2.

5.11 Consider the Morse potential of diatomic molecules:

$$V(r) = D\left(\exp(-2\alpha r) - 2\exp(-\alpha r)\right) \tag{5.33}$$

where $D, \alpha > 0$. Compute the solutions of the problem numerically in one dimension and compare them to the known analytic solutions when $E < 0$:

$$x(t) = \frac{1}{\alpha} \ln \left\{ \frac{D - \sqrt{D(D-|E|)}\sin(\alpha t \sqrt{2|E|/m} + C)}{|E|}, \right\} \tag{5.34}$$

where the integration constant as a function of the initial position and energy is given by

$$C = \sin^{-1}\left[\frac{D - |E|e^{\alpha x_0}}{\sqrt{D(D-|E|)}}\right]. \tag{5.35}$$

We obtain a periodic motion with an energy dependent period $= (\pi/\alpha)\sqrt{2m/|E|}$. For $E > 0$ we obtain

$$x(t) = \frac{1}{\alpha} \ln \left\{ \frac{\sqrt{D(D+E)}\cosh(\alpha t \sqrt{2E/m} + C) - D}{|E|} \right\} \tag{5.36}$$

5.6. PROBLEMS

whereas for $E = 0$

$$x(t) = \frac{1}{\alpha} \ln\left\{\frac{1}{2} + \frac{D\alpha^2}{m}(t+C)^2\right\}. \qquad (5.37)$$

In these equations, the integration constant C is given by a different relation and not by equation (5.35). Compute the motion in phase space $(x, \dot{x})$ and study the transition from open to closed trajectories.

5.12 Consider the effective potential term $V_{eff}(r) = l^2/2mr^2$ ($l \equiv |\vec{L}|$) in the previous problem. Plot the function $V_{tot}(r) = V(r) + V_{eff}(r)$ for $D = 20$, $\alpha = 1$, $m = 1$, $l = 1$, and of course for $r > 0$. Determine the equilibrium position and the ionization energy.

Calculate the solutions $x(t)$, $y(t)$, $y(x)$, $r(t)$ on the plane for $E > 0$, $E = 0$, and $E < 0$ numerically. In the $E < 0$ case consider the scattering problem and calculate the functions $b(\theta)$, $\sigma(\theta)$ and the total cross section σ_{tot}.

5.13 Consider the potential of the molecular model given by the force $\vec{F}(r) = f(r)\hat{r}$ where $f(r) = 24(2/r^{13} - 1/r^7)$. Calculate the potential $V(r)$ and plot the function $V_{tot}(r) = V(r) + V_{eff}(r)$. Determine the equilibrium position and the ionization energy.

Consider the problem of scattering and calculate $b(\theta)$, $\sigma(\theta)$ and σ_{tot} numerically. How much do your results depend on the minimum scattering angle?

5.14 Compute the trajectories of a particle under the influence of a force $\vec{F} = -k/r^3 \hat{r}$. Determine appropriate initial conditions that give a spiral trajectory.

5.15 Compute the total cross section σ_{tot} for the Rutherford scattering both analytically and numerically. What happens to your numerical results as you vary the integration limits?

5.16 Write a program that computes the trajectory of a particle that moves on the plane in the static electric field of N static point charges.

5.17 Solve the three body problem described in the text in the case of three different electric charges by making the appropriate changes to the program in the file rkA_3cb.cpp.

302   CHAPTER 5.   PLANAR MOTION

5.18 Two charged particles of equal mass and charge are moving on the xy plane in a constant magnetic field $\vec{B} = B\hat{z}$. Solve the equations of motion using a 4th order Runge–Kutta Method. Plot the resulting trajectories for the initial conditions that you will choose.

5.19 Three particles of equal mass m are connected by identical springs. The springs' spring constant is equal to k and their equilibrium length is equal to l. The particles move without friction on a horizontal plane. Solve the equations of motion of the system numerically by using a 4th order Runge–Kutta Method. Plot the resulting trajectories for the initial conditions that you will choose. (Hint: Look in the files rkA_3hoc.cpp, rkA_3hoc.csh.)

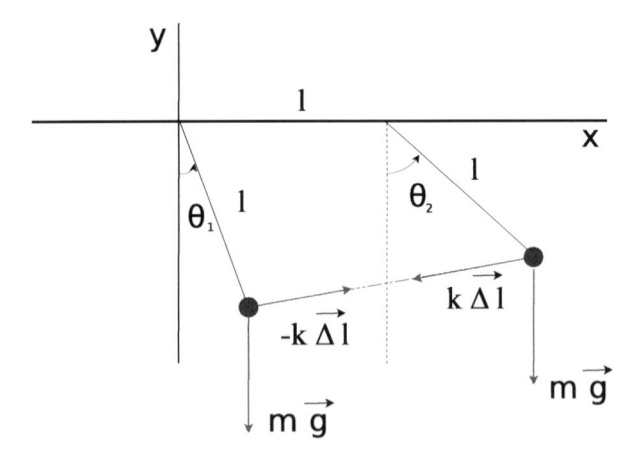

Figure 5.18: Two identical particles are attached to thin weightless rods of length l and they are connected by an ideal weightless spring with spring constant k and equilibrium length l. The rods are hinged to the ceiling at points whose distance is l. (Problem 5.20).

5.20 Two identical particles are attached to thin weightless rods of length l and they are connected by an ideal weightless spring with spring constant k and equilibrium length l. The rods are hinged to the ceiling at points whose distance is l (see figure 5.18). Compute the Lagrangian of the system and the equations of motion for the degrees of freedom θ_1 and θ_2. Solve these equations numerically by using a 4th order Runge–Kutta method. Plot the positions of

5.6. PROBLEMS

the particles in a Cartesian coordinate system and the resulting trajectory. Study the normal modes for small angles $\theta_1 \lesssim 0.1$ and compute the deviation of the solutions from the small oscillation approximation as the angles become larger. (Hint: Look in the files rk_cpend.cpp, rk_cpend.csh)

5.21 Repeat the previous problem when the hinges of the rods slide without friction on the x axis.

5.22 Repeat problem 5.20 by adding a third pendulum to the right at distance l.

Chapter 6

Motion in Space

In this chapter we will study the motion of a particle in space (three dimensions). We will also discuss the case of the relativistic motion, which is important if one wants to consider the motion of particles moving with speeds comparable to the speed of light. This will be an opportunity to use an *adaptive stepsize* Runge-Kutta method for the numerical solution of the equations of motion. We will use the open source code rksuite[1] available at the Netlib[2] repository. Netlib is an open source, high quality repository for numerical analysis software. The software it contains is used by many researchers in their high performance computing programs and it is a good investment of time to learn how to use it. Most of it is code written in Fortran and, in order to use it, you should learn how to link a program written in C++ with functions written in a different programming language.

The main technical skill that you will develop in this chapter is looking for solutions to your numerical problems provided by software written by others. It is important to be able to locate the optimal solution to your problem, find the relevant functions, read the software's documentation carefully and filter out the necessary information in order to call and link the functions to your program.

---

[1] R.W. Brankin, I. Gladwell, and L.F. Shampine, RKSUITE: a suite of Runge-Kutta codes for the initial value problem for ODEs, Softreport 92-S1, Department of Mathematics, Southern Methodist University, Dallas, Texas, U.S.A, 1992.

[2] www.netlib.org

6.1 Adaptive Stepsize Control for RK Methods

The three dimensional equation of motion of a particle is an initial value problem given by the equations (4.6)

$$\frac{dx}{dt} = v_x \qquad \frac{dv_x}{dt} = a_x(t, x, v_x, y, v_y, z, v_z)$$
$$\frac{dy}{dt} = v_y \qquad \frac{dv_z}{dt} = a_y(t, x, v_x, y, v_y, z, v_z)$$
$$\frac{dz}{dt} = v_z \qquad \frac{dv_z}{dt} = a_z(t, x, v_x, y, v_y, z, v_z). \qquad (6.1)$$

For its numerical solution we will use an adaptive stepsize Runge–Kutta algorithm for increased performance and accuracy. Adaptive stepsize is used in cases where one needs to minimize computational effort for given accuracy goal. The method frequently changes the time step during the integration process, so that it is set to be large through smooth intervals and small when there are abrupt changes in the values of the functions. This is achieved by exercising *error control*, either by monitoring a conserved quantity or by computing the same solution using two different methods. In our case, two Runge-Kutta methods are used, one of order p and one of order $p+1$, and the difference of the results is used as an estimate of the truncation error. If the error needs to be reduced, the step size is reduced and if it is satisfactorily small the step size is increased. For the details we refer the reader to [33]. Our goal is not to analyze and understand the details of the algorithm, but to learn how to find and use appropriate and high quality code written by others.

6.1.1 The rksuite Suite of RK Codes

The link http://www.netlib.org/ode/ reads

```
lib     rksuite
alg     Runge-Kutta
for     initial value problem for first order ordinary ←
        differential
           equations. A suite of codes for solving IVPs in ODEs. A
           choice of RK methods, is available. Includes an error
           assessment facility and a sophisticated stiffness checker.
```

6.1. ADAPTIVE STEPSIZE CONTROL FOR RK METHODS

```
      Template programs and example results provided.
      Supersedes RKF45, DDERKF, D02PAF.
ref   RKSUITE, Softreport 92-S1, Dept of Math, SMU, Dallas, ↵
   Texas
by    R.W. Brankin (NAG), I. Gladwell and L.F. Shampine (SMU)
lang  Fortran
prec  double
```

There, we learn that the package provides code for Runge–Kutta methods, whose source is open and written in the Fortran language. We also learn that the code is written for double precision variables, which is suitable for our problem. Last, but not least, we are also happy to learn that it is written by highly reputable people! We download[3] the files rksuite.f, rksuite.doc, details.doc, templates, readme.

In order to link the subroutines provided by the suite to our program we need to read the documentation carefully. In the general case, documentation is available on the web (html, pdf, ...), bundled files with names like README and INSTALL, in whole directories with names like doc/, online help in man and/or info pages and, finally, in good old fashioned printed manuals. Good quality software is also well documented inside the source code files, something that is true for the software at hand.

In order to link the suite's subroutines to our program we need the following basic information:

- INPUT DATA: This is the necessary information that the program needs in order to perform the calculation. In our case, the minimal such information is the initial conditions, the integration time interval and the number of integration steps. The user should also provide the functions on the right hand side of (6.1). It might also be necessary to provide information about the desired accuracy goal, the scale of the problem, the hardware etc.

- OUTPUT DATA: This is the information on how we obtain the results of the calculation for further analysis. Information whether the calculation was successful and error free could also be provided.

---

[3]For the convenience of the reader, these files can be found bundled in the accompanied software in a subdirectory rksuite.

- WORKSPACE: This is information on how we provide the necessary memory space used in the intermediate calculations. Such space needs to be provided by the user in programming languages where dynamical memory allocation is not possible, like in Fortran 77, and the size of workspace depends on the parameters of the calling program.

It is easy to install the software. All the necessary code is in one file rksuite.f. The file rksuite.doc[4] contains the documentation. There we read that we need to inform the program about the hardware dependent accuracy of floating point numbers. We need to set the values of three variables:

```
...
RKSUITE requires three environmental constants OUTCH, MCHEPS,
DWARF. When you use RKSUITE, you may need to know their
values. You can obtain them by calling the subroutine ENVIRN
in the suite:

  CALL ENVIRN(OUTCH,MCHPES,DWARF)

returns values

  OUTCH  - INTEGER
           Standard output channel on the machine being used.
  MCHEPS - DOUBLE PRECISION
           The unit of roundoff, that is, the largest
           positive number such that 1.0D0 + MCHEPS = 1.0D0.
  DWARF  - DOUBLE PRECISION
           The smallest positive number on the machine being
           used.
...
*************************** Installation Details ************

All machine-dependent aspects of the suite have been
 isolated in the subroutine ENVIRN in the rksuite.f file.
...
```

The variables OUTCH, MCHEPS, DWARF are defined in the subroutine ENVIRN. They are given generic default values but the programmer is free to

---

[4]This is a simple text file which you can read with the command less rksuite.doc or with emacs.

6.1. ADAPTIVE STEPSIZE CONTROL FOR RK METHODS

change them by editing ENVIRN. We should identify the routine in the file rksuite.f and read the comments in it[5]:

```
...
      SUBROUTINE ENVIRN(OUTCH,MCHEPS,DWARF)
...
C  The following six statements are to be Commented out
C  after verification that the machine and installation
C  dependent quantities are specified correctly.
...
      WRITE(*,*) ' Before using RKSUITE, you must verify that the  '
      WRITE(*,*) ' machine- and installation-dependent quantities  '
      WRITE(*,*) ' specified in the subroutine ENVIRN are correct, '
      WRITE(*,*) ' and then Comment these WRITE statements and the '
      WRITE(*,*) ' STOP statement out of ENVIRN.                   '
      STOP
...
C  The following values are appropriate to IEEE
C  arithmetic with the typical standard output channel.
C
      OUTCH = 6
      MCHEPS = 1.11D-16
      DWARF = 2.23D-308
```

All we need to do is to comment out the WRITE and STOP commands since we will keep the default values of the OUTCH, MCHEPS, DWARF variables:

```
...
C     WRITE(*,*) ' Before using RKSUITE, you must verify that the  '
C     WRITE(*,*) ' machine- and installation-dependent quantities  '
C     WRITE(*,*) ' specified in the subroutine ENVIRN are correct, '
C     WRITE(*,*) ' and then Comment these WRITE statements and the '
C     WRITE(*,*) ' STOP statement out of ENVIRN.                   '
C     STOP
...
```

In order to check whether the default values are satisfactory, we can use the C++ template numeric_limits, which is part of the C++ Standard Library. In the file numericLimits.cpp, we write a small test program[6]:

```
#include <iostream>
#include <limits>
using namespace std;
```

---

[5] These are lines that begin with a C, as this is old fixed format Fortran code.
[6] The file in the accompanying software, shows you how to compute numeric limits for several types of variables.

```cpp
int main(){
  double MCHEPS,DWARF;

  MCHEPS = numeric_limits<double>::epsilon();
  DWARF  = numeric_limits<double>::min    ();
  cout << "MCHEPS = " << MCHEPS/2.0 << endl;
  cout << "DWARF  = " << DWARF      << endl;
}
```

We compile and run the above program as follows:

```
> g++ numericLimits.cpp -o n
> ./n
MCHEPS = 1.11022e-16
DWARF  = 2.22507e-308
```

We conclude that our choices are satisfactory.

Next, we need to learn how to use the subroutines in the suite. By carefully reading `rksuite.doc` we learn the following: The interface to the adaptive stepsize Runge–Kutta algorithm is the routine UT (UT = "Usual Task"). The routine can use a 2nd-3rd (RK23) order Runge-Kutta pair for error control (METHOD=1), a 4th-5th (RK45) order pair (METHOD=2) or a 7th-8th (RK78) order pair (METHOD=3). We will set METHOD=2 (RK45). The routine SETUP must be called before UT for initialization. The user should provide a function F that calculates the derivatives of the functions we integrate for, i.e. the right hand side of 6.1.

The fastest way to learn how to use the above routines is "by example". The suite include a templates package which can be unpacked by executing the commands in the file `templates` using the sh shell:

```
> sh templates
tmpl1.out
tmpl1a.f
...
```

The file `tmpl1a.f` contains the solution of the simple harmonic oscillator and has many explanatory comments in it. The code is in Fortran, but it is not so hard to read. You may compile it and run it with the commands:

6.1. ADAPTIVE STEPSIZE CONTROL FOR RK METHODS

```
> cd rksuite/templates
> gfortran tmpl1a.f ../rksuite.f -o example1
> ./example1
```

We encourage the reader to study it carefully, run it and test its results.

6.1.2 Interfacing C++ Programs with Fortran

Next, we have to learn how to link the Fortran code in `rksuite.f` to a C++ program. There is a lot of relevant information that you can find with a simple search in the web, we will concentrate on what is relevant to the program that we need to write. The first thing that we need to learn is how to call a function written in Fortran from a C++ program. A simple "Hello World" program can teach us how to do it. Suppose that a Fortran function/subroutine HELLO() is coded in a file `helloF.f90`:

```
SUBROUTINE HELLO()

  PRINT *, 'Hello World!'

END SUBROUTINE HELLO
```

Then, we write in the file `hello.cpp`:

```cpp
#include <iostream>
using namespace std;
extern "C" void hello_();
int main(){
  hello_();
}
```

The first thing that we notice is that we call the function by lowering all letters in its name: HELLO → hello. In Fortran, lowercase and uppercase letters are equivalent, and the compiler creates names with lowercase letters only. Next, we find that we need to append an underscore _ to the function's name[7]: hello → hello_. The Fortran function needs to be declared in the ""C" language linkage":

---

[7] Read your compiler's manual, there could be options that you can use at compile time so that you can avoid it. But the advice is to stick with this convention, so that your code will be portable and ... long lived!

```
extern "C" void hello_();
```

This is something one has to do both for functions written in Fortran, as well as for functions written in plain old C[8].

In order to compile and run the code we have to run the commands:

```
> gfortran -c helloF.f90
> g++ hello.cpp helloF.o -o hello -lgfortran
> ./hello
 Hello World!
```

Compilation is done in two steps: We first need to compile the Fortran program using the Fortran compiler[9] gfortran. The flag -c forces the compiler to perform compilation but not linking. It produces an *object file*, whose extension is .o, helloF.o. These are files which contain compiled code in non-readable text form, but they are not autonomous executable programs. The functions that they contain, can be *linked* to other compiled programs that call them. In the second step, g++ is called to *compile* the C++ source code in hello.cpp and *link* it to the functions in helloF.o. The flag -lgfortran is necessary in order to link to the standard Fortran functions. If you use a different compiler, you should read its manual in order to find the correct linking options.

Another subtle point that needs to be considered is that, in Fortran, variables are passed to functions *by reference* and not *by value*. Therefore, we have to pass variables as pointers or as reference to variables. For example, the following Fortran function[10]

```
REAL (8) FUNCTION SQUAREMYDOUBLE(X)
  REAL(8) x

  X = 2.0*X
```

---

[8] Language linkage encapsulates the set of requirements necessary in order to *link* with a function written in another programming language, and it should be done for all function types, names and variable names. There could be linkage to other programming languages, but only the C and C++ linkage is guaranteed to be available.

[9] Available for free download from gcc.gnu.org/fortran/. On debian based Linux systems, install with sudo apt-get install gfortran.

[10] A Fortran function returns the value stored in a variable with the same name as the name of the function, here SQUAREMYDOUBLE.

6.1. ADAPTIVE STEPSIZE CONTROL FOR RK METHODS

```
SQUAREMYDOUBLE = X*X

END      FUNCTION SQUAREMYDOUBLE
```

must be declared and called as:

```
extern "C" double squaremydouble_(double& x);
...
  double x,x2;
  x2 = squaremydouble_(x);
```

For example, by modifying the hello.cpp program as

```
#include <iostream>
using namespace std;
extern "C" void    hello_();
extern "C" double  squaremydouble_(double& x);
int main(){
  double x,x2;
  hello_();
  x  = 2.0;
  x2 = squaremydouble_(x);
  cout << "x      = "          << 2.0 << endl;
  cout << "2 x  = "            << x   << endl;
  cout << "(2 x)*(2 x) = "     << x2  << endl;
}
```

we obtain the output:

```
> gfortran -c helloF.f90;
> g++ hello.cpp helloF.o -o hello -lgfortran
> ./hello
 Hello World!
x       = 2
2 x  = 4
(2 x)*(2 x) = 16
```

Notice that the value of x is modified in the calling program.

The final issue that we will consider, it how to pass arrays to Fortran functions. One dimensional arrays are quite easy to handle. In order to pass an array double v[N] to a Fortran function, we only need to declare it and pass it as one of its arguments. If the Fortran program is

```
real(8) function make_array1(v,N)
 integer    N
 real(8) v(N)
 ... compute v(1) ... v(N) ...
end        function make_array1
```

then the corresponding C++ program should be:

```
const int N=4;
extern "C" double make_array1_(double v[N], const int& N);
int main(){
  double v[N], x;
  ...
  x = make_array1_(v,N);
  ... use v[0] ... v[N-1] ...
}
```

The only point we need to stress is that the array real(8) v(N) is indexed from 1 to N in the Fortran program, whereas the array double v[N] is indexed from 0 to N-1 in the C++ program. The correspondence of the values stored in memory is v(1) → v[0], ... , v(i) → v[i-1], ... v(N) → v[N-1].

Two dimensional arrays need more attention. In C++, arrays are in *row-major* mode, whereas in Fortran in *column-major* mode. The contents of a two dimensional array are stored linearly in memory. In C++, the elements of an array double A[N][M] are stored in the sequence A[0][0], A[0][1], A[0][2], ... , A[0][M-1], A[1][0], A[1][1], ... , A[1][M-1], ... A[N-1][M-1]. In Fortran, the elements of an array real(8) A(N,M) are stored in the sequence A(1,1), A(2,1), A(3,1), ... , A(1,M), A(2,1), A(2,2), ... , A(2,M), ... , A(N,M). Therefore, when we pass an array from the C++ program to a Fortran function, we have to keep in mind that the Fortran function will use it with its indices transposed. For example, if the Fortran code defines

```
A(i,j) = i + j/10.0
```

which results into A(i,j) = i.j in decimal notation[11] (e.g. A(2,3)= 2.3), then the value of A[i][j] will be (j+1).(i+1) (e.g. A[2][3] =

---

[11]Of course, we assume i,j<10.

6.1. ADAPTIVE STEPSIZE CONTROL FOR RK METHODS 315

4.3, A[2][1] = 2.3)!

All of the above are summarized in the Fortran program in the file CandFortranF.f90:

```fortran
real(8) function make_array1(v,N)
 implicit none
 integer N,i
 real(8) v(N)

 do i=1,N
  v(i) = i
 end do

 make_array1 = -11.0 ! a return value

end function make_array1
!_____
real(8) function make_array2(A,N,M)
 implicit none
 integer N,M,i,j
 real(8) A(M,N) ! Careful: N and M are interchanged!!

 do i=1,M
  do j=1,N
  !A(i,j) = i.j, e.g. A(2,3)=2.3
   A(i,j) = i + j/10.0
  end do
 end do

 make_array2 = -22.0 ! a return value
end function make_array2
```

and the C++ program in the file CandFortranC.cpp:

```cpp
#include <iostream>
#include <fstream>
#include <cstdlib>
#include <string>
#include <cmath>

using namespace std;
```

```
const int N=4, M=3;
extern "C" {
  double make_array1_(double v[N]   , const int& N);
  double make_array2_(double A[N][M], const int& N,
                      const int& M);
}
int main(){
  double A[N][M], v[N];
  double x;

  //Make a 1D array using a fortran function:
  x = make_array1_(v,N);
  cout << "1D array: Return value x= " << x << endl;
  for(int i=0;i<N;i++)
    cout << v[i] << " ";
  cout << "\n-----------------------\n";
  //Make an 2D array using a fortran function:
  x = make_array2_(A,N,M);
  cout << "2D array: Return value x= " << x << endl;
  for(int i=0;i<N;i++){
    for(int j=0;j<M;j++)   //A is ... transposed!
      //A[i][j] = (j+1).(i+1), e.g. A[1][2] = 3.1
      cout << A[i][j] << " ";
    cout << '\n';
  }
}
```

Note that the array A[N][M] is defined as A(M,N) in the Fortran function, and the roles of N and M are interchanged. You can run the code and see the output with the commands:

```
> gfortran -c CandFortranF.f90
> g++ CandFortranC.cpp CandFortranF.o -o CandFortran -lgfortran
> ./CandFortran
1D array: Return value x= -11
1 2 3 4
-----------------------
2D array: Return value x= -22
1.1  2.1  3.1
1.2  2.2  3.2
1.3  2.3  3.3
1.4  2.4  3.4
```

Note that the values of the array A(M,N) are transposed when printed as

6.1.3 The `rksuite` Driver

After we become wise enough, we can write the driver for the integration routine UT (called by ut_ from our C++ program), which can be found in the file rk3.cpp:

```cpp
//================================================================
//Program to solve a 6 ODE system using Runge-Kutta Method
//Output is written in file rk3.dat
//================================================================
// Compile with the commands:
// gfortran -c rksuite/rksuite.f;
// g++ rk3.cpp rk3_g.cpp rksuite.o -o rk3 -lgfortran
#include <iostream>
#include <fstream>
#include <cstdlib>
#include <string>
#include <cmath>
#include "rk3.h"
using namespace std;
//----------------------------------------------------------------
double k1,k2,k3,k4;
double energy(const double& t, double* Y);
void f(double& t,double* Y, double* YP);
extern "C" {
  void setup_(const int& NEQ,
              double& TSTART,double* YSTART,double& TEND,
              double& TOL    ,double* THRES ,
              const int& METHOD, const char& TASK,
              bool  & ERRASS,double& HSTART,double* WORK,
              const int& LENWRK,          bool& MESSAGE);
  void ut_( void f(double& t,double* Y, double* YP),
            double& TWANT, double& TGOT, double* YGOT,
            double* YPGOT, double* YMAX, double* WORK,
            int& UFLAG);
}
//----------------------------------------------------------------
int main(){
  string buf;
  double T0,TF,X10,X20,X30,V10,V20,V30;
  double t,dt,tstep;
```

```
int     STEPS, i;
// rksuite variables:
double  TOL,THRES[NEQ],WORK[LENWRK],HSTART;
double  Y[NEQ],YMAX[NEQ],YP[NEQ],YSTART[NEQ];
bool    ERRASS, MESSAGE;
int     UFLAG;
const char TASK = 'U';
//Input:
cout << "Runge-Kutta Method for 6-ODEs Integration\n";
cout << "Enter coupling constants k1,k2,k3,k4:\n";
cin >> k1 >> k2 >> k3 >> k4;getline(cin,buf);
cout << "Enter STEPS,T0,TF,X10,X20,X30,V10,V20,V30:\n";
cin >> STEPS >> T0   >> TF
    >> X10   >> X20 >> X30
    >> V10   >> V20 >> V30;getline(cin,buf);
cout << "No. Steps= " << STEPS << endl;
cout << "Time: Initial T0 =" << T0
     << "  Final TF="        << TF   << endl;
cout << "          X1(T0)=" << X10
     << " X2(T0)="           << X20
     << " X3(T0)="           << X30 << endl;
cout << "          V1(T0)=" << V10
     << " V2(T0)="           << V20
     << " V3(T0)="           << V30 << endl;
//Initial Conditions:
dt        = (TF-T0)/STEPS;
YSTART[0] = X10;
YSTART[1] = X20;
YSTART[2] = X30;
YSTART[3] = V10;
YSTART[4] = V20;
YSTART[5] = V30;
//Set control parameters:
TOL = 5.0e-6;
for( i = 0; i < NEQ; i++)
  THRES[i] = 1.0e-10;
MESSAGE = true;
ERRASS  = false;
HSTART  = 0.0;
//Initialization:
setup_(NEQ,T0,YSTART,TF,TOL,THRES,METHOD,TASK,
       ERRASS,HSTART,WORK,LENWRK,MESSAGE);
ofstream  myfile("rk3.dat");
myfile.precision(16);
myfile << T0<< " "
```

6.1. ADAPTIVE STEPSIZE CONTROL FOR RK METHODS

```
                << YSTART[0] << " " << YSTART[1] << " "
                << YSTART[2] << " " << YSTART[3] << " "
                << YSTART[4] << " " << YSTART[5] << " "
                << energy(T0,YSTART)<< '\n';
  //The calculation:
  for(i=1;i<=STEPS;i++){
    t = T0 + i*dt;
    ut_(f,t,tstep,Y,YP,YMAX,WORK,UFLAG);
    if(UFLAG > 2) break; //error: break the loop and exit
    myfile << tstep << " "
           << Y[0] << " " << Y[1] << " "
           << Y[2] << " " << Y[3] << " "
           << Y[4] << " " << Y[5] << " "
           << energy(T0,Y) << '\n';
  }
  myfile.close();
}// main()
```

All common parameters and variables are declared in an include file rk3.h. This is done so that they are accessible by the function f which calculates the derivatives:

```
const int NEQ    = 6;
const int LENWRK = 32*NEQ;
const int METHOD = 2;
extern double k1,k2,k3,k4;
```

The number of differential equations is set equal to NEQ=6. The integration method is set by the choice METHOD=2. The variable LENWRK sets the size of the workspace needed by the suite for the intermediate calculations.

The declaration of functions needs some care. The functions energy() and f() are defined in the C++ program and are declared in the global scope (the function f() will also be passed on to the Fortran function UT). The functions setup_() and ut_() are defined in the Fortran program in the file rksuite.f as SETUP() and UT(). Therefore, they are declared within the extern "C" language linkage, defined using lowercase letters and an underscore is appended to their names. All arguments are passed by reference. Scalar doubles are passed as references to double&, the double precision arrays are declared to be pointers double *, the Fortran logical variables are declared to be references to bool&, and the Fortran character* variables[12] as simple references to char&.

The main program starts with the user interface. The initial state of the particle is stored in the array YSTART in the positions $0\ldots 5$. The first three positions are the coordinates of the initial position and the last three the components of the initial velocity. Then, we set some variables that determine the behavior of the integration program (see the file rksuite.doc for details) and call the subroutine SETUP. The main integration loop is:

```
for(i=1;i<=STEPS;i++){
  t = T0 + i*dt;
  ut_(f,t,tstep,Y,YP,YMAX,WORK,UFLAG);
  if(UFLAG > 2) break;  //error: break the loop and exit
  myfile << ... << energy(T0,Y) << '\n';
}
```

The function f calculates the derivatives and it will be programmed by us later. The variable t stores the *desired* moment of time at which we want to calculate the functions. Because of the adaptive stepsize, it can be different than the one returned by the Fortran subroutine UT. The actual value of time that the next step lands[13] on is tstep. The array Y stores the values of the functions. We choose the data structure to be such that $x=$ Y[0], $y=$ Y[1], $z=$ Y[2] and $v_x=$ Y[3], $v_y=$ Y[4], $v_z=$ Y([5] (the same sequence as in the array YSTART). The function energy(t,Y) returns the value of the mechanical energy of the particle and its code will be written in the same file as that of f. Finally, the variable UFLAG indicates the error status of the calculation by UT and if UFLAG> 2 we end the calculation.

Our test code will be on the study of the motion of a projectile in a constant gravitational field, subject also to the influence of a dissipative force $\vec{F}_r = -mk\vec{v}$. The program is in the file rk3_g.cpp. We choose the parameters k1 and k2 so that $\vec{g} =$ -k1 $\hat{k}$ and $k =$ k2.

```
#include <iostream>
#include <fstream>
```

---

[12]The Fortran program uses only their first character, so we don't need to use strings.

[13]When UGLAG ≤ 2, tstep=t and we will not worry about them being different with our program.

6.1. ADAPTIVE STEPSIZE CONTROL FOR RK METHODS

```cpp
#include <cstdlib>
#include <string>
#include <cmath>
#include "rk3.h"
using namespace std;
void f(double& t,double* Y, double* YP){
  double x1,x2,x3,v1,v2,v3;
  x1 = Y[0]; v1 = Y[3];
  x2 = Y[1]; v2 = Y[4];
  x3 = Y[2]; v3 = Y[5];
  // Velocities:   dx_i/dt = v_i
  YP[0] = v1;
  YP[1] = v2;
  YP[2] = v3;
  // Acceleration: dv_i/dt = a_i
  YP[3] = -k2*v1;
  YP[4] = -k2*v2;
  YP[5] = -k2*v3-k1;
}
//------------------------------------------------
double energy(const double& t, double* Y){
  double e;
  double x1,x2,x3,v1,v2,v3;
  x1 = Y[0]; v1 = Y[3];
  x2 = Y[1]; v2 = Y[4];
  x3 = Y[2]; v3 = Y[5];
  // Kinetic   Energy:
  e = 0.5*(v1*v1+v2*v2+v3*v3);
  // Potential Energy:
  e += k1*x3;

  return e;
}
```

For convenience we "translated" the values in the array Y[NEQ] into user-friendly variable names. If the file rksuite.f is in the directory rksuite/, then the compilation, running and visualization of the results can be done with the commands:

```
> gfortran -c rksuite/rksuite.f
> g++ rk3.cpp rk3_g.cpp rksuite.o -o rk3 -lgfortran
> ./rk3
Runge-Kutta Method for 6-ODEs Integration
Enter coupling constants k1,k2,k3,k4:
```

```
10 0 0 0
Enter STEPS,T0,TF,X10,X20,X30,V10,V20,V30:
10000 0 3 0 0 0 1 1 1
No. Steps= 10000
Time: Initial T0 =0 Final TF=3
         X1(T0)=0 X2(T0)=0 X3(T0)=0
         V1(T0)=1 V2(T0)=1 V3(T0)=1
> gnuplot
gnuplot> plot "rk3.dat"  using 1:2 with lines title "x1(t)"
gnuplot> plot "rk3.dat"  using 1:3 with lines title "x2(t)"
gnuplot> plot "rk3.dat"  using 1:4 with lines title "x3(t)"
gnuplot> plot "rk3.dat"  using 1:5 with lines title "v1(t)"
gnuplot> plot "rk3.dat"  using 1:6 with lines title "v2(t)"
gnuplot> plot "rk3.dat"  using 1:7 with lines title "v3(t)"
gnuplot> plot "rk3.dat"  using 1:8 with lines title "E(t)"
gnuplot> set title "trajectory"
gnuplot> splot "rk3.dat" using 2:3:4 with lines  notitle
```

All the above commands can be executed together using the shell script in the file rk3.csh. The script uses the animation script rk3_animate.csh. The following command executes all the commands shown above:

```
./rk3.csh -f 1 -- 10 0. 0 0 0 0 1 1 1 10000 0 3
```

6.2 Motion of a Particle in an EM Field

In this section we study the non-relativistic motion of a charged particle in an electromagnetic (EM) field. The particle is under the influence of the Lorentz force:

$$\vec{F} = q(\vec{E} + \vec{v} \times \vec{B}). \tag{6.2}$$

Consider the constant EM field of the form $\vec{E} = E_x\hat{x} + E_y\hat{y} + E_z\hat{z}$, $\vec{B} = B\hat{z}$. The components of the acceleration of the particle are:

$$\begin{aligned} a_x &= (qE_x/m) + (qB/m)v_y \\ a_y &= (qE_y/m) - (qB/m)v_x \\ a_z &= (qE_z/m). \end{aligned} \tag{6.3}$$

This field is programmed in the file rk3_B.cpp. We set k1 = qB/m, k2 = qE_x/m, k3 = qE_y/m and k4 = qE_z/m:

6.2. MOTION OF A PARTICLE IN AN EM FIELD

```
//================================================
//  Particle in constant magnetic and electric field
//  q B/m = k1 z    q E/m = k2 x + k3 y + k4 z
//================================================
#include "sr.h"
void f(double& t,double* Y, double* YP){
  double x1,x2,x3,v1,v2,v3,p1,p2,p3;
  x1 = Y[0]; p1 = Y[3];
  x2 = Y[1]; p2 = Y[4];
  x3 = Y[2]; p3 = Y[5];
  velocity(p1,p2,p3,v1,v2,v3);
  //now we can use all x1,x2,x3,p1,p2,p3,v1,v2,v3
  YP[0] = v1;
  YP[1] = v2;
  YP[2] = v3;
  //Acceleration:
  YP[3] = k2 + k1 * v2;
  YP[4] = k3 - k1 * v1;
  YP[5] = k4;
}
//------------------------------------------------
double energy(const double& t, double* Y){
  double e;
  double x1,x2,x3,v1,v2,v3,p1,p2,p3,psq;
  x1 = Y[0]; p1 = Y[3];
  x2 = Y[1]; p2 = Y[4];
  x3 = Y[2]; p3 = Y[5];
  psq= p1*p1+p2*p2+p3*p3;
  //Kinetic    Energy:
  e   = sqrt(1.0+psq)-1.0;
  //Potential Energy/m_0
  e += - k2*x1 - k3*x2 - k4*x3;

  return e;
}
```

We can also study space-dependent fields in the same way. The fields must satisfy Maxwell's equations. We can study the confinement of a particle in a region of space by a magnetic field by taking $\vec{B} = B_y \hat{y} + B_z \hat{z}$ with $qB_y/m = -k_2 y$, $qB_z/m = k_1 + k_2 z$ and $qB_y/m = k_3 z$, $qB_z/m = k_1 + k_2 y$. Note that $\vec{\nabla} \cdot \vec{B} = 0$. You may also want to calculate the current density from the equation $\vec{\nabla} \times \vec{B} = \mu_0 \vec{j}$.

The results are shown in figures 6.1–6.4.

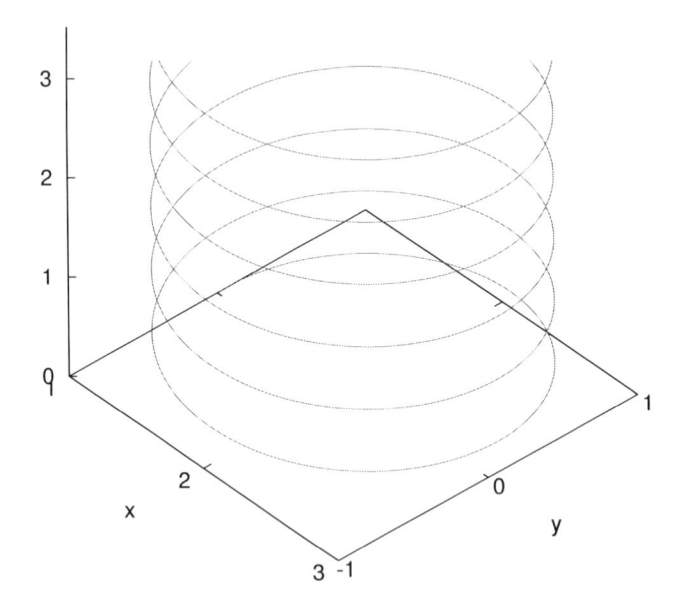

Figure 6.1: The trajectory of a charged particle in a constant magnetic field $\vec{B} = B\hat{z}$, where $qB/m = 1.0$, $\vec{v}(0) = 1.0\hat{y} + 0.1\hat{z}$, $\vec{r}(0) = 1.0\hat{x}$. The integration of the equations of motion is performed using the RK45 method from $t_0 = 0$ to $t_f = 40$ with 1000 steps.

6.3 Relativistic Motion

Consider a particle of non zero rest mass moving with speed comparable to the speed of light. In this case, it is necessary to study its motion using the equations of motion given by special relativity[14]. In the equations below we set $c = 1$. The particle's rest mass is $m_0 > 0$, its mass is $m = m_0/\sqrt{1-v^2}$ (where $v < 1$), its momentum is $\vec{p} = m\vec{v}$ and its energy is $E = m = \sqrt{p^2 + m_0^2}$. Then the equations of motion in a dynamic field $\vec{F}$ are given by:

$$\frac{d\vec{p}}{dt} = \vec{F}. \qquad (6.4)$$

---

[14] Of course for lower speeds, the special relativity equations of motion are a better approximation to the particle's motion, but the corrections to the non relativistic equations of motion are negligible.

6.3. RELATIVISTIC MOTION

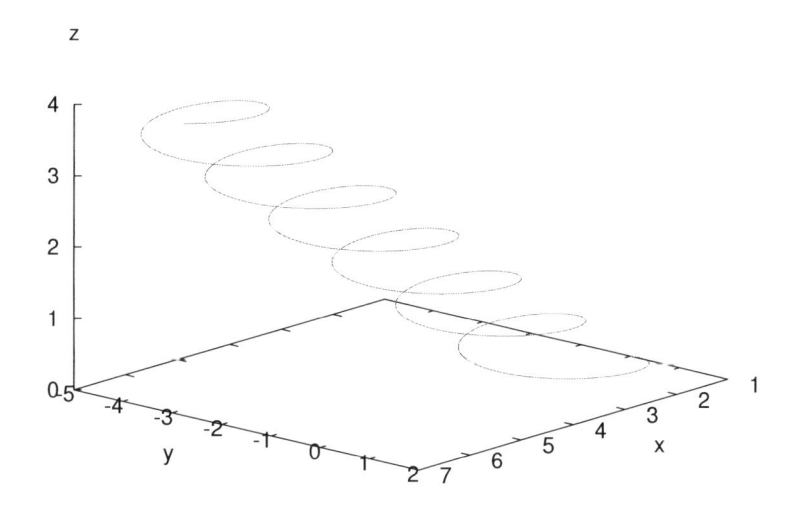

Figure 6.2: The trajectory of a charged particle in a constant magnetic field $\vec{B} = B\hat{z}$, where $qB/m = 1.0$ and a constant electric field $\vec{E} = E_x\hat{x} + E_y\hat{y}$ με $qE_x/m = qE_y/m = 0.1$. $\vec{v}(0) = 1.0\hat{y} + 0.1\hat{z}$, $\vec{r}(0) = 1.0\hat{x}$. The integration of the equations of motion is performed using the RK45 method from $t_0 = 0$ to $t_f = 40$ with 1000 steps. Each axis is on a different scale.

In order to write a system of first order equations, we use the relations

$$\vec{v} = \frac{\vec{p}}{m} = \frac{\vec{p}}{E} = \frac{\vec{p}}{\sqrt{p^2 + m_0^2}}. \tag{6.5}$$

Using $\vec{v} = d\vec{r}/dt$ we obtain

$$\begin{aligned}
\frac{dx}{dt} &= \frac{(p_x/m_0)}{\sqrt{1 + (p/m_0)^2}}, & \frac{d(p_x/m_0)}{dt} &= \frac{F_x}{m_0} \\
\frac{dy}{dt} &= \frac{(p_y/m_0)}{\sqrt{1 + (p/m_0)^2}}, & \frac{d(p_y/m_0)}{dt} &= \frac{F_y}{m_0} \\
\frac{dz}{dt} &= \frac{(p_z/m_0)}{\sqrt{1 + (p/m_0)^2}}, & \frac{d(p_z/m_0)}{dt} &= \frac{F_z}{m_0},
\end{aligned} \tag{6.6}$$

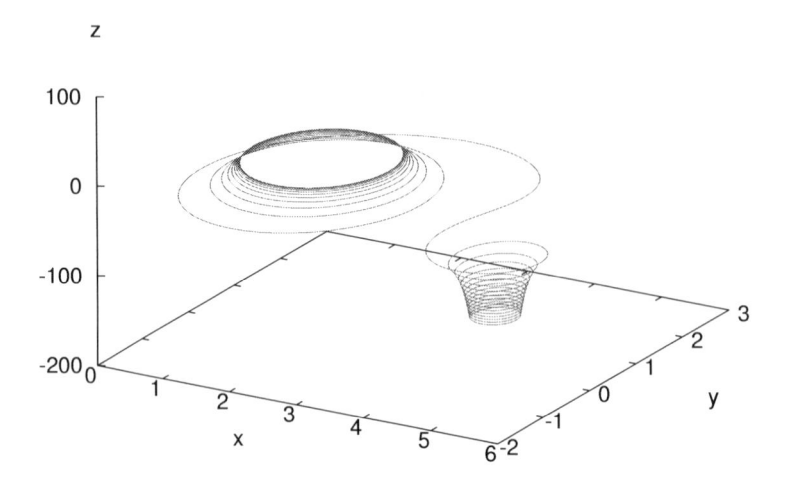

Figure 6.3: The trajectory of a charged particle in a magnetic field $\vec{B} = B_y\hat{y} + B_z\hat{z}$ with $qB_y/m = -0.02y$, $qB_z/m = 1 + 0.02z$, $\vec{v}(0) = 1.0\hat{y} + 0.1\hat{z}$, $\vec{r}(0) = 1.0\hat{x}$. The integration of the equations of motion is performed using the RK45 method from $t_0 = 0$ to $t_f = 500$ with 10000 steps. Each axis is on a different scale.

which is a system of first order differential equations for the functions $(x(t), y(t), z(t), (p_x/m_0)(t), (p_y/m_0)(t), (p_z/m_0)(t))$. Given the initial conditions $(x(0), y(0), z(0), (p_x/m_0)(0), (p_y/m_0)(0), (p_z/m_0)(0))$ their solution is unique and it can be computed numerically using the 4th-5th order Runge–Kutta method according to the discussion of the previous section. By using the relations

$$(p_x/m_0) = \frac{v_x}{\sqrt{1-v^2}} \qquad v_x = \frac{(p_x/m_0)}{\sqrt{1+(p/m_0)^2}}$$

$$(p_y/m_0) = \frac{v_y}{\sqrt{1-v^2}} \qquad v_y = \frac{(p_y/m_0)}{\sqrt{1+(p/m_0)^2}}$$

$$(p_z/m_0) = \frac{v_z}{\sqrt{1-v^2}} \qquad v_z = \frac{(p_z/m_0)}{\sqrt{1+(p/m_0)^2}},$$

(6.7)

6.3. RELATIVISTIC MOTION

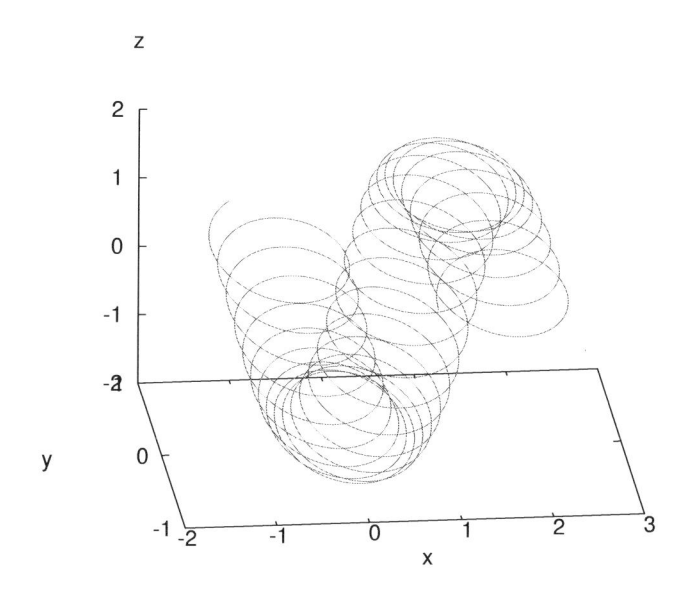

Figure 6.4: The trajectory of a charged particle in a magnetic field $\vec{B} = B_y\hat{y} + B_z\hat{z}$ with $qB_y/m = 0.08z$, $qB_z/m = 1.4 + 0.08y$, $\vec{v}(0) = 1.0\hat{y} + 0.1\hat{z}$, $\vec{r}(0) = 1.0\hat{x}$. The integration of the equations of motion is performed using the RK45 method from $t_0 = 0$ to $t_f = 3000$ with 40000 steps. Each axis is on a different scale.

we can use the initial conditions $(x(0), y(0), z(0), v_x(0), v_y(0), v_z(0))$ instead. Similarly, from the solutions $(x(t), y(t), z(t), (p_x/m_0)(t), (p_y/m_0)(t), (p_z/m_0)(t))$ we can calculate $(x(t), y(t), z(t), v_x(t), v_y(t), v_z(t))$. We always have to check that

$$v^2 = (v_x)^2 + (v_y)^2 + (v_z)^2 < 1. \tag{6.8}$$

Since half of the functions that we integrate for are the momentum instead of the velocity components, we need to make some modifications to the program in the file `rk3.cpp`. The main program can be found in the file `sr.cpp`:

```
//================================================
//Program to solve a 6 ODE system using Runge-Kutta Method
//Output is written in file sr.dat
//================================================
```

```cpp
// Compile with the commands:
// gfortran -c rksuite/rksuite.f
// g++ sr.cpp sr_B.cpp rksuite.o -o rk3 -lgfortran
//----------------------------------------------------------------
#include "sr.h"
double k1,k2,k3,k4;
extern "C" {
  void setup_(const int& NEQ,
              double& TSTART, double* YSTART, double& TEND,
              double& TOL    , double* THRES  ,
              const int& METHOD, const char& TASK,
              bool  & ERRASS, double& HSTART, double* WORK,
              const int& LENWRK,        bool& MESSAGE);
  void ut_( void f(double& t, double* Y, double* YP),
            double& TWANT,  double& TGOT, double* YGOT,
            double* YPGOT,  double* YMAX, double* WORK,
            int& UFLAG);
}
//----------------------------------------------------------------
int main(){
  string buf;
  double T0,TF,X10,X20,X30,V10,V20,V30;
  double P10,P20,P30;
  double P1,P2,P3,V1,V2,V3;
  double t,dt,tstep;
  int    STEPS, i;
  // rksuite variables:
  double TOL,THRES[NEQ],WORK[LENWRK],HSTART;
  double Y[NEQ],YMAX[NEQ],YP[NEQ],YSTART[NEQ];
  bool   ERRASS, MESSAGE;
  int    UFLAG;
  const char TASK = 'U';
  //Input:
  cout << "Runge-Kutta Method for 6-ODEs Integration\n";
  cout << "Special Relativistic Particle:\n";
  cout << "Enter coupling constants k1,k2,k3,k4:\n";
  cin  >> k1 >> k2 >> k3 >> k4;getline(cin,buf);
  cout << "Enter STEPS,T0,TF,X10,X20,X30,V10,V20,V30:\n";
  cin  >> STEPS >> T0 >> TF
       >> X10   >> X20 >> X30
       >> V10   >> V20 >> V30;getline(cin,buf);
  momentum(V10,V20,V30,P10,P20,P30);
  cout << "No. Steps= " << STEPS << endl;
  cout << "Time: Initial T0 =" << T0
       << " Final TF="         << TF   << endl;
```

6.3. RELATIVISTIC MOTION

```
cout <<  "               X1(T0)="  << X10
     <<  " X2(T0)="              << X20
     <<  " X3(T0)="              << X30 << endl;
cout <<  "               V1(T0)="  << V10
     <<  " V2(T0)="              << V20
     <<  " V3(T0)="              << V30 << endl;
cout <<  "               P1(T0)="  << P10
     <<  " P2(T0)="              << P20
     <<  " P3(T0)="              << P30 << endl;
//Initial Conditions:
dt         = (TF-T0)/STEPS;
YSTART[0] = X10;
YSTART[1] = X20;
YSTART[2] = X30;
YSTART[3] = P10;
YSTART[4] = P20;
YSTART[5] = P30;
//Set control parameters:
TOL = 5.0e-6;
for( i = 0; i < NEQ; i++)
  THRES[i] = 1.0e-10;
MESSAGE = true;
ERRASS  = false;
HSTART  = 0.0;
//Initialization:
setup_(NEQ,T0,YSTART,TF,TOL,THRES,METHOD,TASK,
       ERRASS,HSTART,WORK,LENWRK,MESSAGE);
ofstream  myfile("sr.dat");
myfile.precision(16);
myfile << T0            << " "
       << YSTART[0] << " " << YSTART[1] << " "
       << YSTART[2] << " "
       << V1         << " " << V2          << " "
       << V3         << " "
       << energy(T0,YSTART)                << " "
       << YSTART[3] << " " << YSTART[4] << " "
       << YSTART[5] << '\n';
//The calculation:
for(i=1;i<=STEPS;i++){
  t = T0 + i*dt;
  ut_(f,t,tstep,Y,YP,YMAX,WORK,UFLAG);
  if(UFLAG > 2) break; //error: break the loop and exit
  velocity(Y[3],Y[4],Y[5],V1,V2,V3);
  myfile << tstep << " "
         << Y[0]  << " " << Y[1] << " "
```

```
                       << Y[2]    << " "
               << V1    << " "    << V2   << " "
               << V3    << " "
               << energy(T0,Y)         << " "
               << Y[3]  << " "   << Y[4] << " "
               << Y[5]  << " "   << '\n';
    }
    myfile.close();
}// main()
//=========================================================
//momentum -> velocity  transformation
//=========================================================
void velocity(const double& p1,const double& p2,
              const double& p3,
              double& v1,        double& v2,
              double& v3){
    double psq;
    psq = p1*p1+p2*p2+p3*p3;

    v1 = p1/sqrt(1.0+psq);
    v2 = p2/sqrt(1.0+psq);
    v3 = p3/sqrt(1.0+psq);
}
//=========================================================
// velocity -> momentum transformation
//=========================================================
void momentum(const double& v1,const double& v2,
              const double& v3,
              double& p1,        double& p2,
              double& p3){
    double vsq;
    vsq = v1*v1+v2*v2+v3*v3;
    if(vsq >= 1.0){cerr << "momentum: vsq>=1\n";exit(1);}
    p1 = v1/sqrt(1.0-vsq);
    p2 = v2/sqrt(1.0-vsq);
    p3 = v3/sqrt(1.0-vsq);


}
```

The functions momentum and velocity compute the transformations (6.7). In the function momentum we check whether the condition (6.8) is satisfied. These functions are also used in the function F that computes the derivatives of the functions.

Common declarations are now in an include file sr.h:

6.3. RELATIVISTIC MOTION

```
#include <iostream>
#include <fstream>
#include <cstdlib>
#include <string>
#include <cmath>
using namespace std;
//----------------------------------------------------
const int NEQ    = 6;
const int LENWRK = 32*NEQ;
const int METHOD = 2;
extern double k1,k2,k3,k4;
//----------------------------------------------------
double energy(const double& t,        double* Y);
void   f(double& t, double* Y,        double* YP);
void velocity(const double& p1,const double& p2,
              const double& p3,
              double& v1,      double& v2,
              double& v3);
void momentum(const double& v1,const double& v2,
              const double& v3,
              double& p1,      double& p2,
              double& p3);
```

The test drive of the above program is the well known relativistic motion of a charged particle in a constant EM field. The acceleration of the particle is given by equations (6.3). The relativistic kinetic energy of the particle is

$$T = \left(\frac{1}{\sqrt{1-v^2}} - 1\right) m_0 = \left(\sqrt{1 + (p/m_0)^2} - 1\right) m_0 \qquad (6.9)$$

These relations are programmed in the file sr_B.cpp. The contents of the file sr_B.cpp are:

```
//====================================================
//   Particle in constant Magnetic and electric field
//   q B/m = k1 z    q E/m = k2 x + k3 y + k4 z
//====================================================
#include "sr.h"
void f(double& t,double* Y, double* YP){
  double x1,x2,x3,v1,v2,v3,p1,p2,p3;
  x1 = Y[0]; p1 = Y[3];
  x2 = Y[1]; p2 = Y[4];
```

```
  x3 = Y[2]; p3 = Y[5];
  velocity(p1,p2,p3,v1,v2,v3);
  //now we can use all x1,x2,x3,p1,p2,p3,v1,v2,v3
  YP[0] = v1;
  YP[1] = v2;
  YP[2] = v3;
  // Acceleration:
  YP[3] = k2 + k1 * v2;
  YP[4] = k3 - k1 * v1;
  YP[5] = k4;
}
//————————————————————————————————
double energy(const double& t, double* Y){
  double e;
  double x1,x2,x3,v1,v2,v3,p1,p2,p3,psq;
  x1 = Y[0]; p1 = Y[3];
  x2 = Y[1]; p2 = Y[4];
  x3 = Y[2]; p3 = Y[5];
  psq= p1*p1+p2*p2+p3*p3;
  // Kinetic    Energy:
  e   = sqrt(1.0+psq)-1.0;
  // Potential Energy/m_0
  e += - k2*x1 - k3*x2 - k4*x3;

  return e;
}
```

The results are shown in figures 6.5–6.6.

Now we can study a more interesting problem. Consider a simple model of the Van Allen radiation belt. Assume that the electrons are moving within the Earth's magnetic field which is modeled after a magnetic dipole field of the form:

$$\vec{B} = B_0 \left(\frac{R_E}{r}\right)^3 \left[3(\hat{d}\cdot\hat{r})\hat{r} - \hat{d}\right], \qquad (6.10)$$

where $\vec{d} = d\hat{d}$ is the magnetic dipole moment of the Earth's magnetic field and $\vec{r} = r\hat{r}$. The parameter values are approximately equal to $B_0 = 3.5 \times 10^{-5} T$, $r \sim 2R_E$, where R_E is the radius of the Earth. The typical energy of the moving particles is ~ 1 MeV which corresponds to velocities of magnitude $v/c = \sqrt{E^2 - m_0^2}/E \approx \sqrt{1 - 0.512^2}/1 = 0.86$. We choose the coordinate axes so that $\hat{d} = \hat{z}$ and we measure distance in R_E units[15].

---

[15]Since $c = 1$, the unit of time is the time that the light needs to travel distance equal

6.3. RELATIVISTIC MOTION

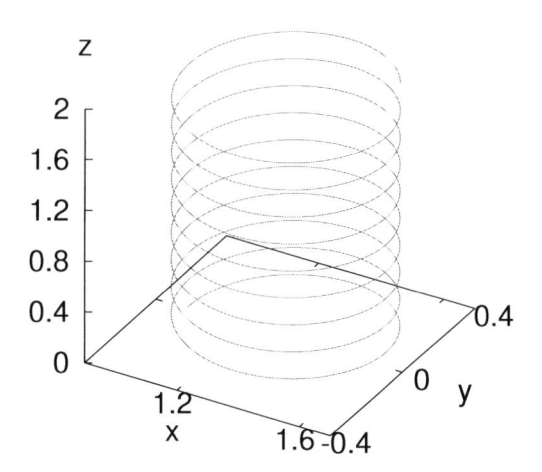

Figure 6.5: The trajectory of a relativistic charged particle in a magnetic field $\vec{B} = B_z \hat{z}$ with $qB_z/m_0 = 10.0$, $\vec{v}(0) = 0.95\hat{y} + 0.10\hat{z}$, $\vec{r}(0) = 1.0\hat{x}$. The integration is performed by using the RK45 method from $t_0 = 0$ to $t_f = 20$ with 1000 steps. Each axis is on a different scale.

Then we obtain:

$$B_x = B_0 \frac{3xz}{r^5}$$
$$B_y = B_0 \frac{3yz}{r^5}$$
$$B_z = B_0 \left(\frac{3zz}{r^5} - \frac{1}{r^3} \right) \tag{6.11}$$

The magnetic dipole field is programmed in the file sr_Bd.cpp:

```
//===========================================================
// Particle in Magnetic dipole field:
// q B_1/m = k1 (3 x1 x3)/r^5
```

---

to R_E in the vacuum.

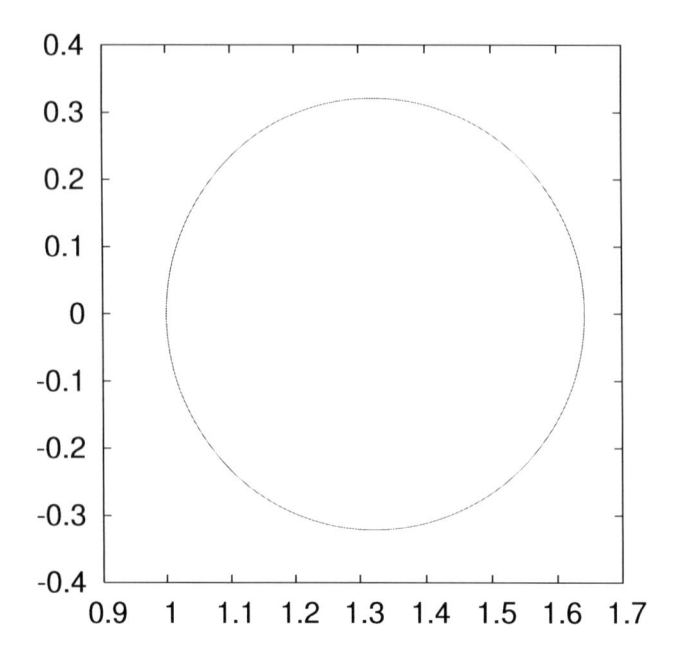

Figure 6.6: Projection of the trajectory of a relativistic charged particle in a magnetic field $\vec{B} = B_z\hat{z}$ with $qB_z/m_0 = 10.0$, on the xy plane. $\vec{v}(0) = 0.95\hat{y} + 0.10\hat{z}$, $\vec{r}(0) = 1.0\hat{x}$. The integration is performed by using the RK45 method from $t_0 = 0$ to $t_f = 20$ with 1000 steps. Each axis is on a different scale.

```
//    q B_2/m = k1 (3 x2 x3)/r^5
//    q B_3/m = k1[(3 x3 x3)/r^5-1/r^3]
//========================================================
#include "sr.h"
void f(double& t,double* Y, double* YP){
  double x1,x2,x3,v1,v2,v3,p1,p2,p3;
  double B1,B2,B3;
  double r,r5,r3;
  x1 = Y[0]; p1 = Y[3];
  x2 = Y[1]; p2 = Y[4];
  x3 = Y[2]; p3 = Y[5];
  velocity(p1,p2,p3,v1,v2,v3);
  //now we can use all x1,x2,x3,p1,p2,p3,v1,v2,v3
  YP[0]    = v1;
  YP[1]    = v2;
  YP[2]    = v3;
  // Acceleration:
```

6.3. RELATIVISTIC MOTION

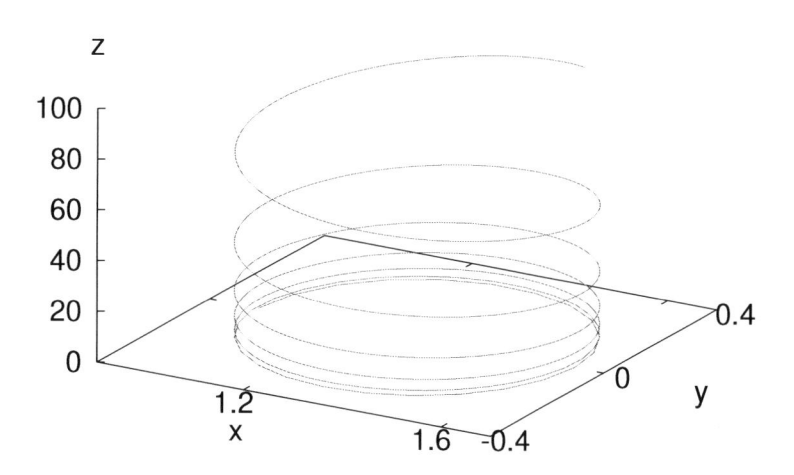

Figure 6.7: The influence of an additional electric field $q\vec{E}/m_0 = 1.0\hat{z}$ on the trajectory shown in figure 6.5.

```
  r       = sqrt(x1*x1+x2*x2+x3*x3);
  r3      = r*r*r;
  r5      = r*r*r3;
  if( r > 0.0){
    B1    = k1*( 3.0*x1*x3)/r5;
    B2    = k1*( 3.0*x2*x3)/r5;
    B3    = k1*((3.0*x3*x3)/r5-1/r3);
    YP[3] = v2*B3-v3*B2;
    YP[4] = v3*B1-v1*B3;
    YP[5] = v1*B2-v2*B1;
  }else{
    YP[3] = 0.0;
    YP[4] = 0.0;
    YP[5] = 0.0;
  }
}
//————————————————————————
double energy(const double& t, double* Y){
  double e;
```

```
    double x1,x2,x3,v1,v2,v3,p1,p2,p3,psq;
    x1 = Y[0]; p1 = Y[3];
    x2 = Y[1]; p2 = Y[4];
    x3 = Y[2]; p3 = Y[5];
    psq= p1*p1+p2*p2+p3*p3;
    // Kinetic   Energy:
    e  = sqrt(1.0+psq)-1.0;

    return e;
}
```

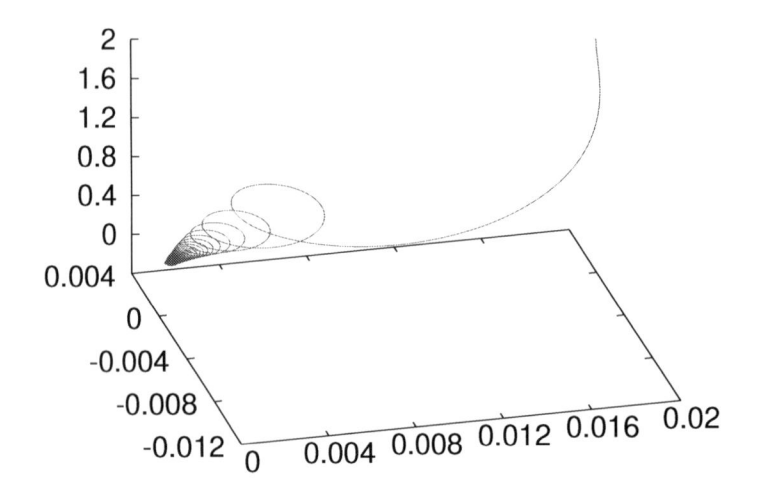

Figure 6.8: The trajectory of a charged particle in a magnetic dipole field given by equation (6.11). We used $B_0 = 1000$, $\vec{r} = 0.02\hat{x} + 2.00\hat{z}$, $\vec{v} = -0.99999\hat{z}$. The integration was done from $t_0 = 0$ to $t_f = 5$ in 10000 steps.

The results are shown in figure 6.8. The parameters have been exaggerated in order to achieve an aesthetically pleasant result. In reality, the electrons are moving in very thin spirals and the reader is encouraged to use more realistic values for the parameters $\vec{v}_0$, B_0, $\vec{r}_0$. The problem of why the effect is not seen near the equator is left as an exercise.

6.4 Problems

6.1 Compute the trajectory of a projectile moving in space in a constant gravitational field and under the influence of an air resistance proportional to the square of its speed.

6.2 Two point charges are moving with non relativistic speeds in a constant magnetic field $\vec{B} = B\hat{z}$. Assume that their interaction is given by the Coulomb force only. Write a program that computes their trajectory numerically using the RK45 method.

6.3 Write a program that computes the trajectory of the anisotropic harmonic oscillator $\vec{F} = -k_x x\hat{x} - k_y y\hat{y} - k_z z\hat{z}$. Compute the three dimensional Lissajous curves which appear for appropriate values of the angular frequencies $\omega_x = \sqrt{k_x/m}$, $\omega_y = \sqrt{k_y/m}$, $\omega_z = \sqrt{k_z/m}$.

6.4 Two particles of mass M are at the fixed positions $\vec{r}_1 = a\hat{z}$ and $\vec{r}_2 = -a\hat{z}$. A third particle of mass m interacts with them via a Newtonian gravitational force and moves at non relativistic speeds. Compute the particle's trajectory and find initial conditions that result in a planar motion.

6.5 Solve problem 5.19 of page 302 using the RK45 method. Choose initial conditions so that the system executes only translational motion. Next, choose initial conditions so that the system executes small vibrations and its center of mass remains stationary. Find the normal modes of the system and choose appropriate initial conditions that put the system in each one of them.

6.6 Solve the previous problem by putting the system in a box $|x| \leq L$ and $|y| \leq L$.

6.7 Solve the problem 5.20 in page 302 by using the RK45 method.

6.8 Solve the problem 5.21 in page 303 by using the RK45 method.

6.9 The electric field of an electric dipole $\vec{p} = p\hat{z}$ is given by:

$$\vec{E} = E_\rho \hat{\rho} + E_z \hat{z}$$
$$E_\rho = \frac{1}{4\pi\epsilon_0} \frac{3p\sin\theta\cos\theta}{r^3}$$
$$E_z = \frac{1}{4\pi\epsilon_0} \frac{p(3\cos^2\theta - 1)}{r^3} \quad (6.12)$$

where $\rho = \sqrt{x^2 + y^2} = r\sin\theta$, $E_x = E_\rho \cos\phi$, $E_y = E_\rho \sin\phi$ and (r, θ, ϕ) are the polar coordinates of the point where the electric field is calculated. Calculate the trajectory of a test charge moving in this field at non relativistic speeds. Calculate the deviation between the relativistic and the non relativistic trajectories when the initial speed is $0.01c, 0.1c, 0.5c, 0.9c$ respectively (ignore radiation effects).

6.10 Consider a linear charge distribution with constant linear charge density λ. The electric field is given by

$$\vec{E} = E_\rho \hat{\rho} = \frac{1}{4\pi\epsilon_0} \frac{2\lambda}{\rho} \hat{\rho}$$

Calculate the trajectories of two equal negative test charges that move at non relativistic speeds in this field. Consider only the electrostatic Coulomb forces and ignore anything else.

6.11 Consider a linear charge distribution on four straight lines parallel to the z axis. The linear charge density is λ and it is constant. The four straight lines intersect the xy plane at the points $(0,0)$, $(0,a)$, $(a,0)$, (a,a). Calculate the trajectory of a non relativistic charge in this field. Next, compute the relativistic trajectories (ignore all radiation effects).

6.12 Three particles of mass m interact via their Newtonian gravitational force. Compute their (non relativistic) trajectories in space.

6.13 There is a C++ "translation" of rksuite. Download it from netlib.org/ ode/ rksuite and teach yourself how to use it. The documentation is not as explicit as for the Fortran version, part of it is in the source code file rksuite.cpp. You can teach yourself how to use it by reading the example file RksuiteTest.cpp and the methods of the class RKSUITE in the file rksuite.h. Write a program to study the motion of the non relativistic electron in a constant magnetic field. Then repeat for the relativistic electron.

Bibliography

[Textbooks]

[1] www.physics.ntua.gr/~konstant/ComputationalPhysics/. The site of this book. The accompanying software and additional material can be found there. You may also find contact information about the author for sending corrections and/or suggestions. Fan mail accepted too!

[2] H. Gould, J. Tobochnik and H. Christian, *"Computer Simulation Methods, Application to Physical Systems"*, Third Edition, Addison Wesley (2007). A great introductory book in computational physics. Java is the programming language of choice and a complete computing environment is provided for using and creating interacting and visual physics applications. The software is open source and can be downloaded from opensourcephysics.org. The book has open access and can be downloaded freely.

[3] R. Landau, M. J. Páez and C. C. Bordeianu, *"Computational Physics: Problem Solving with Computers"*, Wiley-VCH, 2 ed. (2007).

[4] M. E. J. Newman and G. T. Barkema, *"Monte Carlo Methods in Statistical Physics"*, Clarendon Press, Oxford (2002). Excellent book for an introductory and intermediate level course in Monte Carlo methods in physics.

[5] B. A. Berg, *"Markov Chain Monte Carlo Simulations and Their Statistical Analysis. With Web-Based Fortran Code"*, World Scientific, 2004. Graduate level Monte Carlo from a great expert in the field. Covers many advanced Monte Carlo methods.

[6] D. P. Landau and K. Binder, "*A Guide to Monte Carlo Simulations in Statistical Physics*", Cambridge University Press, 3rd Edition, 2009.

[7] K. Binder and D. W. Heermann, "*Monte Carlo Simulation in Statistical Physics*", Fifth Edition, Springer (2010).

[8] W. H. Press, S. A. Teukolsky, W. T. Vetterling and B. P. Flanney, "*Numerical Recipes, The Art of Scientific Computing*", Third Edition, Cambridge University Press (2007), www.nr.com. Well, this is *the* handbook of every physicist in numerical methods.

[Chapter 1]

[9] www.cplusplus.com, C++ Tutorials.

[10] cppreference.com, C++ reference.

[11] N. M. Josuttis, "*The C++ standard library: a tutorial and reference*", Pearson, 2012.

[12] S. Oualline, "*Practical C++ Programming*", 2nd Ed., O' Reilly, 2002.

[13] B. Stroustrup, "*The C++ Programming Language*", 3rd Ed., Addison-Wesley, 1997.

[14] D. Goldberg, "*What Every Computer Scientist Should Know About Floating-Point Arithmetic*", Computing Surveys, 1991, http://docs.oracle.com/cd/E19957-01/806-3568/ncg_goldberg.html. See also Chapter 1 in [8].

[15] Gnuplot official site http://gnuplot.info/

[16] P. K. Janert, "*Gnuplot in Action: Understanding Data with Graphs*", 2nd Ed., Manning Publications (2012).

[17] L. Phillips, "*gnuplot Cookbook*", Packt Publishing, 2012.

[18] tcsh homepage: http://www.tcsh.org/Home

[19] P. DuBois, "*Using csh & tcsh*", O'Reilly and Associates (1995), www.kitebird.com/csh-tcsh-book/.

BIBLIOGRAPHY

[20] M. J. Currie, *"C-shell Cookbook"*, http://www.astro.soton.ac.uk/unixtut/sc4.pdf.

[21] Wiki book: *"C Shell Scripting"*, http://en.wikibooks.org/wiki/C_Shell_Scripting.

[22] G. Anderson and P. Anderson, *"The Unix C Shell Field Guide"*, Prentice Hall (1986).

[Chapter 3]

[23] R. M. May, *"Simple Mathematical Models with Very Complicated Dynamics"*, Nature **261** (1976) 459. The first pedagogical and relatively brief introduction to the logistic map and its basic properties.

[24] C. Efthimiou, *"Introduction to Functional Equations: Theory and Problem-Solving Strategies for Mathematical Competitions and Beyond"*, MSRI Mathematical Circles Library (2010). Section 16.7 presents a brief and simple presentation of the mathematical properties of the logistic map.

[25] P. Cvitanović, R. Artuso, R. Mainieri, G. Tanner and G. Vattay, *"Chaos: Classical and Quantum"*, ChaosBook.org, Niels Bohr Institute (2012). An excellent book on the subject. Can be freely downloaded from the site of the book.

[26] L. Smith, *"Chaos: A Very Short Introduction"*, Oxford University Press (2007).

[27] M. Schroeder, *"Fractals, Chaos, Power Laws: Minutes from an Infinite Paradise"*, W.H. Freeman (1991).

[28] S. H. Strogatz, *"Non Linear Dynamics and Chaos"*, Addison-Wesley (1994).

[29] Wikipedia: "Chaos Theory", "Logistic Map", "Bifurcation Diagram", "Liapunov Exponents", "Fractal Dimension", "Feigenbaum constants".

[30] Wikipedia: "List of chaotic maps".

[31] Wikipedia: "Newton's method".

[32] M. Jakobson, *"Absolutely continuous invariant measures for one-parameter families of one-dimensional maps"*, Commun. Math. Phys. **81** (1981) 39.

[Chapter 4]

[33] *"Numerical Recipes"* [8]. See chapters on the Runge–Kutta methods.

[34] E. W. Weisstein, *"Runge-Kutta Method"*, from MathWorld–A Wolfram Web Resource.
http://mathworld.wolfram.com/Runge-KuttaMethod.html.

[35] J. H. E. Cartwright and O. Piro, *"The dynamics of Runge-Kutta methods"*, Int. J. Bifurcation and Chaos **2**, (1992) 427-449.

[36] J. H. Mathews, K. Fink, *"Numerical Methods Using Matlab"*, Prentice Hall (2003), Chapter 9.

[37] J. H. Mathews, *"Numerical Analysis - Numerical Methods Project"*, http://math.fullerton.edu/mathews/numerical.html.

[38] I. Percival and D. Richards, *"Introduction to Dynamics"*, Cambridge University Press (1982). See also [40].

[39] J. B. McLaughlin, *"Period Doubling bifurcations and chaotic motion for a parametrically forced pendulum"*, J. Stat. Phys. **24** (1981) 375–388.

[Chapter 5]

[40] J. V. José and E. J. Saletan, *"Classical Dynamics, a Contemporary Approach"*, Cambridge University Press, 1998. A great book on Classical Mechanics. You will find a lot of information on non linear dynamical systems exhibiting chaotic behavior. See also the chapters on scattering and planetary motion.

[Chapter 6]

[41] R. W. Brankin, I. Gladwell, and L. F. Shampine, *"RKSUITE: a suite of Runge-Kutta codes for the initial value problem for ODEs"*, Softreport 92-S1, Department of Mathematics, Southern Methodist University, Dallas, Texas, U.S.A (1992). Available at www.netlib.org/ode/rksuite and in the accompanying software of the book.

[Chapter ??]

[42] See the Mathematica Notebooks of Peter West http://young.physics.ucsc.edu/115/.

[43] U. Wolff, B. Bunk, F. Knechtli, *"Computational Physics I"*, http://www.physik.hu-berlin.de/com/ teachingandseminars/previous_CPI_CPII.

[44] F. T. Hioe and E. W. Montroll, *"Quantum theory of anharmonic oscillators. I. Energy levels of oscillators with positive quartic anharmonicity"*, J. Math. Phys. **16** (1975) 1945, http://dx.doi.org/10.1063/1.522747

[Chapter ??]

[45] L. Kadanoff, *"Statistical Physics – Statics, Dynamics and Renormalization"*, World Scientific (2000). A great book in advanced statistical physics by one of the greatest in the field!

[46] J. Ambjørn, B. Durhuus and T. Jonsson, *"Quantum Geometry"*, Cambridge Monographs on Mathematical Physics, Cambridge University Press (1997). More in depth discussion of random walks in field theory and statistical mechanics.

[47] C. Itzykson and J. M. Drouffe, *"Statistical Field Theory"*, Volume 1, Cambridge Monographs on Mathematical Physics, Cambridge University Press (1989). Random walks and Euclidean quantum field theory.

[48] D. E. Knuth, *"Seminumerical Algorithms"*, Vol. 2 of *"The Art of Computer Programming"*, Addison-Wesley (1981).

[49] K.G. Savvidy, *"The MIXMAX random number generator"*, Comp. Phys. Commun. **196** (2015) 161, [arXiv:1403.5355]; K.G. Savvidy and G.K. Savvidy, *"Spectrum and entropy of C-systems. MIXMAX random number generator"*, Chaos, Solitons & Fractals **91** (2016) 11, [arXiv:1510.06274]; MIXMAX site: mixmax.hepforge.org; Wikipedia: "MIXMAX generator".

[50] M. Lüscher, *Comput. Phys. Commun.* **79** (1994) 100; F. James, *Comput. Phys. Commun.* **79** (1994) 111; *Erratum*

97 (1996) 357. The code is available at the journal's site http://cpc.cs.qub.ac.uk/summaries/ACPR_v1_0.html as well as from CERN at
http://wwwasd.web.cern.ch/wwwasd/cernlib/
download/2001_wnt/src/mathlib/gen/v/ranlux.F.

[51] L. Schrage, *"A More Portable Fortran Random Number Generator"*, ACM Transactions on Mathematical Software, **5** (1979) 132-138; P. Bratley, B. L. Fox and L. Schrage, *"A Guide to Simulation"*, Springer-Verlag, 1983.

[52] G. Marsaglia and A. Zaman, *Ann. Appl. Prob.* **1** (1991) 462.

[53] B. Li, N. Madras and A. D. Sokal, *"Critical Exponents, Hyperscaling and Universal Amplitude Ratios for Two- and Three-Dimensional Self-Avoiding Walks"*, J.Statist.Phys. **80** (1995) 661-754 [arXiv:hep-lat/9409003]; G. Slade, *"The self-avoiding walk: A brief survey"*, Surveys in Stochastic Processes, pp. 181-199, eds. J. Blath, P. Imkeller and S. Roelly, European Mathematical Society, Zurich, (2011), http://www.math.ubc.ca/~slade/spa_proceedings.pdf.

[Chapter ??]

[54] J. J. Binney, N. J. Dowrick, A. J. Fisher and M. E. J. Newman, *"The Theory of Critical Phenomena"*, Clarenton Press (1992). A simple introduction to critical phenomena and the renormalization group.

[55] R. K. Pathria and P. D. Beale, *"Statistical Mechanics"*, Third Edition, Elsevier (2011). A classic in statistical physics.

[56] F. Mandl, *"Statistical Physics"*, Second Edition, Wiley (1988).

[57] R. J. Baxter, *"Exactly Solved Models in Statistical Mechanics"*, Dover Publications (2008).

[Chapter ??]

[58] E. Ising, *"Beitrag zur Theorie des Ferromagnetizmus"*, Z. Phys. **31** (1925) 253–258.

[59] L. Onsager, *"Crystal Statistics. I. A Two–Dimensional Model with an Order–Disorder Transition"*, Phys. Rev. **65** (1944) 117–119.

[60] K. Huang, *"Statistical Mechanics"*, John Wiley & Sons, New York, (1987). A detailed presentation of the Onsager solution.

[61] C. N. Yang, *Phys. Rev.* **85** (1952) 809.

[62] N. Metropolis, A. W. Rosenbluth, M. N. Rosenbluth, A. H. Teller and E. J. Teller, *Chem. Phys.* **21** (1953) 1087.

[63] M. P. Nightingale and H. W. J. Blöte, *Phys. Rev. Lett.* **76** (1996) 4548.

[64] H. Müller-Krumbhaar and K. Binder, *J. Stat. Phys.* **8** (1973) 1.

[65] B. Efron *SIAM Review* **21** (1979) 460; *Ann. Statist.* **7** (1979) 1; B. Efron and R. Tibshirani, *Statistical Science* **1** (1986) 54. Freely available from projecteuclid.org.

[Chapter ??]

[66] R. H. Swendsen and J.-S. Wang, *Phys. Rev. Lett.* **58** (1987) 86.

[67] U. Wolff, *Phys. Rev. Lett.* **62** (1989) 361.

[68] A. Pelisseto and E. Vicari, *"Critical Phenomena and Renormalization–Group Theory"*, *Phys. Reports* **368** (2002) 549.

[69] F. Y. Wu, *"The Potts Model"*, *Rev. Mod. Phys.* **54** (1982) 235.

[70] P. D. Coddington and C. F. Baillie, *Phys. Rev. Lett.* **68** (1992) 962.

[71] H. Rieger, *Phys. Rev.* **B 52** (1995) 6659.

[72] E. J. Newman and G. T. Barkema, *Phys. Rev.* **E 53** (1996) 393.

[73] A. E. Ferdinand and M. E. Fisher, *Phys. Rev.* **185** (1969) 832; N. Sh. Izmailian and C.-K. Hu, *Phys. Rev.* **E 65** (2002) 036103; J. Salas, *J. Phys.* **A 34** (2001) 1311; W. Janke and R. Kenna, *Nucl. Phys. (Proc. Suppl.)* **106** (2002) 929.

[74] J. Ambjørn and K. N. Anagnostopoulos, *Nucl. Phys.* **B 497** (1997) 445.

[75] K. Binder, *Phys. Rev. Lett.* **47** (1981) 693.

[76] K. Binder, *Z. Phys.* **B 43** (1981) 119; G. Kamieniarz and H. W. J. Blöte, *J. Phys.* **A 26** (1993) 201.

[77] J. Cardy, *"Scaling and Renormalization in Statistical Physics"*, 1st Edition, Cambridge University Press (1996).

[78] A. M. Ferrenberg and D. P. Landau, *Phys. Rev.* **B44** (1991) 5081.

[79] M. S. S. Challa, D. P. Landau and K. Binder, *Phys. Rev.* **B34** (1986) 1841.

[80] H. E. Stanley, *"Introduction to Phase Transitions and Critical Phenomena"*, Oxford (1971).

[81] R. Creswick and S.-Y. Kim, *J. Phys. A: Math.Gen.* **30** (1997) 8785.

[82] C. Holm and W. Janke, *Phys. Rev.* **B 48** (1993) 936 [arXiv:hep-lat/9301002].

[83] M. Hasenbusch and S. Meyer, *Phys. Lett.* **B 241** (1990) 238.

[84] M. Kolesik and M. Suzuki, *Physica* **A 215** (1995) 138 [arXiv:cond-mat/9411109].

[85] M. Kolesik and M. Suzuki, *Physica* **A 216** (1995) 469.

Index

++ increment operator, 43
. (current directory), 5
.. (parent directory), 5
; (separate commands), 11
$PATH, 12
- (switch, options), 11
/, 5
< (redirection), 14
>> (redirection), 14
>& (redirection), 14
> (redirection), 13
NF, 21
#include, 38
$path, 12
& (background a process), 23
a.out, 76
awk, 21, 60
    BEGIN, 21, 241
    END, 21
    NR, 21
    $1, $2, ..., 21
cat, 18, 60
cd, 6
chmod, 8
cin, 44
cout, 39
cp, 9
date, 60
double, 109
echo, 60

emacs, 23
endl, 73
float, 109
g++, 40, 76
getline, 73
gfortran, 312
grep, 20
head, 18
info, 15
iostream, 38
less, 18
ls, 7
main(), 38
man, 15
mkdir, 6
mv, 9
pwd, 6
rk2.csh, 263
rmdir, 6, 11
rm, 9
setenv, 12
set, 12
sort, 19
stderr, 13
stdin, 13
stdout, 13
tail, 18
whatis, 15
where, 16
which, 12, 16

| (piping), 14

absolute path, 3
attractor, 152, 235

basin of attraction, 164, 191, 242
bifurcation, 152, 159, 171

C++, 37
    ++ increment operator, 43
    Hello World, 38
    #include, 38
    cerr, 74
    cin, 44
    cout, 39
    double, 109
    endl, 73, 74
    exit(), 74
    float, 109
    getline, 73
    iomanip, 110
    iostream, 38
    main(), 38
    array, static, 42
    comment, 38
    comments, 38
    compile, 40
    double precision, 309
    epsilon, 309
    for, 42
    Fortran programs, calling, 311
    fstream, 44
    function, 38
    function definition, 45
    Input/Output to files, 44
    instantiation, 44
    link to Fortran and C, 311
    linkage, 311
    memory allocation, 288

    new, 288
    numeric limits, 309
    precision, floating point, 75
    preprocessor, 38
    string, 74
    void, 47
chaos, 149, 154, 240
    period doubling, 240
circle map, 196
cobweb plot, 156
command completion, 17
command substitution, 59
compile
    object file .o, 312
completion
    command, 17
    filename, 17
cpp, 38
cross section, 273
    differential, 273, 275
    total, 273
current directory, 4

derivative
    numerical, 170
directory, 4
    home, 3, 5
    parent, 4
Duffing map, 198
dynamic memory allocation, 287

eccentricity, 268
Emacs, 22
    abort command, 65, 67
    auto completion, 35
    commands, 65
    Ctrl key, C-, 23
    cut/paste, 29, 66

INDEX

edit a buffer, 28
frames, 31
help, 35, 65
info, 35, 67
kill a buffer, 32
mark, 27
Meta key, M-, 23
minibuffer, 67
minibuffer, M-x, 23
modes, 33
    LaTeX, 33
    auto fill, 34
    C, 33
    C++, 33
    font lock (coloring), 34
    overwrite, 34
point, 27
read a file, 28, 32, 65
recover a buffer, 32
recover file, 65
region, 27
replace, 66
save a buffer, 28, 32, 65
search, 65
spelling, 67
undo, 29, 65
window, split, 31, 66
windows, 31, 66
entropy, 184
error
    integration, 114
estimator, 217
Euler method, 203
Euler-Verlet method, 204, 243

Feigenbaum constant, 161
file
    owner, 7
    permissions, 8
filename completion, 17
filesystem, 3
fit, 178
focus,foci, 268
Fortran, 312
    gfortran, 312
    rksuite, 306

Gauss map, 195
Gnuplot, 50
    <, 54, 262
    1/0 (undefined number), 262
    animation, 81, 99
    atan2, 80
    comment, 52
    fit, 178
    functions, 215
    hidden3d, 55
    load, 82, 122
    log plots, 177
    parametric plot, 56
    plot, 52
    plot 3d, 55, 99
    plot command output, 54, 262
    plot expressions, 53, 80, 262
    pm3d, 55
    replot, 53, 54
    reread, 262
    save plots, 54
    splot, 55, 99
    using, 53, 80
    variables, 92, 215
    with, 53

Hénon map, 197
hard sphere, 271
home directory, 3, 5

impact parameter, 274, 275

Kepler's law, 268

leapfrog method, 246
Liapunov exponent, 175
linking
    object file .o, 312
logistic map, 150
    2^n cycles, 154
    attractor, 152
    bifurcation, 152, 159, 171
    cobweb plot, 156
    entropy, 184
    fixed points, 152
        stability, 152
    onset of chaos, 154, 183
    special solutions, 151
    strong chaos, 183
    transient behavior, 159
    weak chaos, 183

man pages, 15
memory
    allocation, dynamic, 287
    allocation, static, 287
minibuffer, 23
mouse map, 195

Netlib, 306
    rksuite, 306
Newton's law of gravity, 267
Newton-Raphson method, 163, 167
numerical
    derivative, 170

options, 11

parent directory, 4
path
    absolute, 3
    command, 12
    file, 3
    relative, 3
period doubling, 154
piping, 14
Poincaré diagram, 240
preprocessor, 38

redirection, 13
relative path, 3
relativity
    special, 324
rksuite, 306
root, 4
Runge-Kutta method, 216, 246, 254
    adaptive stepsize, 306
Rutherford scattering, 275

scattering, 271, 277
    rate, 273
    Rutherford, 275
shell
    argv, 60
    array variable, 58
    arrays, 60
    command substitution, 59
    foreach, 60
    here document, 57, 60
    if, 60
    input $<$, 60
    script, 56, 60
    set, 58
    tcsh, 56
    variable, 58
sine map, 194
splinter, 151
standard error, 13

standard input, 13
standard output, 13
subdirectory, 4

tent map, 195
Tinkerbell map, 198
transient behavior, 159, 234

user interface, 73

variables
    environment, 12
    shell, 12

working directory, 4

Printed in Great Britain
by Amazon